## ABOUT THE AUTHOR

MADELEINE ALBRIGHT served as America's sixty-fourth secretary of state from 1997 to 2001. Her distinguished career also includes positions on Capitol Hill, on the National Security Council, and as U.S. ambassador to the United Nations. She is a resident of Washington, D.C., and Virginia.

# PRAGUE WINTER

*A Personal Story
of Remembrance and War,
1937–1948*

## MADELEINE ALBRIGHT

### WITH BILL WOODWARD

———◆———

HARPER ● PERENNIAL

NEW YORK ● LONDON ● TORONTO ● SYDNEY ● NEW DELHI ● AUCKLAND

HARPER  PERENNIAL

A hardcover edition of this book was published in 2012
by HarperCollins Publishers.

PRAGUE WINTER. Copyright © 2012 by Madeleine Albright. All rights reserved. Printed in the United States of America. No part of this book may
be used or reproduced in any manner whatsoever without written permission
except in the case of brief quotations embodied in critical articles and reviews.
For information address HarperCollins Publishers, 10 East 53rd Street, New
York, NY 10022.

HarperCollins books may be purchased for educational, business, or sales
promotional use. For information please write: Special Markets Department,
HarperCollins Publishers, 10 East 53rd Street, New York, NY 10022.

FIRST HARPER PERENNIAL EDITION PUBLISHED 2013.

*Designed by Fritz Metsch*

The Library of Congress has catalogued the hardcover edition as follows:

Albright, Madeleine Korbel.
Prague winter : a personal story of remembrance and war, 1937–1948 /
Madeleine Albright.—FIRST EDITION.
p. cm
Includes bibliographical references and index.
ISBN 978-0-06-203031-3 (hardback)
1. Czechoslovakia—History—1938–1945. 2. World War, 1939–1945—
Czechoslovakia. 3. Albright, Madeleine Korbel—Family. 4. Albright,
Madeleine Korbel—Childhood and youth. 5. Jewish families—Czech
Republic—Prague—Biography. 6. Prague (Czech Republic)—Biography.
7. Prague (Czech Republic)—History—20th century. 8. World War,
1939–1945—Czech Republic—Prague. I. Title.
DB2207.A43 2012
943.71'2033092—dc23
2011049416

ISBN 978-0-06-203034-4 (pbk.)

13 14 15 16 17  OV/RRD  10 9 8 7 6 5 4 3 2 1

TO THOSE WHO
DID NOT SURVIVE
BUT TAUGHT US HOW TO LIVE—
AND WHY

## MEMORIES OF PRAGUE

❧

*How long since I last saw*
*The sun sink low behind Petřín Hill?*
*With tearful eyes I gazed at you, Prague,*
*Enveloped in your evening shadows.*
*How long since I last heard the pleasant rush of water*
*Over the weir in the Vltava river?*
*I have long since forgotten the bustling life of Wenceslas Square.*
*Those unknown corners in the Old Town,*
*Those shady nooks and sleepy canals,*
*How are they? They cannot be grieving for me*
*As I do for them . . .*
*Prague, you fairy tale in stone, how well I remember!*

PETR GINZ (1928–1944)
*Terezín*

# CONTENTS

# PRAGUE WINTER

# Setting Out

I was fifty-nine when I began serving as U.S. secretary of state. I thought by then that I knew all there was to know about my past, who "my people" were, and the history of my native land. I was sure enough that I did not feel a need to ask questions. Others might be insecure about their identities; I was not and never had been. I knew.

Only I didn't. I had no idea that my family heritage was Jewish or that more than twenty of my relatives had died in the Holocaust. I had been brought up to believe in a history of my Czechoslovak homeland that was less tangled and more straightforward than the reality. I had much still to learn about the complex moral choices that my parents and others in their generation had been called on to make—choices that were still shaping my life and also that of the world.

I had been raised a Roman Catholic and upon marriage converted to the Episcopalian faith. I had—I was sure—a Slavic soul. My grandparents had died before I was old enough to remember their faces or call them by name. I had a cousin in Prague; we had recently been in touch and as children had been close, but I no longer knew her well; the Iron Curtain had kept us apart.

From my parents I had received a priceless inheritance: a set of deeply held convictions regarding liberty, individual rights, and the rule of law. I inherited, as well, a love for two countries. The United States had welcomed my family and enabled me to grow up in freedom; I was proud to call myself an American. The Czechoslovak Republic had been a beacon of humane government until snuffed out by Adolf Hitler and then—after a brief period of postwar revival—extinguished again by the disciples of Josef Stalin. In 1989, the Velvet Revolution, led by Václav Havel, my hero and later my cherished friend, engendered new hope. All my life I had believed in the virtues of democratic government, the need to stand up to evil, and the age-old motto of the Czech people: *"Pravda vítězí,"* or "Truth shall prevail."

FROM 1993 UNTIL 1997, I had the honor of representing the United States as ambassador to the United Nations. Because I was in the news and because of Central Europe's liberation after the fall of the Berlin Wall, I began to receive mail about my family. Some of these letters had the facts wrong; others were barely legible; a few requested money; still others never reached me because staff members—strangers to the language—could not distinguish between correspondence on personal as opposed to public issues. By late in President Bill Clinton's first term, I had seen several missives from people who had known my parents, who had the names and dates approximately right, and who indicated that my ancestors had been of Jewish origin. One letter, from a seventy-four-year-old woman, arrived in early December 1996; she wrote that her family had been in business with my maternal grandparents, who had been victimized by anti-Jewish discrimination during the war. I compared memories with my sister, Kathy, and brother, John, and also shared the information with my daughters, Anne, Alice, and Katie. Since I was in the process of being vetted for secretary of state, I told President Clinton and his senior staff. In January 1997, before we had time to explore further, a hardworking *Washington Post* reporter, Michael Dobbs, uncovered news that stunned us all: according to his research, three of my grandparents and numerous other family members had died in the Holocaust.

In February 1997, Kathy, John, and John's wife, Pamela, visited the Czech Republic; they confirmed much of what had been in the *Post* story and identified a few errors. That summer, I was able to make two similar though briefer trips. For me, the moment of highest emotion came inside Prague's Pinkas Synagogue, where the names of our family members were among the eighty thousand inscribed on the walls as a memoriam. I had been to the synagogue before but—having no cause—had never thought to search for their names.

That episode is recounted in my memoir *Madam Secretary* and will not be elaborated on here. The core revelation, however, is central because it provided the impetus for this book. I was shocked and, to be honest, embarrassed to discover that I had not known my family history better; my sister and brother shared this emotion. Nor was I entirely reassured by the many people who spoke or wrote to me of having had comparable experiences concerning secrets kept by their own parents. I could accept without being satisfied that there was nothing inexplicable or unique about the gap that existed in my knowledge; still, I regretted not having asked the right questions. I also felt driven to learn more about the grandparents whom I had been too young to know—especially since by then I had become a grandparent myself.

Having decided to delve more deeply into my family's history, I soon realized that I could not do so without placing my parents within the context of the times in which they had lived and especially 1937–1948, the era encompassing World War II—also the first dozen years of my life.

IN THE LATE 1930s, the global spotlight was drawn to Czechoslovakia, a faraway place that few people in such capitals as London and Washington had visited or even knew how to spell. The country was familiar, if at all, in the guise of Bohemia—a land of magic, marionettes, Franz Kafka, and Good King Wenceslas. But to those knowledgeable about Central Europe, the nation was respected for its thousand-year history and valued for its location as a crossroads between West and East. It was also the scene of a long and at times bitter rivalry between Czechs and Germans. In that struggle's climactic chapter, Adolf Hitler demanded that the government surrender its sovereignty by opening its borders to his troops, thus creating for all of Europe a moment of hard reckoning. To the major powers in the West, Czechoslovakia was not thought to be worth fighting for, so it was sacrificed in the quest for peace; yet still the war came—and with it the near total destruction of European

Jewry and ultimately a realignment of the international political order.

My family spent World War II in England, arriving just as the population of that island nation was awakening from two decades of complacency. We were there when Winston Churchill rallied his countrymen to unite against the Nazi darkness, endure the Blitz, find space for the continent's refugee children, and play host to the Czechoslovak government in exile, whose cause my father served. My earliest memories are of London and the British countryside, of bomb shelters and blackout curtains, and of being taken by my parents to the seashore despite the massive steel barriers erected to foil enemy attempts to invade.

From the day the United States entered the war, my parents and their friends were confident the Allies would win. As democrats from Central Europe, they prayed that it would be the United States—not the Soviet Union—that would possess the decisive postwar influence in our region. It was not to be. With the Nazis defeated, Czechoslovakia was once more to become a pivotal battleground where the forces of totalitarianism would prevail, driving my family again into exile, this time finding a permanent refuge in the United States.

Nothing could be more adult than the decisions people were compelled to make during this turbulent era, yet the issues involved would be familiar to any child: How can I be safe? Whom can I trust? What can I believe? And (in the words of the Czech national anthem) "Where Is My Home?"

A child of my generation born in Prague would almost certainly be familiar with the novel *The Grandmother*. Written in 1855, the book was one of the first serious works of literature to be published in the Czech language. The story has a special place in my heart because of the heroine's name: "Madaline." One of the supporting characters is a striking young woman, who—bewitched and "ruined for marriage" by a passing soldier—retires to a forest cave, going barefoot even in winter, surviving on berries, roots, and the occasional handout. When asked by a child how the young woman can endure such harsh conditions, the grandmother replies that it is because the poor creature never enters a room warmed by fire, "so she is not as sensitive to cold as we."

While I was growing up, tens of millions of people were denied the chance, metaphorically, to enter a room warmed by fire. Instead, they were forced to adapt to the hardships of war: occupation by enemy troops, separation from home and loved ones, shortages of food and heat, and the constant presence of suspicion, fear, danger, and death. Without the chance to evolve gradually, amid familiar people and places, they were thrown back on their primal instincts and forced to make practical and moral judgments from a short menu of bad options.

In many cases, the choices made were brave, in some purely pragmatic, in others accompanied by the shame of betrayal or cowardice. Often the path selected was a crooked one, as caution, then courage, pointed the way. Sometimes an action chosen in response to immediate circumstances had long-term impacts that could not be foreseen. In this environment, hurried decisions—whether made by national leaders, enemy combatants, harried bureaucrats, next-door neighbors, or even parents—could have fatal or lifesaving consequences.

In the end, no one who lived through the years of 1937 to 1948 was a stranger to profound sadness. Millions of innocents did not survive, and their deaths must never be forgotten. Today we lack the power to reclaim lost lives, but we have a duty to learn all that we can about what happened and why—not to judge with the benefit of hindsight but to prevent the worst of that history from playing out again.

RESEARCHING THIS BOOK began, as many family-centered explorations do, with a stack of boxes stowed in the garage. My father had published half a dozen works of nonfiction and, when taking notes, used a Dictaphone to record his thoughts. I have a basketful of recordings to which I had never listened for fear that his voice would prompt too painful a sense of deprivation. I felt a similar anxiety about those boxes. When in government, I had been too busy to sort through them; in the years since, a profusion of other projects had allowed me to persuade myself that the time was still not right. But I had waited long enough.

Gathering my courage, I pulled down a few cartons and began my journey. Inside, I discovered a bounty of papers separated by rusted paper clips and held together by rubber bands so brittle that they snapped when they should have stretched. Much of the material was routine, but mixed in were more interesting findings. Here were the original drafts of talks that my father had given about the figures he admired most: T. G. Masaryk, the founder of modern Czechoslovakia, and his son, Jan, who had been my father's boss. I came across books written by people I had met as a child, including a multivolume set by Prokop Drtina, with whom we had shared an apartment house in wartime London. Inside one of his books, a page had been turned down at the corner, noting the place. I soon learned that our long-ago neighbor had thought to include a description of a little girl named Madlenka, the first time anyone had written about me. It could only have been my mother who had marked the page.

In recent years, I have taught a course at Georgetown University titled "The National Security Toolbox." I found an article written four decades earlier by my father—a piece I had never known existed—called "The Tools of Foreign Policy." In another folder was a pile of some 120 pages, neatly typed and divided into chapters. At one point in the past, my father had confided that he was attempting to write a novel. I asked, "Concerning what?" He replied, "A man returning to Czechoslovakia at the end of World War II." This must be it. Eagerly, I plunged in; before long, my eyes had filled with tears. In pages to come, my father's words will have a prominent place.

So, too, my mother's. In 1977, shortly after my father's death, she had written an eleven-page letter that provides the sole firsthand information I have about dramatic moments in my parents' lives, including our escape from Prague following the German invasion. For several weeks, I searched for the essay but could not find it. Growing nervous, I asked my sister and brother if they knew where it might be: no luck. I turned my office upside down, then looked for the tenth time through my desk. There in the drawer with the papers I care about most was my mother's writing, squashed and pushed to the side. Smoothing the edges of the lined yellow pages, I began to read:

> On a high mountain near Denver is a little cemetery and there on the wall of a mausoleum is a plaque with the name: Josef Korbel 1909–1977. Maybe one day somebody will wonder who was this

man with such an unusual spelling and why was he buried in the mountain in Colorado.

Well I would like to write something about him, because his life was even more unusual than the spelling. He is buried in the mountain because he loved nature, because he loved fishing, because it was in Colorado where he spent many happy years after an active life in many different fields and countries. He used to say often, "I was in many glorious jobs, but to be a college teacher in a free country is what I enjoy best."

Joe was born in Czechoslovakia in a little town where his father had at that time a small shop with building materials. There was not even a high school in this place so he had at the age of twelve to live in a neighboring town. It was in this high school where we met and fell in love . . .

There it was in summary: the beginning of the story and the end; but surely there was more to learn about all that transpired between the high school and the mountain.

*Josef and Mandula Korbel*

———————

SOME PEOPLE PURSUE enlightenment by sitting quietly and probing their inner consciousness; I make plane reservations. On a Saturday morning in September 2010, I rang the buzzer of a modest flat in London. This was the apartment house where I had spent the early days of World War II. Responding to the bell was Isobel Alicia Czarska, a charming woman who even in the middle of preparing for a trip happily gave me a whirlwind tour. For the first time in almost seventy years, I walked down the stairway to the cellar where I had once taken shelter from the bombs of the Luftwaffe. Isobel explained that the basement had never been refurbished—a fact confirmed by me as soon as I saw the ceiling, painted the same unattractive green I remembered. As we stood in the cramped space, I explained my quest; Isobel kindly offered to do research into the building's wartime history and send along what she learned—a commitment she would faithfully keep.

Before departing London, I attended a symposium titled "Ties That Bind," commemorating the seventieth anniversary of the Czechoslovak government in exile. Hosted by Michael Žantovský, the Czech ambassador to the United Kingdom, the conference served as a forum for revisiting past controversies in light of newly available information. I was struck again by how pivotal a period in history this had been and by how wide an array of opinions scholars can have about the same set of events. By the end of the day, some of us had been moved to cheer, some to weep, and some virtually to come to blows.

I also went to Prague, where a variety of friends, old and new, aided my research. Tomáš Kraus, the executive director of the Federation of Jewish Communities in the Czech Republic, answered my questions about the history of Prague's Jewish settlements, which extend as far back as the seventh century. Daniel Herman of the Institute for the Study of Totalitarian Regimes provided me with a file that the postwar Communist government had kept on my family. Not all the papers were legible, but the evidence was clear that my father had had powerful enemies in the Marxist regime. At the Czech Foreign Ministry, I was given documents related to my father's career, among them a police report on my paternal grandfather, who was apparently not the most careful of drivers—in 1937, he was ordered to pay compensation for running over a hen. I visited, as well, the Terezín fortress and prison; our last stop was the cemetery where the victims from a long history of conflicts are laid to rest. Czechs, Germans, Hungarians, Jews, Poles, Russians, Serbs, Slovaks, and others are remembered together even if, in life, they were often at one another's throats.

In the course of my travels to Prague, I spent much time with my cousin Dáša, who always welcomed me with a plate of plum dumplings. For more than two years, we were in steady communication, exchanging memories, sharing pictures, working together to translate letters and other writings. No one alive at that point had known me earlier than she; it was often her arms that had cradled me in the bomb shelter. But her parents could not leave Czechoslovakia when mine did, and later, when the Communists took over, she chose to remain and marry her sweetheart instead of leaving for the West. We had led far different lives yet seemed to draw on the same bottomless reserve of energy. Her contributions to my research were immeasurable. When last I saw her, in April 2011, she had a schedule filled with activities, including a program for children aimed at Holocaust education and remembrance. Early in July, returning from a trip to England, she complained of pain in her neck and spine. Less than two weeks later, she passed away. I will always be grateful that this project helped bring us back together.

Mortality claimed a second person close to this book when, on December 18, 2011, President Havel succumbed to respiratory ailments. I had seen him most recently at his seventy-fifth birthday celebration two months before his death. My gift to him was a compass that had been used by a U.S. soldier in World War I, the conflict that had first brought freedom to Czechoslovakia. In my note, I cited the irony of giving a compass to a man who served as the moral North Star for a generation. The twentieth century produced only a handful of authentic democratic heroes; he was one of them.

In October 2010, we had shared a meal at the Café Savoy, a favorite smoke-filled haunt of Havel's from the days of the Velvet Revolution. When I explained to my friend what I was planning, he immediately pledged his help. I asked about his experiences as a boy and invited him to reflect on the choices made by leaders during the war. In any discussion with Havel, issues of public policy arise and then, inevitably, also morality. He had often told me of his idea that God could be compared to the sun—a big eye in the sky that can see what we are doing when no one else is around. I had always been spooked by that image but agree that conscience is the quality no scientist has quite figured out. The nostrum "Let your conscience be your guide" was drummed into America's baby boom generation by Walt Disney. Life is more complicated than that, but sitting with Havel, I worried that we sometimes look past what is plain. Two decades earlier, I had listened while he had delivered a speech that had thoroughly perplexed the U.S. Congress. Czechoslovakia had just regained its freedom, and the legislators were anticipating a howl of Cold War triumph. Instead, they heard a plea on behalf of the Family of Man and a declaration that the real battle—for moral responsibility toward the earth and our neighbors on it—had barely begun.

What fascinates me—and what serves as a central theme of this book—is why we make the choices we do. What separates us from the world we have and the kind of ethical universe envisioned by someone like Havel? What prompts one person to act boldly in a moment of crisis and a second to seek shelter in the crowd? Why do some people become stronger in the face of adversity while others quickly lose heart? What separates the bully from the protector? Is it education, spiritual belief, our parents, our friends, the circumstances of our birth, traumatic events, or more likely some combination that spells the difference? More succinctly, do our hopes for the future hinge on a desirable unfolding of external events or some mysterious process within?

My search for an answer to these questions begins with a look back—to the time and place of my earliest years.

# PART I

✢

# *Before March 15, 1939*

*Did not Sibilla prophesy that much misery was to come
to Bohemia, that there would be wars, famines and
plagues, but that the worst time would come when . . .
the given word or promise would not be held sacred; that
then the Bohemian land should be carried over the earth
upon the hooves of horses?*

—BOŽENA NĚMCOVÁ,
*The Grandmother: A Story of Country Life in Bohemia, 1852*

———◆———

## An Unwelcome Guest

On a hill in Prague there is a castle that has stood for a thousand years. From its windows one can see a forest of gilded cupolas and baroque towers, slate roofs and sacred spires. Visible too are the stone bridges spanning the broad and winding Vltava River as its waters flow northward at a leisurely pace. Through the centuries, the beauty of Prague has been enriched by the labor of artisans from a plethora of nationalities and creeds; it is a Czech city with a variety of accents, at its best in spring when the fragrant blossoms of the lindens burst forth, the forsythia turns gold, and the skies seem an impossible blue. The people, known for their diligence, resilience, and pragmatism, look forward each winter to when the days lengthen, the breezes soften, the trees regain their covering, and the river banks issue a silent summons to play.

On the morning of March 15, 1939, that promise of spring had never seemed so distant. Snow lay thick on the castle grounds; winds blew fiercely from the northeast; the heavens were a leaden gray. At the U.S. Legation, two disheveled men cornered a diplomat en route to his office and begged desperately for asylum. They had been Czechoslovak spies in Germany and were known to the Gestapo. The diplomat, a young Foreign Service officer named George Kennan, turned them away; there was nothing he could do.

Czechs had awoken that morning to a startling announcement: "Today at six o'clock German troops crossed our borders and are proceeding to Prague by all routes. Stay calm." The light of dawn was still searching for cracks in the clouds when the first convoy of jeeps and trucks roared by, heading toward the castle. The vehicles, plastered with ice, were driven by red-faced soldiers wearing steel helmets and wool coats. Before long, the people of Prague had had their coffee and it was time to go to work. The sidewalks filled with men and women stopping to gape at the alien procession, defiantly waving their fists, crying, or staring in stony silence.

*German troops occupy Prague.*

In Wenceslas Square, voices were raised in a spontaneous rendering of patriotic songs. On and on the mechanized battalions came, penetrating every neighborhood of the ancient city. At the train station, artillery pieces and tanks were unloaded. By midmorning, heavy-footed Germans were striding purposefully into government ministries, the town hall, prisons, police offices, and barracks. They seized the airfields, deployed field guns on the snow-covered slopes of Petřín Hill, draped flags and banners across the fronts of buildings, and attached loudspeakers to lampposts and trees. Martial law was declared and a 9 p.m. curfew announced.

In the early-evening darkness, a motorcade arrived from the north. Its passengers were ushered through the deserted streets, across the river, and up the curving byways to the castle mount; and so that night, the fabled home of Bohemian kings served as headquarters for the ruler of Germany's Third Reich. Adolf Hitler and his top aides, Hermann Göring and Joachim von Ribbentrop, were in an exultant mood. "The Czechs may squeal," the führer had told his military commanders, "but we will have our hands on their throats before they can shout. And anyway, who will come to help them?" Ever mindful of a statement attributed to Bismarck that "he who controls Bohemia controls Europe," Hitler had long planned for this day. He thought the Czechs, because of their cleverness, to be the most dangerous of Slavs; he coveted their air bases and munitions factories; he knew that he could satisfy his ambitions in the rest of Europe only when the Czech homeland had been crushed. Now his triumphal march had begun. The German language was dominant within the castle walls, above which the German flag had been hoisted. Ordinarily a vegetarian teetotaler, Hitler treated himself to a victor's communion: a bottle of Pilsener and a slice of Prague ham.

The next day, Ribbentrop commandeered the main radio stations to proclaim that Czechoslovakia had ceased to exist. Bohemia and Moravia would be incorporated into greater Germany, and their government, now a protectorate, would take orders from Berlin. Citizens should await instructions. Hitler, meanwhile, was receiving visitors. First Emil Hácha, the Czech president, pledged his cooperation, then the minister of defense, then the mayor; no one wanted a bloodbath.

Around noon, a crowd of German-speaking civilians and soldiers gathered to cheer the führer when he appeared in a third-floor window. The resulting image so pleased the Nazis that they put it on a postage stamp.

In succeeding days, the snow stopped but the air remained bitter and cold. German soldiers occupied the local army barracks; Nazi administrators made themselves at home in the finest residences and hotels. Each morning before dawn, men in long coats moved swiftly about the city; they carried nightsticks and lists of names. My parents sent me to stay with my grandmother and did their best to do what their beloved country had done: disappear.

# 2

# Tales of Bohemia

I am not sure how old I was—though certainly very young—when I first heard the stories of Čech the founder and the wise and fearless Libuše. My mother read to me often, and she loved the old Bohemian tales. As in many cultures, these combined myth and reality in a blend of rousing adventures, epic quests, magic swords, and inventive explanations concerning the origin of things. Over time real heroes and villains appeared to take their place alongside the make-believe ones and together created the saga of a nation. The historian's job is to sift through such narratives and separate truth from fiction. Frequently, however, facts are redesigned to fit a pattern that conforms to the author's sensibility at the time the writing is done. That is why the past seems constantly to change. "A scholar," wrote my father, "inescapably reads the historical record in much the same way as he would look in a mirror—what is most clear to him is the image of his own values [and] sense of . . . identity."

I never had an academic course in Czech history; instead I absorbed information piecemeal from random bits of conversation, research while in college, and the books that my mother read and my father wrote. Over time, I became conditioned to think of my homeland as exceptional, a country filled with humane and democratic people who had struggled constantly to survive despite foreign oppression. The nation's finest moments had been marked by a willingness to defend itself against more powerful foes; the saddest by a failure to fight back when betrayed by supposed allies and friends. Its purest expression could be found in the period between the two world wars, when the Czechoslovak Republic served as a model of twentieth-century democracy within an otherwise dismal Europe.

I was confident of this history, so much so that when defending my PhD dissertation, I was taken aback to be challenged by professors with family ties from elsewhere in Central Europe who didn't understand why I thought the Czechoslovak experience so distinctive. At that stage in my life, I was not about to abandon the historical narrative with which I was most comfortable, a version that had the advantage of simplicity and clear partitions between right and wrong. The professors were just jealous, I thought, of my homeland's democratic institutions and values. To appreciate the country, they needed to know more about its heroes and myths, its fight to establish an identity, and the singular characteristics of its people.

THE EARLIEST SETTLERS of the lands that lie within the heart of Europe between the Carpathian Mountains and the Danube were the Boii, a Celtic tribe on the run from northern floods. Those pioneers were gradually pushed out by Germanic warriors, who were then suppressed by the legions of imperial Rome. The Romans called the land "Bohemia" after the Boii, which means that the territory was named by Italians in honor of the Irish, demonstrating—if nothing else—that globalization is not new.

When Rome crumbled, the Germans returned, joined in the eighth century AD by Slavs who migrated from the Central Asian steppes. According to legend, the patriarch Čech led his people on the arduous journey west across three great rivers until they came to a hill of a most peculiar shape: round at the top with inordinately steep sides. From the summit, Čech announced to his weary companions that they had reached at last the "Promised Land . . . [of] vast forests and sparkling rivers, green meadows and blue lakes, a land filled with game and birds and wet with sweet milk and honey."

A daughter of Čech's successor, the prophetess Libuše, is described in the odd way of ancient chroniclers as "the pride and glory of the female sex, doing wise and manly deeds." It was she who envisioned the creation of a city—Prague—"whose glory shall touch the stars." The story may be fantastic, but there was nothing fictional about the city and its fame. By the end of the tenth century, control of the Czech lands had been consolidated by the Přemyslids, an indigenous clan whose dynasty brought the nation into being. During their reign, grand cathedrals, monasteries, and synagogues were built; the castle district was fortified; and commerce flourished on both sides of the river.

Among the nation's early rulers was Václav (in English, Wenceslas), a devout Christian who incurred resentment among the pagan nobility due to his kindness toward the poor. In search of allies, Václav made peace with German Saxony and, in return for protection, paid an annual tribute of silver and oxen. The king was beloved by his people but envied by his treacherous brother Boleslav, whose minions murdered the young monarch while he was on his way to mass. Every nation needs its martyrs, and Wenceslas became Bohemia's first.

*King Wenceslas*

The Czech lands thrived under the Přemyslid kings with Prague becoming a model of diversity: Czechs, Germans, Jews, Poles, Roma, and Italians lived in the city's crowded buildings and haggled each day over the furs, scarves, saddles, shields, and other goods for sale in kiosks along its busy streets.

Toward the end of the thirteenth century, the kingdom extended its sway briefly as far south as the Adriatic Sea—just long enough for Shakespeare to set a scene in *The Winter's Tale* on the otherwise hard-to-imagine "seacoast of Bohemia."

One of the few medieval leaders to leave a lasting legacy was Charles IV (1316–1378), the first king of Bohemia to rule also in Germany and as sovereign of the Holy Roman Empire. A forward thinker, the monarch went through several wives, one French, the next three German. The fourth, Elisabeth of Pomerania, entertained dinner guests by ripping chain metal to shreds and bending horseshoes with her bare hands. There was no fifth wife.

Among the many highlights of Charles's reign was the founding of a university in Prague that attracted students from as far away as England, Scandinavia, and the Balkans. This was in 1348, before printed books and at a time when scientific inquiry was still limited to what the Church would allow. The king also ordered the construction of a sixteen-arch stone bridge over the Vltava. His architects recommended that a special ingredient—eggs—be mixed with the mortar to ensure its strength.* Supplying the builders was too big a job for the hens of Prague, so a decree went out summoning cartloads of the special ingredient from around the country. The royal masons were bemused when wagons from one northern town rolled up bearing an impressive quantity—all hard-boiled.

Charles, though cosmopolitan in his personal tastes, zealously promoted Bohemia's national myths. He confirmed the region's autonomy and designated Czech (along with German and Latin) as an official language within the empire. To honor Saint Wenceslas, he commissioned

---

* This story has always had its skeptics, but a scientific scan of Charles Bridge, conducted in 2008, confirmed the presence of egg protein in the mortar.

a crown of pure gold encrusted with precious stones topped by a cross and sapphire cameo said to contain a thorn from the crown of Christ. Today, the royal diadem and other coronation jewels are hidden away within an iron safe behind a door with seven locks in a special chamber of the towering Saint Vitus Cathedral. According to popular wisdom, if a false ruler dons the crown, death will claim him within a year.

THE MARTYR WENCESLAS was the political icon of the Bohemian nation; a second martyr, Jan Hus, became the spiritual one. Born in 1372, Hus launched his career modestly enough, as an expert in spelling. Short and plump, he developed into a popular preacher and, in 1409, was named rector of Charles University. The Czech motto, "Truth shall prevail," derives from Hus's refusal to accept fully the authority of the Church. Instead of Latin, he insisted on preaching in

*Crown of Saint Wenceslas*

the local tongue, thus making the words and message of the Gospel more accessible. He advocated a host of doctrines that presaged the Protestant Reformation, including the idea that Jesus, not the pope, was the true head of the Church; that the Communion wafer and wine were merely symbolic; and that encouraging sinners to buy their way to salvation had no scriptural sanction. Liturgical issues were amplified by economic ones: the Church owned half of Bohemia's arable land. According to Hus, such wealth was the dowry of Satan. These teachings brought him into conflict with the archbishop of Prague, who accused him of heresy.

In 1415, when Catholic leaders assembled in the German city of Constance, the fate of Jan Hus was on their agenda. Despite being given a promise of safe passage, the troublesome rector was confined in chains to a jail next to a cesspool. When confronted by his accusers, he refused to recant, prompting the Church delegates to condemn him. The prisoner was stripped of his vestments, shorn of his hair, crowned with a paper hat bearing three images of the devil, and burned at the stake. Not wanting to leave relics, his executioners took care to incinerate every part of his body and all articles of clothing. This scheme to erase memory, however, had precisely the opposite effect.

*The martyrdom of Jan Hus*

Within weeks of the martyr's death, a Hussite movement was up-ending the religious and economic order in Prague. Prominent priests were ousted from their pulpits and replaced by advocates of the new way. Hussite peasants wanted lower rents, while nobles, eyeing the estates of their Catholic neighbors, wished to make "Satan's dowry" their own. Meanwhile, the entrenched Church and its patrons struggled to retain their privileges. For half a decade, the rivalry between the two sides simmered; it boiled over when, in July 1420, Hussite warriors routed the Catholic forces assembled by the Holy Roman Emperor.

The rebel commander, Jan Žižka, was a fierce and inventive fighter who had lost his right eye early in his career but remained, at the age of sixty, a brilliant military strategist. In this campaign he transformed an unlikely array of farmers and peasants into an intimidating force that fashioned weapons out of farm implements, turned wagons into mobile fortresses, and triumphed over heavily armed cavalry. Military victories, especially against high odds, provide a firm foundation for national mythology, and Žižka, despite his eventual death from sepsis, has enjoyed a long career as a Czech hero. He was the standard-bearer who fought back and won against foreign enemies, a leader who chose the sword over acquiescence or martyrdom.*

Žižka's uprising helped set the battle lines that would bedevil Europe for the next two hundred years. His prowess enabled the Czech aristocracy to seize vast landholdings from the Catholics, while also fostering the development of the national language and a populist culture noted for its devotion to universal literacy. "This wicked people," admitted Pope Pius II in the fifteenth century, "has one good quality—it is fond of learning. Even their women have a better knowledge of scripture than Italian bishops."

In the years that followed, the religious rivalry quieted and the Hussite (or Protestant) nobility was content to accept Habsburg rule, based

---

* The Hussite cause was viewed as a threat by Catholics throughout Europe. Among them was the eighteen-year-old Joan of Arc, who, in 1430, wrote a letter addressed to the "heretics of Bohemia." "If I wasn't busy with the English wars," she warned, "I would have come to see you long before now; but if I don't find out that you have reformed yourselves I might leave the English behind and go against you."

*Jan Žižka*

in Vienna and led by German-speaking Catholics. This arrangement was based on the understanding that their religious and property rights would be respected. For a time all was well; but then, in 1618, Protestant leaders submitted a list of grievances to the Habsburg crown, demanding a fuller measure of self-government. The response was dismissive. Angered, the Protestants marched to the castle, where on May 23 they confronted the king's representatives. The interview went poorly and, to vent their dissatisfaction, the intruders propelled two of the royal counselors and a scribe out the window—several stories up. The bureaucrats survived the experience, a miracle attributed by the Catholics to divine intervention and by the Protestants to the victims' landing on a dunghill.

For almost two centuries, Bohemian aristocrats of different faiths had lived and prospered together; now they allowed irritations to fester into anger and violence. The Battle of White Mountain, fought on the foggy forenoon of November 8, 1620, is remembered by Czechs as a day of national infamy. However, the two sides that clashed on that date were separated not by ethnicity but religion. Ferdinand, the new Habsburg emperor, had recruited a coalition of Catholics from Spain, Italy, Bavaria, and Poland. The opposing alliance included Protestant sympathizers from throughout Europe and was led by the young Prince Friedrich of Germany. Because the wealthy of neither faith wished to give rifles to peasants, popular passions were not engaged and most of the soldiers were hired mercenaries.

On the day of battle, the Protestants, though outnumbered, controlled the approach to the mountain, which was actually a hill on the outskirts of Prague. In ninety minutes of hard fighting, more than two thousand men were killed. The Catholics appeared to hold the upper hand, but the Protestants remained in a position to defend the city. At that moment of crisis, they turned for leadership to Friedrich, their chosen prince, only to find that he had fled. Deserted and betrayed, they promptly surrendered, allowing the imperial army to march into the capital.

To Protestant Bohemia, White Mountain felt like the end of history. Defeated nobles were executed or banished, their religion prohibited, and their estates parceled out among the emperor's Spanish and Austrian allies. The Czech people survived but as a nation of peasants without an upper or middle class. For a time, Prague experienced a building boom as Catholic nobles commissioned grandiose projects that contributed much to the capital's architectural glory but deepened the alienation of most Czechs. Their language, replaced by German, was no longer spoken in administrative offices or princely courts. Amid the dazzle of the Age of Royalty, the Bohemian people, if thought of at all, were dismissed as backward and of little account.

IN THEIR STUDIES of Czech history, my father and his colleagues discerned two opposing dimensions: the fighters, such as Žižka, and the scholars. Foremost among the latter was Jan Ámos Komenský, best remembered for his writings while in exile. The bishop of the Hus-inspired Unity of Czech Brethren, Komenský was among those forced to flee in the aftermath of the Battle of White Mountain. He survived by eating nuts and escaped his pursuers by hiding in the trunk of a linden tree.

With no choice but to begin a new life, Komenský soon proved himself to be an educator of astonishing humanity and vision. In keeping with his Bohemian ideals, he stressed universal literacy and access to free schools for girls and boys alike. He pioneered role playing in contrast to rote teaching methods, invented the illustrated children's book, and wrote an essay on language that was reportedly used by Native American students at Harvard. Having seen his laboriously compiled Czech-language dictionary burned by foreign peasants, he advocated the creation of a universal tongue that would help bring humankind together; he did not think civilized people should allow mere language to divide them. In Amsterdam for the last years of his life, he lamented his inability to return to his homeland: "My whole life is merely the visit of a guest." Although religious martyrs and warrior generals have places in my personal pantheon, Komenský is the early thinker whom I most admire.

TO THE EAST of the Czech lands is Slovakia, the home of fellow Slavs whose history is mingled with that of the Bohemians. The two peoples were united under the Great Moravian Empire, which had, in the ninth century, exercised a loose sovereignty over much of Central Europe. The empire's downfall after eighty years stemmed from an invasion by the Magyar alliance, a dynasty that founded the kingdom of Hungary and ruled the Slovaks for most of the next millennium. Despite the political separation, Czechs and Slovaks continued to travel back and forth for purposes of evangelism, commerce, and study.

Slovakia's principal city, Bratislava, sits astride the Danube River, a 250-mile journey from Prague. The mountainous land features gorgeous peaks and dense forests, lakes formed during the Ice Age, and mineral-rich soil. The picturesque scenery has served as a backdrop for thousands of folk songs, native dances, tall tales, and a true-to-life story centered on an eighteenth-century adventurer, Juro Jánošík, who deserted the imperial army and formed a band of thieves. Jánošík's highwaymen made their home in a forest, befriended a local priest, stole only from the wealthy, and shared their booty with the poor. This Slovak Robin Hood's passion for economic justice was a hint of events to come, for Central Europe had reached the threshold of far-reaching social change.

EMPEROR JOSEPH II, who reigned from 1780 to 1790, thought of himself as a modern man, and a good one. He issued grants of food and medicine to the indigent, founded hospitals, asylums, and orphanages, and opened public parks and gardens. By his decree, "No man shall be compelled in future to profess the religion of the state." This "tolerance patent" meant that, after 150 years, Czechs were once again free to practice the Protestant and Christian Orthodox faiths. Joseph also endeavored to integrate Bohemia's Jewish community—at the time the largest in the world—by lifting restrictions on employment, eliminating special taxes, and requiring the use of German in education. Those changes, which greatly accelerated the exposure of Jews to German language and culture, were resisted by some but welcomed by others as a way to expand their participation in society.

In that preindustrial era, the majority of Czechs still pursued a rural life, tilling the soil, tending livestock, sewing clothes, and working as millers, gamekeepers, blacksmiths, cabinetmakers, and shepherds. Most lay down at night in huts decorated with religious icons. Medicinal needs were addressed by collecting herbs or by purchasing the special balms that itinerant peddlers promised would soothe tired muscles and aching teeth. Men generally had mustaches, wore baggy trousers, carried snuffboxes, and smoked pipes; the women in their aprons labored at baking, washing, and food gathering; the children were kept in line with tales about an old crone who stuffed bad girls and boys into her shoulder bag and carried them away. Before the Christmas holidays, entire villages gathered to devour sweets, have poultry-plucking parties, and swap stories about water sprites and ghosts. People believed what they had been brought up to believe: a mixture of scriptural tenets, pagan myth, and good manners. A sleeping child was best roused by a tap on the forehead, causing the soul to awaken first. For reasons both physical and spiritual, bread—the gift of God—was treated with reverence. To step on so much as a crumb was to make the souls in purgatory shed tears. Friends and strangers alike were greeted with a slice from a brown or black loaf smeared with grease and sprinkled with salt.

In such an environment, everyone knew everybody else and each knew his place; the division of the population into social classes was a given. Joseph II enlarged the freedom of his subjects, but his goal was to preserve an empire, not build a democracy. Ever mindful of defense needs, he wanted to create an army that would love its emperor and speak a single tongue. To guard against the incursion of enemies from the north, he built an eight-sided military fortress that he named in honor of his mother, Empress Maria Theresa; the garrison was called Theresienstadt—or, in Czech, Terezín.

# 3

# The Competition

Sir Arthur Conan Doyle's first Sherlock Holmes story—"A Scandal in Bohemia"—begins with a knock on the door of 221B Baker Street. The identity of the mysterious stranger is quickly deduced by the great detective, who recognizes the hereditary king of Bohemia by his German accent. It is a story designed to make Czech nationalists grind their teeth.

But by 1891, the time of Doyle's writing, the cultural balance was already shifting. Any assumption that a gentleman of Bohemia must be a German speaker was increasingly precarious. The Enlightenment, the French and American revolutions, and industrialization had prompted a political awakening across Europe. Workers and peasants began to believe that their lives could be freer and more varied than those of their ancestors, causing the feudal system that had enriched the Austrian and Magyar nobility to break apart. Social activists churned out pamphlets advocating autonomy and equal treatment for Czechs within the Austrian Empire. Slovaks transmitted similar requests to the leaders of Hungary. These reformers were not so bold as to seek national independence but instead petitioned for prerogatives within the empire, such as the right to form political parties, elect representatives to parliament, exercise more control over local government, and operate their own schools.

After many false starts and some bloodshed, the agitation had an impact, albeit an uneven one. In 1867, the court in Vienna recognized its cousin in Budapest as an equal partner, thus giving birth to the Austro-Hungarian Empire. However, a dual monarchy meant that there were two systems of government. In Hungary, all who lived within the borders were considered Hungarian; there were no minorities and hence, for Slovaks, no minority protections. In Austria, the new constitution acknowledged the right of each national group to preserve its language and culture.

The revival of national identification in the Czech lands was spurred by intellectual theories concerning the role of the nation in history and the centrality of language in forging a people. If such ideas had arisen in an earlier era, they would not have spread far; but the nineteenth century was a time of broadening horizons as newspapers and political journals multiplied and books other than the Bible found their way into homes. Especially for people migrating from country to city, the idea of the nation served as a star by which to navigate in a world where the old signposts of religion and social class were losing authority.

*Austria-Hungary, including Czech lands, 1867*

Even though many of the early Czech nationalists wrote in German, they urged the development of Bohemian literature and cheered the inauguration of Czech opera, most notably Bedřich Smetana's *Libuše* and *The Bartered Bride*. They also championed the national theater, the philharmonic orchestra, the Sokol gymnastics organization, an academy of arts and sciences, and, in 1882, the division of Charles University into separate German and Czech branches. They began pondering, as well, what it meant to be Czech.

According to the era's leading journalist, Karel Havlíček, "A Czech does not rely on others [but] . . . sets out to do his work and will overcome everything." Havlíček suggested that the destruction of Bohemian nobility had given to the Czech people a uniquely democratic character: unpretentious, practical, and steeped in humanitarian values. Whereas others were divided between a rich minority and the poor majority, the Czechs were egalitarian, rejecting fancy titles and addressing their countrymen as brothers and sisters. In his view, the people's commitment to decency and fair play was an asset to all of Europe and a welcome departure from the backbiting so characteristic of neighboring nationalities. Of course, Bohemians also admitted to a tendency to drag down anyone who rose too high. "When a Czech owns a goat," so went the saying, "his neighbor does not yearn for a goat of his own; he wants the neighbor's goat to die." There were, in addition, more ominous assessments of the local character. The German historian and Nobel Laureate Theodor Mommsen commented darkly, "The Czech skull is impervious to reason, but is susceptible to blows."

Throughout the nineteenth century, Czech and German nationalists competed with one another, seeming not to notice that in attempting to prove how different their peoples were, they expressed similar aspirations and lay claim to comparable virtues. In unison and with equal vehemence, they demanded that parents raise their young as patriots. Božena Němcová could have spoken for either side when, in "To the Bohemian Women," she urged that:

> *With the first tender, flattering word,*
> *With the first sweet kiss*

*Let's pour the Czech sound into their souls*
*With burning love of country.*
*Czech women, Czech Mothers!*
*We have but one Joy:*
*Raising our children*
*For the glorious, dear country.*

Such bromides were not for everyone. Many inhabitants of the region cared little about national distinctions, which were, in any event, difficult to discern. The original Slav and Teutonic tribes had long since passed into history, and their descendants had shared the same land for centuries, during which intermarriage had been commonplace. Czech and German names were scrambled up, as were physical characteristics, and many people were bilingual. This meant that purity of blood was more often than not an illusion, albeit a seductive one.

Ironically, the burgeoning rivalry between Czechs and Germans was reinforced by the Austrian Empire's commitment to minority rights. To honor that pledge, authorities had to know who belonged to which nationality. This imposed upon a fluid and imprecise social reality one of the most rigid of human inventions—bureaucracy. Agents of the empire arrived in every town and village with forms to be filled out. Citizens were to choose one label or another. The larger the group, the more schools it could have, the more votes it would be entitled to in parliament, and the more local officials its members could elect. Hence a declaration of nationality, once a personal and voluntary option, became both a legal mandate and a political act.

For many families, the choice was based on well-established ethnic and linguistic affinity, but for others the designation was more a question of what was practical. If there were not enough Czechs in a town to merit their own school, it was convenient for a family to be German. If a town were largely Czech, it was prudent for a German shopkeeper to transact business in that tongue. Impoverished parents were tempted by offers of free lunches and school supplies in return for sending their children to the "right" school or athletic club. The mix-and-match nature of the process reflected the fact that many families had relatives on both sides of the divide. My paternal grandfather, Arnošt Körbel, settled in the Czech-speaking heart of the country; some of his siblings did so in German areas. Earlier generations had, most probably, lived neither in Germany nor Bohemia but in what is now part of Poland.

Such complexities only caused activists to become more insistent. To their way of thinking, national identity was not some article of clothing to be shopped for and then put on and taken off; it was the key determinant of who one was. People had an obligation to choose and, having chosen, to conform. A German should vote for German politicians, patronize German stores, eat German food, dress in German clothes, join German clubs, and give one's heart to a German mate. The same catechism applied to Czechs. This elevated national identification to an absurd level. Some partisans claimed qualities for their people that were grossly exaggerated; others focused on magnifying the faults of their neighbors. Still others were angered by families—referred to derisively as "hermaphrodites"—who neglected to choose a side or, even worse, chose the wrong one. According to a 1910 Czech newspaper editorial, "If every Czech person could double their hatred and contempt for the renegades . . . enough people would think twice before Germanizing themselves and their children."

As Czech nationalism took hold, the frustration of being confined within the Austro-Hungarian Empire increased. The Czechs had minority rights, but these did not translate into political and social equality. Whether at the imperial court in Vienna or in the typical Bohemian town, German speakers still held most of the leading positions. The Czechs in 1910 believed themselves to be less free than their ancestors had been in 1610, a sense of grievance that prompted some to look abroad for allies. A number of writers envisioned a future of unity for all the Slavic peoples, from Russians in the east to Bohemians in the west. The fly in this ointment was that the Czech intellectuals who traveled to other Slavic lands did not like what they saw. Neither the Polish nobility nor the czarist courts appealed to populist thinkers, while the idea of a pan-Slav brotherhood seemed far-fetched after hearing Poles describe Russians as Mongols and Russians dismiss Poles as a race of backward peasants. The consensus among Czech nationalists, then, was that their best option was to assert their identity within rather than outside the empire. Perhaps with time and the emergence of the right leader, circumstances would change, and the black-and-yellow Habsburg banners could be replaced by the Czech colors of red and white, with possibly a touch of Slovak blue.

ALL THIS SQUABBLING was no bar to prosperity. By 1900, 80 percent of the empire's industrial production was based in the historic Czech lands of Bohemia, Moravia, and Silesia. The literacy rate was 96 percent, twice that of Hungary and higher, even, than the German. The economy was expanding more rapidly than that of England or France. The Czechs were leaders in rail service, coal mining, iron and steel production, chemicals, paper, textiles, glass, armaments, and industrial machinery. Guided by the motto "In work and knowledge is our salvation," they developed novel techniques for processing ham and fermenting beer, made a popular liquor from beets, invented a convenient way to market sugar (in cubes), introduced the assembly-line production of shoes, and were among the first to install electric rails and trams. Lecturers attracted to Charles University included the Austrian sound wave pioneer Christian Doppler, the shock wave expert Ernst Mach, and a young German professor working on a theory of physics, Albert Einstein. The credit for introducing the safety helmet into workplaces belongs to a bilingual insurance employee from Prague; a writer in his spare time, his name was Franz Kafka.

The improved legal status of Jews did not always mesh well with the intensification of national feelings. People of Jewish background achieved extraordinary success in business, the professions, and the arts, but their position within society resisted easy summary. Slovak Jews were more rural and tended to be conservative; the opposite end of the spectrum could be seen among the emerging intelligentsia in and around Prague. For some Jews, the rising sense of nationalism was translated into Zionism or into a deeper study of ethical and scholastic traditions. For others, it meant a growing association with the movement for Czech rights; but this desire to be part of the Bohemian national movement was not always welcomed.

Siegfried Kapper, a Czech Jew, composed patriotic verse while vigorously asserting his dual heritage. Among his works was an 1846 poem titled, "Do Not Say I Am Not Czech." Karel Havlíček, the journalist, responded by arguing precisely that; it was impossible, he insisted, to be both Semitic and Czech. This theory, widely held, posed an obstacle for Jews seeking to associate with the patriotic feelings of the place in which their families had lived for hundreds of years. Did blood (to the extent it could be determined) define nationality, or was it a mix of geography, language, customs, and personal preferences? An endless argument. Regrettably, even where vitriolic anti-Semitism was rare, the more casual variety was widespread. The brilliant Jan Neruda, often compared to Anton Chekhov, was typical. His fictional Jewish characters consisted almost entirely of greedy moneylenders whose race was castigated as cruel and thirsting for power. Neruda didn't bother to furnish evidence; he simply assumed that his readers would agree. In this atmosphere, many Jews were unsure in which direction to turn. Dr. Theodor Herzl, the father of Zionism, captured the dilemma:

> Poor Jews, where should they stand? Some tried to be Czechs; these were assaulted by the Germans. Others wanted to be Germans, and both the Czechs and the Germans attacked them. What a situation!

ODDLY, THE INDIVIDUAL who would do most for Czech independence—and also much to fight anti-Semitism—was the son of a Roman Catholic Slovak coachman. Tomáš Masaryk was born on the seventh of March 1850; he grew up speaking the local dialects but was instructed by his mother, a Moravian, to count and pray in German. As a youngster he was trained briefly as a locksmith, then a blacksmith. Years later, he recalled the skills demanded of a nineteenth-century boy: how to whistle, run, swim, walk on his hands, ride a horse, climb a tree, catch beetles, kindle a fire, toboggan, walk on stilts, throw snowballs, skip stones, whittle, tie knots of horsehair, use a jackknife, and fight "all kinds of ways," adding, "I can't say what sort of life the girls lived, since we had nothing to do with them."

When young Masaryk was not otherwise occupied, he was studying. A local priest taught him Latin and recommended that the boy be sent to school. While earning his way as a tutor, he ascended the academic ladder. In 1872, he graduated from the University of Vienna; four years later, he earned a PhD in philosophy and moved to Leipzig, where he attended lectures on theology. Having met one challenge through meticulous study, he moved on to the next, borrowing from the library a stack of books about the psychology of women. Thus prepared, he met Charlotte Garrigue, a young American blessed with fine auburn hair, a talent for music, and an independent mind. At first, she responded reticently to his courtship and left to vacation at a spa. Masaryk followed her there in a fourth-class railway car, took her for long walks, and soon won her over. The couple married in March 1878 in Charlotte's hometown of Brooklyn, establishing not only a matrimonial connection but an international one between the people of the Czech lands and the United States. In a sign of respect rare then and since, Masaryk adopted Charlotte's last name as his middle one. They had four children, the youngest a boy named Jan.

T. G. Masaryk began teaching at the university in Prague and quickly developed a reputation as a freethinker. No one could ever claim that he lacked convictions or the spine to defend them. As an academic, he startled senior faculty with his frank lectures on such matters as sex education and prostitution. Upon becoming a member of parliament, he denounced the Austro-Hungarian Empire's occupation of Bosnia-Herzegovina. Along the way, he incurred the enmity of the Catholic Church by lauding Jan Hus and by transferring his allegiance to a fervent but anticlerical Protestantism. Then, as a journalist, he ran headlong into the locomotive of Czech nationalism.

In 1817, two supposedly ancient Czech manuscripts were discovered in Zelená Hora, a town in the Plzeň district in western Bohemia. The documents purported to show that the nation's literature predated that of the Germans and that the old Bohemians had attained a higher standard of education and culture. For decades, Czech propagandists used the writings as a starting point in discussing their people's history; artists, meanwhile, employed them as a source for patriotic works.

In February 1886, an article endorsed by Masaryk offered convincing evidence that the manuscripts were fraudulent. This puncturing of the nationalist balloon was not happily received. Masaryk recalled that, a few days after the article appeared, a local businessman engaged him in a heated conversation:

> He didn't know who I was, and started going on about me, saying I'd been bribed by the Germans to drag the Czech past through the mud and so on. . . . Another time I joined some people on a tram cursing that traitor Masaryk. I found it amusing. What made me angry was to see people defending the manuscripts when they did not believe in them but were afraid to admit it.

As a child, Masaryk had been told by his mother that Jews used Christian blood in their rites. The adult Masaryk had no use for such superstitions, but not all his countrymen felt the same. In 1899, a nineteen-year-old seamstress was found in a forest with her throat slashed and clothes torn. A rumor spread that a ritual murder had taken place. Police had no good suspects and so—egged on by local sentiment—arrested Leopold Hilsner, an itinerant Jew who had been seen around the forest. In a trial that traumatized the country's Jewish population and attracted attention across Europe, Hilsner was convicted on the basis of circumstantial evidence. Masaryk successfully appealed the verdict, prompting a second trial, and wrote pamphlets denouncing bigotry and raising questions about the facts.* The episode gave Masaryk's enemies fresh ammunition; he was accused of accepting payments from Jews and forced by his university to suspend classes until the protests against him died down.

Masaryk was a product of the Victorian Age, but his intellect and sensibility were thoroughly up-to-date. He inquired into almost everything and wrote with insight (if not always accuracy) about suicide,

---

* Hilsner was also convicted in the second trial, but his sentence was commuted from death to life imprisonment. In 1918, he was pardoned by the emperor of Austria and released.

the Soviet Union, Greek philosophy, hypnotism, evolution, the virtues of physical exercise, and the tug-of-war between science and religious faith. He had what might even now be considered advanced ideas about the equality of women and the connections between a clean body and a long life. He was impatient with dogma and had a special contempt for the kind of partial education that caused people to believe that they knew more than they did.

Masaryk's view of nationalism was especially relevant as the twentieth century began. The professor prized patriotism for furnishing an incentive to undertake productive work but stressed that "love of nation does not imply hatred towards another." He insisted that racial purity in the modern world was neither desirable nor possible and that no group should consider itself without fault. He cited pointedly the times when, as a child, he had fought with boys from the next town. "Every Sunday," he said, "we'd come to blows with the Podvorov gang over who would ring the church bells. There you have nationalism in a nutshell."

*Tomáš Masaryk*

Masaryk saw a world in which the settled verities of religious conviction, political order, and economic status were under attack. Modernization was essential but also dangerous because it could leave people without a way to anchor themselves either intellectually or emotionally. The solution, in his eyes, was to embrace religion without the straitjacket of the Church, social revolution without the excesses of Bolshevism, and national pride without bigotry. He believed in democracy and the capacity of people to learn and to grow. His dream was to build a Czech society that could take its place alongside the Western countries he admired.

## 4

# The Linden Tree

My father was five and my mother four when, in June 1914, shots were fired in Sarajevo, mortally wounding Archduke Ferdinand, the heir apparent to the Habsburg throne. The assassination ignited World War I, or the Great War, in which Austria-Hungary, Germany, and Turkey were aligned against the leading Western countries, including czarist Russia and later the United States. The gargantuan conflict brought low three once mighty empires: the Habsburg, Ottoman, and Romanov. In their place would arise a newly combustible European mix featuring the first Communist state, a sullen Germany, a weary England, a wary France, and seven freshly spawned independent states, including the Czechoslovak Republic.

These outcomes had not been foreseen. The less adventuresome Czech nationalists hoped to curry favor with the Austrian Crown by supporting the war; they sought in that way to improve the prospects for autonomy. Tomáš Masaryk led a bolder contingent whose members saw the conflagration as an opportunity to break completely free. In April 1915, he prepared a lengthy memorandum that characterized Austria-Hungary as an "artificial state" and pledged the creation of a "constitutional and democratic Bohemia." In July, on the five hundredth anniversary of the martyrdom of Jan Hus, he identified himself publicly as an opponent of the empire. During the next three years, he traveled to friendly capitals throughout Europe and the United States in support of his nation's independence.

As subjects of Austria-Hungary, Czechs and Slovaks were required to serve in its military. However, many were loath to risk their lives on behalf of a German-speaking coalition against a Russian army consisting of fellow Slavs. This clash between duty and desire was wittily captured in Jaroslav Hašek's stories about the good soldier Švejk, a Bohemian everyman who, when called by his draft board, showed up in a wheelchair. "Death to the enemy!" he shouted while waving two crutches above his head. Inducted nonetheless, Švejk is asked by his lieutenant how it feels to serve in the imperial army. "Humbly report, sir, I'm awfully happy," comes the reply. "It'll be really marvelous when we both fall dead."

Hašek was among the thousands of Czech and Slovak soldiers who switched sides during the war either by deserting or—as in his case—being recruited out of a Russian prisoner of war camp. In 1917, the men were organized into a Czech and Slovak Legion, a ragged but intrepid band that fought bravely and well against the Germans. Matters became more complicated when the Bolshevik Revolution turned Russia upside down, causing the country to withdraw from the war and stranding the legionaries thousands of miles from home. They were left with the choice of surrendering or of trying to escape to the east by dodging warlords, bandits, and hostile Bolsheviks all the way to the Pacific. Masaryk did his best to help by securing a secret promise from the Communist leader, Vladimir Lenin, to grant safe passage. However, the deal soon broke down in a dispute over weapons, and the men had to fight from station to station along the five thousand miles of the Trans-Siberian Railroad.

Upon reaching the coast, the men discovered to their shock that the Western Allies would not allow them to return home. Instead, the tired fighters were ordered to retrace their steps and lead a hastily conceived and poorly coordinated attempt to oust the Bolsheviks. By that time, the weather had turned frigid and the war in Europe had been won. For an additional year, the legion was enmeshed in a multisided conflict over the fate of Russia in which it had no immediate stake. Allies came and went while one Russian side and then another gained the advantage. Finally, with help from the U.S. military, the legionaries were able to depart, but not before many had had to walk the last several hundred miles to Vladivostok.

Due to the timely presence of newspaper correspondents, the legion's exploits had been reported widely in the United States and became a significant diplomatic asset for Masaryk. Arriving for an event in New York City, he was greeted by a giant map on display in front of the main public library, permitting observers to monitor the legion's progress as it fought its way to the Pacific.

In America, the campaign for Czechoslovak independence had been launched by an immigrant-run newspaper in Omaha, Nebraska. "It is up to us living outside Austria to take the first step," wrote the editor of *Osvěta* on August 12, 1914. "Long live the United States of Bohemia, Moravia, Silesia, and Slovakia!" This summons was taken up in Cleveland, Cedar Rapids, Chicago, Philadelphia, and other cities, culminating in Pittsburgh, where Masaryk joined local organizations in a pact demanding immediate freedom. Such demonstrations garnered publicity but had no legal effect. Masaryk's goal was to change the policy of the United States. In this, he was simultaneously encouraged by President Woodrow Wilson's support for the principle of national self-determination and stymied by Washington's desire to separate Austria-Hungary from Germany in order to shorten the war. Throughout 1917 and the early months of 1918, the State Department opposed dismembering the empire in the hope that Vienna would agree to a separate peace. This pragmatic policy was hard to sustain because of its contrast with Wilson's idealistic words. As the negotiations with Austria dragged on, the administration expressed sympathy for the Czechoslovak cause but withheld formal recognition.

In June, Masaryk met Wilson at the White House. As a child, I had been taught to believe that the two presidents warmed to each other immediately, but there is always a risk of friction when two professors are given the chance to compare brains. Wilson admitted to Masaryk that, as a descendant of Scots Presbyterians, he had a tendency to be stubborn. Masaryk found the U.S. president "somewhat touchy." Both wished to talk more than to listen. Masaryk outlined the case for independence; Wilson discussed the Czechoslovak Legion's ongoing battle with the Russian Bolsheviks. Whether or not the two men enjoyed each other's company, the results from Masaryk's perspective were satisfactory. Within days, the State Department declared that "all branches of the Slav race should be completely freed from Austrian rule," and in September, the United States formally recognized Masaryk's National Council as a belligerent in the war. These steps, coupled with Wilson's image as the instigator of a new and more honorable global order, would make the American president a hero throughout Czechoslovakia and add unprecedented luster to his country's international reputation.

To be certain that there would be no backsliding in the final days of the war, Masaryk chose, on October 18, to issue a declaration of independence. The document was released in Washington with a dateline in Paris, where the rebel government was based. The decisive action, however, was on the battlefield, where Allied forces routed what remained of the enemy armies, and in Prague, where Czech politicians invited their Austrian overseers to leave and ushered the new state into existence.

This was the day, October 28, 1918—the Czech equivalent of July 4 or Bastille Day—that those on the scene would not forget. In my garage, among my father's papers, I came across an account written exactly half a century later:

> I was just nine years old. On the preceding night, I was awakened by patriotic songs coming from the lips of a happy group who were on the way to the railroad station of our little town, Kyšperk, to tear down the emblems of the Austro-Hungarian Empire. I watched them from the window with a sense of self-importance, feeling that I was participating in something extraordinarily significant.
>
> The following morning, Mother dressed me in my Sunday suit, gave me a double portion of butter for breakfast, a rarity in wartime, and sent me to school. The whole village, some 2000 people, was in an uproar. They embraced each other, sang and shouted, put up Czech flags and cleaned in front of their houses. In school, the teachers rejoiced and the principal delivered a

speech about the greatness of Czech history, the dissolution of the hated Habsburg monarchy, the victorious struggle for freedom, and the promising future that lay ahead. In the afternoon, we marched to a park to plant a linden—a linden of freedom.

Shortly before Christmas, Masaryk returned from his triumphal diplomatic mission to face new responsibilities in Prague. Riding from the train station to the castle in an open-topped car, he was serenaded by the republic's newly organized army band and showered with cheers; the aging president with the snowy beard and old-fashioned spectacles responded with a jovial wave. After centuries of subjugation, his country had won its freedom; it even had a national anthem, or rather two ("Where Is My Home?" for the Czechs, and for the Slovaks, "Lightning Flashes atop the Tatras"). The dream of independence had come true; the question that nagged was: What next?

THE BORDERS OF Czechoslovakia were delineated at the 1919 Paris Peace Conference but only after a prolonged wrangle. Masaryk and Edvard Beneš, his thirty-five-year-old foreign minister, entered the negotiations with the obvious advantage that Germany and Austria-Hungary had lost the war. This, combined with the saga of the Czechoslovak Legion and Masaryk's personal stature, assured them a fair hearing. According to the British historian Margaret MacMillan:

> Beneš and Masaryk were unfailingly cooperative, reasonable and persuasive as they stressed the Czechs' deep-seated democratic traditions and their aversion to militarism, oligarchy, high finance, indeed all that the old Germany and Austria-Hungary had stood for.

On February 5, 1919, Beneš rose to present his case concerning the country's northern border. He had been preceded by the loquacious delegate from Poland, who had spoken for five hours, beginning, in the words of an American observer, "at eleven o'clock in the morning and in the fourteenth century." Beneš then took the chair, "began a century earlier," and talked an hour longer. As he was at pains to make clear, Europe would be stable only if Czechoslovakia had defensible borders. He found a receptive audience, especially among the French, who wished to create as many constraints as possible on German power. Sadly, many Central Europeans live in what mapmakers find are inconvenient locations. Thus, when the new republic's southern border was drawn along the Danube, three-quarters of a million Hungarians were included in the country's Slovak region. Further to the east, Carpathian Ruthenia was added, contributing half a million Ukrainians. In the north, a bitter compromise was worked out around the coal-rich railway hub of Těšín, leaving Czechoslovakia with less land than it wanted but also jurisdiction over 100,000 unhappy Poles.

In the end, the tireless Beneš achieved most of what he had sought: mountains, forests, and rivers would separate the 54,000 square miles of the new and fragile republic from its neighbors. Still, the Czechoslovak borders would be hard to defend because of the state's tadpolelike profile, running from west to east: Bohemia, Moravia, Slovakia, and Ruthenia. Six hundred miles long, the country was but 150 miles wide in Bohemia and barely half that in Slovakia and Ruthenia. Enemy tanks would have a rugged time penetrating the wooded northern hills; however, if they did, they would then have little trouble slicing the country in two. Worse, the nation was hemmed in by its historic rivals—in the north and west by Germany and in the less protected south by Austria and Hungary. Because of the dispute over Těšín, relations with Poland would also be strained.

The thorniest question was how to incorporate three million ethnic Germans who were concentrated in the Sudetenland (or Southland)* region of Bohemia and Moravia, thoroughly mixed with the Czechs— this out of a total population of roughly thirteen million. German attempts to declare the region independent or part of Austria received no international support and were promptly and in one instance brutally

---

* The Sudetenland was to the south in relation to Germany; it encompassed parts of the northern, western, and southern border regions of Czechoslovakia.

*Czechoslovak Republic, 1919–1938*

suppressed by the Czechoslovak army. The leaders of the Sudeten population responded by refusing to participate in drafting a constitution or forming a parliament. Masaryk did not help matters when, in his inaugural address, he characterized "our Germans" as people "who originally entered the country as immigrants and colonists." Surely it was wrong to assign second-class status to a people that had lived on the land for centuries. Stung by criticism, the new president promised respect for the rights of all who demonstrated loyalty to the state. He considered it vital that the minorities participate in building a united and prosperous country. This vision was embodied in the 1920 constitution, which guaranteed women's suffrage, freedom of assembly and speech, and the equality of citizens before the law.* Still, Masaryk was realistic, telling advisers that to build a true democracy would require fifty years of undisturbed peace. All he could hope, in what remained of his life, was to make a good beginning.

And he did. Much about that first decade was promising. Political parties were formed in which Czechs, Slovaks, and Germans were all able to participate. Ethnic Germans were included as cabinet ministers, making Czechoslovakia the only country in Europe where a minority

---

* Despite American democracy's 130-year head start, the Czechoslovak Republic enshrined women's suffrage six months before the United States reached that milestone.

was so represented. Elections were held at every level, with the balloting open and fair. Because the press was independent, citizens could vent their opinions without fear. There were no political prisons, torture, or officially sponsored disappearances. Legislative power was exercised by parliament and guided by the Committee of Five, an informal body consisting of the heads of the leading parties. There were so many factions that none could dominate; this encouraged moderation. Communists, fascists, and separatists enjoyed legal status but operated at the margins of public life.

The new regime began by translating into policy the egalitarian spirit of Czech tradition. Under the old empire, three wealthy families had owned as much property as the 600,000 poorest. In the republic, German and Hungarian aristocrats were stripped of their titles, imperial estates were carved up, and the size of property holdings limited. Nationalized lands were sold to independent farmers at a nominal price.

Meanwhile, in urban areas, workers benefited from the introduction of modern social legislation, including an eight-hour day, disability payments, health insurance, and retirement pensions. Masaryk, the old professor, emphasized education from first grade to university, especially in regions, such as Slovakia, that had previously been underserved. Economically, Czechoslovakia was a success. The currency was stable, budgets were balanced, and exports of textiles and glassware flourished. The innovative spirit that had emerged in the nineteenth century continued to blossom. By 1930, the country ranked tenth among the world's industrialized powers. Well-known Czechoslovak brands included Škoda automobiles, Pilsener beer, Prague ham, and Bat'a shoes—whose corporate headquarters in Zlin was 250 feet high and featured an air-conditioned elevator complete with washbasin.

The Czechoslovaks were also respected in world affairs, having earned a reputation for supporting disarmament, international law, and peace. It was said of Masaryk that, if there were such a post, he could easily have been elected president of Europe. With his erect bearing, handsome features, and silver beard, he certainly looked the part. His energy astonished as well—he played tennis, rode horseback, and swam long distances. One afternoon, the president invited the swash-buckling actor Douglas Fairbanks to join him for tea in the garden outside his country home. Masaryk challenged Fairbanks to demonstrate his athleticism. Fairbanks briefly surveyed the scene before rising from his chair, hoisting his cup of tea, and leaping over the table without spilling a drop. "Very good," said Masaryk. "That's not something I can do with a cup of tea—but as for jumping the table, just watch." And at seventy-seven years old, he was as good as his word.

IT WAS IN that heady environment of national optimism and pride that my parents came of age.

My father, Josef Körbel, was the youngest of three children. He had been born on September 20, 1909, in the farming community of Kyšperk (now Letohrad) about ninety miles east of Prague. Early in 1997, after learning of our family's Jewish heritage, my sister, my brother, and his wife visited the town. In August, joined by my sister and daughters, I retraced their steps. We were shown the trim row house across a maple-lined street from the train station where my father had grown up. The mayor and some of the older residents helped to fill us in on the history. The Kyšperk of 1909 had been a village of two to three thousand, predominantly Czech but with a smattering of German speakers. Although merchants had previously advertised in both languages, the trend at that time was to favor only Czech. My grandfather, Arnošt Körbel, operated a modest building supply shop out of the first floor of his house. Among the firm's customers was the local match factory, which Arnošt had helped found and which employed many in the village. Typical for a Körbel man, he was of average height, with a handsome roundish face and cleft chin. He was personable and viewed by the community as thoughtful and kind. In 1928, he and his wife, Olga, moved closer to Prague, where he was a manager in a firm that built some of the city's most ambitious construction projects, including the cross-Vltava Jirásek Bridge.

In Kyšperk, which had no synagogue, the Körbels did not attend religious services; they did participate in Saint Nicholas Day, Easter celebrations, and other community-wide festivities. Such disregard for cultural boundaries was typical of many nonobservant Jews. For them, as for many Czech Christians, these occasions—with their songs, pageants, decorations, and special foods—had more social than religious significance. Easter was as much a rite of spring as a testament to resurrection, and Christmas trees were not only for Christians. When I was young, my father often told the story of how he and his brother had fought over and broken a washbowl one Easter morning; the message of the story was entirely secular: if I misbehaved, then I—like them— would be punished.

Kyšperk was not large enough to have its own secondary school, so at the age of twelve, my father began attending class in the nearby town of Kostelec nad Orlicí. He was an excellent student who took a prominent role in school plays, chafed at boring teachers, and found himself in trouble for shooting the hat off a stranger's head with an air gun. According to a letter he would later write to members of his high school class, he loved to stride along the trails that snaked their way through the foothills of the Orlické Mountains. He also began spending time in the town square, which had broad sidewalks lined with roses of all colors and dense forests of red and white carnations.

One of the local boys' favorite games was to spy on the village policeman, a man so fat that his "trousers hung like an accordion." The moment he looked away, the youths would leap from the sidewalk to a flower bed, plucking a rose or carnation to give to a girlfriend. In my father's case, the flowers he liked best were near the town's leading wholesale food store. There he was able to catch sight of Anna Spiegelová, the young woman who would become my mother.* The Spiegel family ran a company that sold flour, barley, spices, jellies, and other foodstuffs to shops throughout the region. Anna's parents, Růžena and Alfred, were proud of their products, especially a sweet homemade liquor called Asko and freshly roasted coffee beans that Růžena insisted were the most flavorsome in the land. In later years, after a large meal, my father would say that the reason we had so much food was that my mother had come from a wholesale family.

---

* In Czech, the suffix "-ová" is added to the names of women and girls.

Interviewed in the 1970s, my mother recalled:*

I had a very nice childhood. In the spring and summer my older sister and I would go out to the forest and pick mushrooms, blueberries, and wild strawberries. On a rainy day we went to see a silent movie if our teachers let us. In the winter we went skiing and sleigh riding or, when older, cross-country skiing. I enjoyed reading the books my sister had read but did not care so much for having to wear her hand-me-down dresses.

Between my parents, it may not have been love at first sight, but pretty close. My father, who was never shy, simply walked up to Anna and introduced himself: "Good afternoon, I am Josef Körbel and you are the most talkative girl in Bohemia," whereupon she slapped him. Anna's nickname was Andula, but from her time in high school she was known as Mandula, a contraction of "my Andula," an endearment supplied by my father. She called him Jožka and said yes when, in 1928, he proposed. He was nineteen, she a year younger. Her parents counseled patience and shipped her off to a school in Switzerland to learn French, secretarial skills, and other social necessities.

If they thought that distance would put an end to romance, they were wrong. My mother, writing after my father's death, recalled, "Joe was certainly a man worthwhile waiting for for seven years." She then added—and crossed out—"but I was not always so passioned. Couple times I was thinking of leaving this." (Even after decades in America, my mother's English remained heavily accented and governed by her own rules of grammar.) She continued: "Very often I was wondering what was I admiring most in his personality. Was it his perseverance which he probably inherited from his father . . . [or] did I love him because of his good heart, gentleness, unselfishness and loyalty to his family, which he inherited from his lovely mother?"

Mandula Spiegelová, later Körbelová, was pretty and petite; she

---

* The interview was conducted by my daughter Katie Albright for an elementary school project; she received an A.

wore her brown hair short, flapper style, and had hazel-green eyes and dimpled cheeks. My father referred to her in a letter as "a person of rather unruly inclinations," by which he meant that she had a mind of her own and was not afraid to speak it. As for my father, he had a strong, serious face and wavy hair; my mother said he grew more handsome as the years went by. What they shared from the beginning was an exuberant desire to explore the possibilities of life. In my father's case, that meant completing his education as rapidly as he could with an eye to becoming a journalist or, in the manner of Masaryk and Beneš, a diplomat. To acquire the necessary language skills, he studied German and French and later spent a year in Paris. At the age of twenty-three, he received a doctorate of jurisprudence from Charles University. He worked briefly for one law firm, then another, on either side of a period of obligatory military service. In November 1934, he achieved his goal of joining the Czechoslovak Ministry of Foreign Affairs. The time had finally come to take Mandula up on her promise.

The wedding, a civil one, was held at the Old Town Hall on April 20, 1935. As at any Czech wedding, there would have been a lot of singing, led no doubt by the groom, who had a wonderful tenor voice and knew all the traditional songs. On the marriage certificate, my parents were identified as *bez vyznání*: without religious confession.

*Wedding, Josef and Mandula Körbel*

My mother, typical of women in that era, lacked a university degree. However, she fully supported my father's professional ambitions and was pleased to accompany him from the countryside to the sophisticated capital of Prague. My father's recollections show their happiness:

> As other European countries went through political and social upheavals, unstable finances, and one by one succumbed to fascism, Czechoslovakia was a fortress of peace, democracy, and progress. We . . . gulped the elixir of liberty. We read avidly national and foreign literature and newspapers, attended every opening night in the National Theatre and Opera; and wouldn't miss a single concert of the Prague Philharmonic Orchestra.

In that first year together, young and unburdened by children, Josef and Mandula lived in an apartment done up in art deco style, all black and white. Eagerly, they took their place in café society, patronizing the restaurants and strolling through the parks and squares. Several times a week, women of my mother's generation went down to the Old Town market, which bustled with peddlers selling meat, vegetables, sweets, baked goods, and fruit. Large canvas umbrellas in various hues provided protection from the sun and rain. Purchases were wrapped in newspapers and collected in large net bags. Especially on Saturday, the market was scented with the mingled fragrances of flowers, fruit, and fowl. Street musicians, almost invariably men, competed for attention and tips. On special occasions, they were joined by dancers in the Czech national dress, their torsos twisting and skirts flowing as their feet pounded the pavement.

Old Town had always been filled with markets. New Town, so called because it was not settled until the fourteenth century, was more residential and greener, with ample space for forested groves and parks. Prague's past is written in the statues, synagogue walls, and church spires that are visible from everywhere except its darkest and narrowest streets, but in my parents' time the preoccupations were distinctly modern. As I learned when doing research for my PhD on the role of the press in Czechoslovakia, Prague at the time had 925,000 residents and no fewer than ten important newspapers, most owned by political parties. City shops also sold the leading journals from throughout Europe—French, English, and Russian as well as German and Czech—giving cafés the appearance of reading rooms. Public policy was a subject of constant discussion. My father was among the officers of Přítomnost (The Present), a civic affairs and debating society centered in Prague that attracted ambitious young professionals working in government, journalism, and academia. It was through this club that he came to know Prokop Drtina, its president and a man whose career and life would intersect with my family's throughout the decade to come.

The cultural and political dynamism of the capital owed much to the intellectual energy of a man born in an earlier era whose impending departure few were ready to accept. Since the republic's founding, residents had grown accustomed to the sight of T. G. Masaryk riding on a stallion through the streets, nothing separating him from the crowds. In the winter of 1936, my father met the patriarch for the only time when the Foreign Ministry asked him to accompany a group of Yugoslav scholars who had sought an audience. To my father, it was like meeting George Washington: "There he was, Tomáš Garrigue Masaryk, 86 years old, tall and slim in a dark suit, in a simply furnished room, surrounded by his library, a bouquet of roses over the mantelpiece." On his desk two books stood like opposing duelists: *Faust,* by Johann Wolfgang von Goethe, and Adolf Hitler's *Mein Kampf.*

WHEN MY FATHER joined its ranks, the Czechoslovak Foreign Ministry was not the sprawling bureaucracy one might imagine. It consisted instead of a small cadre of officials who served as ambassadors to foreign countries and about a hundred more employees providing staff support. The budget was far from lavish; several months elapsed before my father was even paid. The unchallenged leader and chief strategist of the operation was Edvard Beneš, Masaryk's closest adviser and foreign minister since the republic's founding.

Born in 1884, Beneš was the tenth and youngest child of a peasant family with roots in the northwest corner of Bohemia. Animated from the start by nationalism, the precocious lad wrote an ode to Hus while still in knee pants and frequently traded punches with German children. He was also a systematic and ambitious thinker. An altar boy at ten, he was agnostic by twelve. The following year, he smoked his last cigarette, and a year later swore off liquor. At sixteen, he was "captivated by radicalism and socialism and was celebrating the first of May with a red flower in his buttonhole." By eighteen, he had moved on from radical ideology to seeking truth through science. At twenty-one, he vowed to prepare for a career in politics and, to that end, enrolled in three universities simultaneously. In Paris to study, he met and kept company with Anna Vlčková, the daughter of a Czech railway man. Through a friend, he learned that the young woman had fallen for him, a frightening prospect; also impractical. The next day, he invited her for a walk and explained that because further romance would be a hindrance to his career, they should go their separate ways. She agreed. He went to London, she to Prague, but love eventually won out. Four years later, Edvard and Anna (now Hana)* embarked on a marriage that, although frequently surrounded by turmoil, would remain solid for the rest of their lives.

While at university, Beneš came to the attention of Charlotte Masaryk, who prevailed upon her husband to employ the impecunious student as a German-Czech translator. The young man attended Masaryk's lectures and soon became a disciple. When World War I broke out, the two agreed to work together. Although not imposing physically, Beneš had been a gifted schoolboy soccer player and was no coward. Throughout the war, he traveled by train from one European capital to the next, transmitting messages to agents of the Czech underground and carrying code books in double-bottomed suitcases. This was no game. If apprehended, he would have been hanged or shot. Ironically, Beneš was arrested three times by the British and twice by the French on suspicion of spying—not for the Czechs but for Austria.

Among fellow diplomats, Masaryk's protégé was known for his

---

* Beneš's first romance, also with a girl named Anna, had ended badly, so he asked Miss Vlčeková to change her name to Hana, which she did.

intelligence, strategic vision, absence of humor, and enthusiasm for debating even minor issues. He also possessed an aptitude for organization and was incorruptible. At the Paris Peace Conference, he was approached by a longtime friend who proposed creating a special fund from which the foreign minister might make discretionary withdrawals. Such a practice, though of questionable ethics, would hardly have been unusual. Beneš was careful to obtain his friend's room number and promptly had him arrested.

In designing his country's foreign policy, Beneš began by accepting the inescapable: because Czechoslovakia was small and therefore dependent for help on others, it could thrive only within a climate of regional peace. Thus he built a network of alliances beginning with the Little Entente, a partnership with Yugoslavia and Romania to provide a barrier against Hungary. For more powerful friends, he turned to the West, completing in 1925 a mutual defense treaty with France. Ten years later he balanced this with a similar but more limited agreement with the Soviet Union. Under the arrangement, Czechoslovakia and the USSR would be obliged to assist each other in the event of an attack only if France were already fighting on the same side. That sounds complicated, but it made sense to Beneš, who did not want his country dragged into a war between Germany and Russia.

The 1935 treaty with Moscow would turn out to be less useful than Beneš hoped, but he considered its negotiation a highlight of his career, in part because of the warm welcome he received in the Soviet capital. At the train station, there had been a red carpet and many more flags than had been present for earlier visits by dignitaries from Great Britain and France. Beneš enjoyed the full VIP tour of Russian treasures, including the opera, the new subway system, and the Lenin Mausoleum. At the farewell reception, he had to fend off numerous attempts to make him drink, a show of discipline not matched by Kliment Voroshilov, the Kremlin's commissar for defense. Voroshilov assured Beneš that his country, in the event of a German attack, would fight back, or, as he phrased it, "rip the enemy apart." He also promised that the Soviet Union would not leave the Czechoslovaks to fight alone. This pledge prompted the question "But how will you do that? After all, our countries are not neighbors. Would you really cross the territory of other nations in order to come to our aid?" "Of course," replied Voroshilov. "We take it for granted."

WHEN, AT THE age of nine, my father had joined his fourth-grade class in planting a linden tree in honor of the new Czechoslovakia, his teacher had predicted that the tree would grow tall and strong— "able to withstand high winds." As the 1930s began, the gales buffeting the new republic promised to put that forecast to the test. The Great Depression tossed hundreds of thousands of workers onto the unemployment rolls; by 1933, one in six was without a job. Export-dependent industries in the predominantly German Sudetenland, especially the textile sector, were badly damaged. Suddenly it seemed that skills and discipline were not enough; a strong work ethic mattered little if there was no employment to be had. Such economic frustrations have a way of fueling unrest. In Czechoslovakia's neighborhood, this posed a particular danger.

On January 30, 1933, Adolf Hitler ascended the steps of the Presidential Palace in Berlin to receive from President Paul von Hindenburg a formal invitation to become chancellor of Germany. Rarely has a transfer of power been more plainly a handoff from one generation to the next. The brittle-boned Hindenburg had begun his military career an epoch earlier—during the Austro-Prussian War in 1866. He was the standard-bearer of German military tradition and only a few months previously had contemptuously dismissed the forty-three-year-old Hitler. "That man for chancellor?" he fumed. "I'll make him a postmaster and he can lick stamps with my head on them."

The Austrian-born führer had prevailed nonetheless, and from the moment he took office, Germany's military rise and moral descent became the central story in Europe. With startling swiftness, he transformed his country from a tottering democracy into a tightly organized dictatorship with a skyrocketing military budget and an aggressive international agenda. During the peace conference at the end of the Great War, the dominant image in world affairs had been that of old men in starched collars holding polite conversations in ornate rooms.

The new image was that of a brown-shirted mob breaking bank and shop windows while screaming, "Judah perish!" To foreign observers, Hitler seemed an ill-bred nuisance, spawned by the Depression, giving shrill voice to resentments flowing from the Treaty of Versailles. That pact had included steep financial penalties that could not be paid and were never collected. The defaults irritated the Allies; the penalties reminded the Germans of their humiliation.

The French had insisted after the war that German power be checked at the Rhine, thinking that the river would become a moat behind which they could mount their fortifications. The area within Germany that was west of the river (known as the Rhineland) was to remain demilitarized and under international supervision. In March 1936, Hitler ordered his army to reoccupy the region. The French were powerful enough at the time to drive the Germans back and had a legal right, under the treaty, to do so. Instead they consulted with the British, who deferred to them; resolute statements were issued, and nothing happened. The Rhineland, it seemed, was a small amount to pay for peace. The Nazis, however, soon began to fortify their new front line—closer to France and further from Berlin. A German invasion of France, unthinkable until then, could now be thought of. The price of peace would grow.

## 5

## A Favorable Impression

Tomáš Masaryk's body was wearing down. In May 1934, he suffered a stroke that impaired his eyesight. In December 1935, he resigned the presidency and was succeeded by Beneš. In September 1937, at the age of eighty-seven, he died.

Mourners came from across the country and throughout Europe. For hours they shuffled past the casket on display outside the castle entrance. On the day of the funeral, the cortege retraced the route that had brought the new president from the train station nineteen years earlier. Twenty-five thousand veterans of the Czechoslovak Legion were among the marchers. A million people formed jagged lines along the streets, standing on chairs and perched on shoulders, each straining to glimpse the flag-covered coffin; eventually it passed, flanked by soldiers from each of the country's major nationalities: a Czech, a Slovak, a German, a Hungarian, a Ruthenian, a Pole. Peter Demetz, aged fifteen, recalled, "You heard only the muffled sound of the horses' hooves, the clink of the wheels and weapons, the infantry boots on cobbled streets, and quiet sobbing." In Beneš's farewell tribute, he called Masaryk "the awakener" and urged his countrymen to put aside their conflicts in order to build a democracy in which all citizens could find their rightful place. The great leader's body was transported to a small cemetery in the village of Lány; there it was laid to rest in a leafy plot near the family's country home.

Interviewed shortly before his death, Masaryk marveled that, in all his time as president, he had never found it necessary to sacrifice his principles. He said that he had been guided in office by the same beliefs that had steered him as a student, teacher, and apprentice politician. Further, he felt that his faith in democracy had been validated. "My satisfaction," he explained, comes "from having seen the . . . ideals I professed prove themselves and stand firm through trial after trial." One can but wonder if Masaryk would have retained his sense of satisfaction had he been granted a single more year of life.

FOUR MONTHS BEFORE Masaryk died, I was born on a warm spring day in Prague. The date was May 15, 1937; I was named Marie Jana out of respect for my mother's sister, but that appellation didn't stick. Grandmother Růžena dubbed me "Madla" after a character in a popular stage drama, *Madla from the Brick Factory*. My mother, with her unique pronunciation, modified the name to "Madlen." From there it was a short hop to "Madlenka," which is what I was called growing up. For several weeks, I was agreeably fussed over in Prague, then went on my initial foreign trip—to Yugoslavia, where my father had been appointed press attaché to the Czechoslovak legation. I spent most of my first year in Belgrade.

*The author flanked by grandmothers Růžena Spiegelová (left) and Olga Körbelová*

My parents were enthralled by the Yugoslav capital and worked dili-
gently to add Serbo-Croatian to the list of languages in which they were
fluent. Although the black clouds hanging over Europe could not be
ignored, it was human nature to hope that the worst might be avoided.
The memory of the Great War was still fresh—surely the world's leaders could prevent a repetition? The Czechoslovaks had placed their faith
in the League of Nations, an alliance with France, a partnership with
Soviet Russia, and measures to accommodate the demands of their res-
tive German minority. Yes, Hitler was appalling, and yes, his swagger
was having an unnerving effect throughout Europe, but for my parents
and for their generation, the full reality of Nazism was, in the local
expression, "still behind the mountains."

As a representative of Czechoslovak democracy, my father was
intrigued by Yugoslavia's democratic opposition, which was at odds
with the country's conservative monarchy. He met with his freedom-
loving friends frequently and sometimes secretly, while also arrang-
ing events that publicized Czechoslovak history and culture. He soon
learned that democratic enthusiasm and professional diplomacy could
make for an awkward fit. In April 1937, he was preparing for a visit to
Belgrade by President Beneš when a group of prodemocracy students
came to the legation. "We love Beneš," they exclaimed. "Please tell us
when he is coming so that we can put him on our shoulders and carry
him through the streets." My father replied as a diplomat is trained
to do, cordially and at considerable length, saying nothing. Word of
the visitor's itinerary still escaped, and the students, possessing more
courage than common sense, rushed Beneš's car. Fortunately or not,
they were pushed back by a cordon of police. At the insistence of the
Yugoslav government, all public events were removed from the president's schedule.

A few months later, the French foreign minister came to Belgrade.
Again, demonstrators took to the streets to express their prodemocracy
yearnings. My father, startled by the noise outside his office, ventured
to the balcony for a look. Below him were about five hundred people
chanting and holding signs that condemned Yugoslavia's reactionary regime. He hesitated, wanting neither to discourage the idealistic

gathering by turning away nor to signal support for the downfall of the government to which he was accredited. Pondering these alternatives, he stood there unmoving for several minutes, until finally the police rolled up and dispersed the crowd. My father's restraint was not sufficient, however, to overcome the hostility of Yugoslavia's ruling elite. In 1938, he was accused by the foreign minister of writing articles for the Czechoslovak press regarding internal developments in Belgrade and of having Bolshevik sympathies, neither of which was true. With support from his own government, he continued doing his job the way he thought it should be done.

To my parents, the social aspects of diplomatic life came naturally. My mother was not the only one who was talkative. My father chatted with people on all sides of every issue, and learned firsthand about the ethnic rivalries that would resurface so tragically in the Balkans when I was secretary of state. Among the friendships my parents forged were with Vladimir Ribnikar, a Serb newspaper publisher, and his wife, Jara, a Czech. The couple had small children, with whom I was now old enough to play such exciting games as standing up, then falling down. Our families met weekly for dinner, and my father and Mr. Ribnikar exchanged phone calls daily. The Ribnikars were among the last people to whom we said good-bye upon leaving Belgrade—and the first we would try to greet, under harrowing conditions, upon our return.

TO BUILD AN empire, Germany needed an industrial base that extended well beyond its traditional borders. Even before seizing power, the führer confided to his advisers:

> We shall never be able to make grand politics without a steelhard power center of 80 or 100 million Germans living in an enclosed area! My first duty therefore will be to create this center which will not only make us invincible but ensure for us once and for all the decisive ascendancy over European nations. . . . In these areas there is today a large majority of alien tribes, and if we want to put our Great Power on a permanent basis, it will be our duty to remove them. . . . The Bohemian-Moravian Basin . . . will

be colonized by German farmers. The Czechs will be ... expelled from Central Europe.

Small countries can survive hostile neighbors, but the odds lengthen when a significant national minority identifies with the enemy. This happened in Czechoslovakia not as an inevitable consequence of ethnic diversity but because of a tragic convergence of events: Hitler's rise, the Depression, Tomáš Masaryk's declining health, and the failure of governments in and outside Central Europe to comprehend the scope of the danger they faced.

During the republic's first decade, the majority of its German population was reconciled to life within the state. Nationalist feelings, though present, were expressed peacefully. In the elections of 1920, 1925, and 1929, parties advocating separatism received at most 26 percent of the German vote. International conditions were also favorable, as relations between Prague and the Weimar Republic were cordial. A leader of the Czechoslovak German Agrarian party declared in mid-decade:

> We have lived with the Czech[s] for a thousand years, and through economic, social, cultural and even racial ties, we are so closely connected with them that we really form one people. To use a homely metaphor: we form different strands in the same carpet.

Clearly, coexistence was possible, but Hitler quickly changed the psychology on both sides of the border. The Germany that had been demoralized after World War I was now a resurgent land with a grossly inflated view of its rights and few internal constraints on what it would do to secure them. The renewed promise of a powerful Germany fed the desire of Sudeten nationalists to recapture their past primacy. The Nazi party was banned in Czechoslovakia, but members of the Sudeten German Heimat Front, a party founded by Konrad Henlein in 1933, were Nazis by another name. They professed loyalty to Prague, but their ambitions were linked to those of the Third Reich. The interaction of economic insecurity and Teutonic solidarity inflamed political attitudes. Henlein's party outpolled all others in the 1935 parliamentary election—a shocking upset.

Henlein was an intelligent, nearsighted, somewhat paunchy former gymnastics instructor with a receding hairline. There was little alarming about his demeanor, but to counter fears of extremism, he denied any affiliation with Hitler or animus toward Jews. He also disclaimed interest in pushing Czechoslovakia's foreign policy in a pro-German direction. His only concern was to protect the rights of his people, which were, he alleged, being routinely abused by the chauvinistic government in Prague.

Henlein was motivated less by Nazi ideology than by the lure of power and fame. His skill as a politician stemmed from his gift for lying with apparent sincerity. Few, having heard him describe the racial superiority of the German race, suspected that his own mother was Czech. He did not fool the authorities in Prague, but that didn't matter as the men he sought most to influence spoke neither German nor Czech but English. Late in 1935, the first secretary to the British legation sent a dispatch regarding him to London: "To judge by his personality, as well as by his speeches, he seems to be a moderate and man of his word." A month later, the envoy added, "One wonders why Dr. Beneš makes no attempt to take advantage of the moderation shown by Henlein before it is too late."

The Sudeten champion made several trips to London, where he was introduced by sympathetic Britons to well-placed friends. In those encounters, he voiced his bewilderment at the refusal of Czechoslovak officials to see reason. He was, he insisted, fully committed to the republic and warned that if his modest requests were rejected, a figure more radical than he would surely emerge. In July 1936, one of the most seasoned English diplomats, Sir Robert Vansittart, took his measure. "He makes a most favorable impression," the nobleman reported. "I should say he was . . . honest and clear-sighted. . . . [He] said that he had always been the advocate and leader of the effort for reconciliation with the government."

Henlein received a less rapturous welcome when, accompanied by a hulking Gestapo agent, he paid a call on the Czechoslovak ambassador to London, Jan Masaryk. The younger Masaryk lacked his father's self-discipline but compensated with an irreverent sense of humor and a unique personal style. On this occasion, he ushered the Sudeten leader into his study and was surprised when the bodyguard came too. "He goes wherever I do," Henlein explained. Masaryk nodded agreeably and arranged four chairs around his table. "You won't mind, then," asked Jan, "if I am joined by my assistant, who goes wherever I go?" The ambassador opened the door and gave a whistle. In bounded Gillie, his pet Aberdeen terrier, who leaped into place on the appropriate chair. Notwithstanding this incident, Henlein's personal diplomacy had a telling effect. Repeatedly, he accused Beneš of playing a dangerous game by failing to satisfy Hitler. If the British wanted to stop worrying about Central Europe, he argued, they would have to persuade the Czechs to back down. As the months passed, he was able to convince even experienced British opinion makers that peace hinged entirely on whether this demand was met.

In truth, Sudeten German grievances, though legitimate to a degree, were hardly of a magnitude to warrant an international crisis. Yes, Sudetens were underrepresented in such institutions as the postal service and military. Out-of-work Germans resented the awarding of government contracts in their region to firms that employed Czech workers. Because the area was heavily industrialized, the Depression hit harder and its effects lasted longer than elsewhere in the country. Still, Sudeten families had easy access to German-language schools, an equitable ratio of teachers to students, and a fair share of social services. They had politicians to speak for them, newspapers and magazines to argue on their behalf, and more liberty to express dissenting opinions than did their ethnic brethren in Hamburg, Frankfurt, and Berlin.

At the center of many an argument were statistics. Germans cited data from the 1910 census, when the region had been ruled by the Austro-Hungarian Empire and it was often more beneficial to be German than Czech. The Czechoslovak government relied on the 1930 survey, which tilted the other way. Sudetens complained that the percentage of schools that were German had declined from 43 percent to 22 percent. The government pointed out that Germans made up only 21 percent of the population. However, the federal authorities could fairly be criticized for worrying more about the rights of Czechs living in German regions than of Germans living in Czech areas.

A more basic problem was that the country's system of ethnically distinct schools and social organizations was an obstacle to building a unified Czechoslovakia. In the 1920s, there was a sharp divide among Sudeten Germans over whether to teach their young to accept or reject integration in the state. The rise of the fascists pushed schools rapidly in the direction of separation despite the opposition of more liberal elements, including Jews. While the embattled moderates continued to defend a traditional approach to education, Henlein and his followers looked to the pedagogical model on display next door.

In the Third Reich, children were taught to believe that they were part of something far greater than just another country. They were members of an exceptional race, one hundred million strong, chosen by God and led by a führer who was God's prophet. From an early age, boys and girls were trained as warriors, conditioned to hate Jews, and led to feel contemptuous toward Slavs, who were dismissed as dirty and slow-witted. Thomas Mann was one of many German intellectuals to flee Germany and to live for a time in Prague.* In May 1938, he described the Nazi educational program as

> an inexorable first draft of what the German of the future is to be. . . . The result is that education is never for its own sake; its content is never confined to training, culture, knowledge, [and] the furtherance of human advancement through instruction.

---

* The Czechoslovak Republic did not require visas from travelers who had German passports. Refugees were registered in Prague, granted residence permits, and given stateless passports. No other country in Europe accorded such generous treatment to refugees.

Instead, it has sole reference . . . to the idea of national preeminence and warlike preparedness.

Mann observed that "the glory of the German nation has always lain in a freedom which is the opposite of patriotic narrow-mindedness, and in a special and objective relation to mind. 'Patriotism corrupts history.' It was Goethe who said that." It was Hitler who said that "education must have the sole object of stamping the conviction into the child that his own people and his own race are superior to all others."

In Czechoslovakia, fascist doctrines were bravely resisted by profederalist Germans, many of whom had close ties to the Catholic Church. Like their neighbors, their families had lived in the Sudeten region for generations. Their leaders had advocated on behalf of provincial interests within the republic and had little sympathy for Czech nationalism, but their sense of what it meant to belong to their race did not coincide with Hitler's. They judged the warmth and humanism that characterized German Romanticism to be a sign of strength, not of sentimentality or softness. Throughout the confrontation between fascism and democracy, a group of Sudeten parliamentarians stood their ground, declaring support for freedom and the rule of law. In so doing, they underwent the challenge faced by moderates in any political maelstrom, which is to make their voices heard amid the roar emanating from the extremes. To Sudeten fascists, the moderates were traitors; to Czech nationalists, German moderates were still German.

As the symptoms of crisis began to manifest themselves in open discrimination and localized violence, the obvious, albeit painful, option for German democrats and especially for Jews was to seek refuge elsewhere within Czechoslovakia. Many fled to Prague, where they joined like-minded people who had emigrated earlier from Germany proper. For a time, the city was the European capital of humanist discourse. Among the topics most avidly discussed was the question of identity. According to the laws of the republic, Jews had the right, but not an obligation, to declare Jewish nationality. Roughly one half did, while the remainder identified themselves as Czechoslovak, German, Hungarian, Polish, or other. Although the Jewish population made up less than 3 percent of the country, it accounted for more than a third of capital investment and 10 percent of students at university. It was hardly a monolithic group; the rate of marriage outside the faith was the highest in Central Europe, and there were constant debates about worship obligations, ethics, language, social customs, dietary restrictions, and politics. With Hitler next door, many Jews with relatives living elsewhere used those contacts to emigrate. Several thousand moved to Palestine. Still others sought, often in vain, to obtain visas for travel to the West. Thinking to improve their chances of obtaining passage, some converted to Christianity or obtained forged certificates of baptism—which were readily available from the growing (and ecumenical) antifascist underground.

German-speaking Jews who stayed in Czechoslovakia were embraced, at least by the more liberal elements of society, who, in the tradition of Masaryk, supported human rights. Between 1935 and 1937, some nine hundred German émigrés, many of whom were Jews, were granted citizenship. In 1937, a statue of Moses the lawgiver was unveiled in Prague's Old Town and a street was named in honor of Louis Brandeis, a Jewish-American jurist with Bohemian ancestry. Ordered by Hitler to report on the treatment of Jews, the German ambassador to Prague replied that he could find no evidence of discrimination.

"HISTORY IS REPLETE with examples of men who have risen to power by employing stern, grim, and even frightful methods, but who, nevertheless, when their life is revealed as a whole, have been regarded as great figures whose lives have enriched the story of mankind. So may it be with Hitler." This assessment, authored by an Englishman in 1935, was not just another example of British naïveté. On the contrary, it was the testimony of a man who had been among the first to warn of Germany's plans to rearm and to denounce Hitler's persecution of Jews and democrats: Winston S. Churchill.

Looking back, we tend to see only the ranting Hitler of scratchy films, the screaming dictator who appeared to command the right arms of an entire German generation and whose agenda was a litany of hate. We wonder how any person of intelligence—let alone an observer as tough-minded as Churchill—could have arrived at a more positive assessment. *Mein Kampf,* even in the bowdlerized versions then available in the West, made clear Hitler's animosity toward France, his fantasies about Aryan superiority, and his desire to bend every part of Europe to his will. But were these writings mere puffery designed to lift the spirits of downtrodden Germany? And wasn't it natural that Germany would reassert itself as a leading power? "Those who have met Hitler face to face," continued Churchill, "have found a highly competent, cool, well-informed functionary with an agreeable manner [and] a disarming smile. . . . Thus the world lives on hopes that the worst is over, and that we may yet live to see Hitler a gentler figure in a happier age."

A gentler figure in a happier age. Those impressed by Hitler included Arnold Toynbee, the prestigious if far from the most discerning British historian of the era. David Lloyd George, who had led England through the Great War, lamented after meeting the führer that "I only wish we had a man of his supreme quality as the head of affairs in our country." A third who fell under the spell was Edward Albert Christian George Andrew Patrick David Windsor, formerly King Edward VIII. In 1937, shortly after his controversial marriage to the American socialite Wallis Simpson, he paid a friendly call on Hitler, greeting him snappily with the Nazi salute.

Then there was Lord Halifax, previously viceroy of India and for many years a close associate of Prime Minister Neville Chamberlain. Edward Wood was born with an atrophied left arm and three sickly older brothers, each of whom died before the age of nine. This left Wood heir to the title and a vast estate in Yorkshire. Like many of his social standing, the slender and hawk-faced Halifax measured politicians primarily by the intensity of their distaste for Bolshevism. In 1936, he visited Germany for the first time and pronounced Hitler's regime— which had locked up every Communist it could find—"fantastic." In November of the following year, he went back, at the request of Chamberlain, ostensibly to attend a hunting exhibition. When the sport was concluded, he made his way to Hitler's villa in the mountainous retreat of Berchtesgaden. Exiting his car, he was about to hand his coat to the footman when he was alerted by an aide's urgent whisper, "Der Führer!

Der Führer!" Halifax scrutinized the "footman" more closely, held on to his coat, and greeted his host.

During their three-hour meeting, Halifax informed Hitler that British opinion was offended by some aspects of the Nazi administration but that his government desired nonetheless to work with him on behalf of European peace. To that end, the Englishman cited three potential trouble spots: Austria, Czechoslovakia, and the Polish port city of Danzig. "On all these matters," he told Hitler, "we are not necessarily concerned with preserving the status quo, but are concerned to avoid such treatment of them as would be likely to cause trouble." British policy, then, was to tolerate changes to the European order, but with the earnest hope that any adjustments could occur without confrontation among the major powers.

This head-in-the-sand thinking stemmed from Great Britain's weakened condition. For three hundred years, the country had been a leading arbiter in world affairs. By the nineteenth century, its empire stretched from Canada and the Caribbean to South Africa, India, and Australia, with many outposts in between. Due to the breakup of the Ottoman Empire, it had gained valuable mandates in the oil-rich Middle East. The British had no doubt of the civilizing benefits of their rule: educating the backward, enlightening the heathen, training administrators, and stepping in with billy clubs raised when needed to maintain order. From their public schools to the House of Commons, from the great London banks to the pages of the *Times*, they treasured their institutions. They had reached the mountaintop and found the view magnificent. They were discovering, however, that from that lofty vantage point, all trails led down.

The First World War had been a shock, the Allied victory hard won; the feeling persisted that too many had died for too little. The romantic notion of battle, the exhilarating vision of King Henry V at Agincourt, had run aground in the trenches of Verdun and the Somme. Fifteen or twenty years later, portraits of slain fathers, brothers, and sons were arranged carefully on many a mantelpiece; middle-aged men with limbs missing were still a common sight on English streets; greedy weapons manufacturers remained objects of contempt;

and any future conflict was expected to be far worse. Poison gas was everywhere feared, and the warning of one parliamentarian had been accepted as gospel: "It is well . . . for the man in the street to realize that there is no power on Earth that can protect him from being bombed. Whatever people may tell him, the bomber will always get through."

Economic considerations also weighed heavily. Administering an empire in the twentieth century generated more headaches and less income than in earlier times; the people of India, led by the charismatic Mahatma Gandhi, seemed a particular nuisance. Long a creditor nation, Great Britain had become a debtor, with a trade ledger tipped in the wrong direction. Pressure was intense to balance the budget through arms reductions, which, in the era of the League of Nations, were thought essential for world peace. The fourth of Woodrow Wilson's famous Fourteen Points had called for the reduction of weaponry "to the lowest level consistent with domestic safety." This objective was echoed in a series of disarmament conferences held in Geneva in the 1920s. Efforts to promote pacifism and to end war by enacting laws against it enjoyed a modest vogue. Words alone, however, mean little. Of the major powers, Great Britain was the only one to actually cut spending for defense.

The Nazis' entrance upon the European stage did not, at first, alarm the British. After all, under the Versailles treaty, the size of the German army and navy was limited and the defeated country was forbidden to maintain an air force. The wake-up bell began sounding only when, in March 1935, Hitler renounced the treaty and declared that his country would indeed rebuild its military. The following year, when Germany reoccupied the Rhineland, Britons were unsettled to learn that his army was already three times the legal size and that his air force, or Luftwaffe, would soon surpass their own.

His Majesty's government began then to rethink its needs, especially in the air and on the sea. Rearmament proceeded sluggishly, however, and paid scant attention to the army, which remained well below authorized strength and was not intended in any case for deployment on the European continent. An American who observed all this while writing his senior-year college thesis compared the English attitude to that of a gentleman who requires a new suit but concludes that what he needs more is a good dinner. John F. Kennedy, a son of the U.S. ambassador to London, titled his study *Why England Slept*. "It takes time to change men's minds," he wrote, "and ... violent shocks to change an entire nation's psychology." The emergence of Nazism was a disturbance, not a shock. To many Britons, including those high up in government, fascism seemed a phase that the Germans would grow out of once their more legitimate needs were satisfied.

Neville Chamberlain had assumed the responsibilities of prime minister in May 1937. His policy, which combined appeasement with rearmament, aimed to restore confidence in European security. Sixty-eight when he took the job, Chamberlain had spent much of his career in the shadow of his father, a wealthy industrialist, and of his brother, who had served as foreign secretary. Now, late in his life, he rose higher than either. He was fortunate to live in an era when one could thrive in politics without liking people; his primary passions were music, gardening, and the relentless pursuit of fish.

Chamberlain was a practical, business-oriented man, supremely confident in his judgments and disdainful of critics. He did not believe that war was a solution to any problem and felt sure that all intelligent men would conclude the same. He had the ability, usually valuable but in this case treacherous, of being able to put himself into the shoes of another. He could readily understand Hitler's resentment of the peace treaty and his accompanying desire to restore a measure of German might. Chamberlain could also be philosophical about the führer's coarse rhetoric and bullying, which he ascribed to poor breeding. However, the prime minister could not imagine anyone intentionally causing a second world war. In Chamberlain's universe, people might be flawed, but they worried about their souls and did not set out to do monstrous things.

## 6

## *Out from Behind the Mountains*

Nineteen thirty-seven was the year Europe approached the cliff's edge but could not quite see beyond. In Czechoslovakia, the economy was starting to recover. Exports were up, the budget was in surplus, unemployment had dropped by two-thirds, and a state-of-the-art subway system was planned for Prague. Concert halls and theaters were filled to capacity, and citizens of all ethnicities cheered the nation's sports teams, which in ice hockey and soccer were among the world's best.

Hitler was an inescapable presence, but the Nazis were still a novel phenomenon and, from a distance, vaguely preposterous. The notion that the summit of the human race was represented by the homely Austrian and his pear-shaped colleagues was laughable—and people did laugh.

At the Liberated Theater in Prague, audiences roared at the satiric sketches of the popular cabaret team of Voskovec* and Werich:

> *Before civilization dawned*
> *Everything was fine*
> *The brontosaurus swept the streets,*

---

* Jiří Voskovec later became a film actor in America. His performance as a patriotic immigrant in the jury room drama *Twelve Angry Men* was cited by U.S. Supreme Court justice Sonia Sotomayor as strengthening her resolve to pursue a career in law.

*Cannibals devoured their eats;*
*They all slept in the shadow of the pine.*

*But then came the heel*
*Who invented the wheel*
*From the wheel came coins*
*From coins, inflation,*
*A Screaming Ass,*
*An angry nation.*

*And now civilization's here,*
*Poverty's in style.*
*The Ass is braying in the sun,*
*Everybody starts to run,*
*And all the other Asses shout: Sieg Heil!*

The dream of a united and democratic Czechoslovakia still lived. Even in 1937 and 1938, many Czech families sent their children to the Sudetenland for summer vacation; Germans returned the favor by depositing their children in Czech towns. The structures that had maintained peace in Europe since the end of World War I were starting to crumble, but the conviction remained firm that disaster could be avoided. However, this did not stop the government in Prague from preparing for the worst.

Since the republic's founding, its leaders had been mindful of the nation's vulnerability. T. G. Masaryk had stressed the need not just to oppose evil but to campaign actively against it. He wanted Czechoslovakia's soul to reflect "Jesus, not Caesar" but refused to go so far as to turn the other cheek. "I'd dedicate the combined power of my brain and my love of country and humanity to keep the peace," he declared, "but also, if attacked, to fight a war."

He and later Beneš backed this vow by asking French advisers to assist in military training and by devoting ample resources to national defense. To minimize dependence, they nurtured a robust armaments industry that became known as the arsenal of Central Europe. By 1938, the country had in hand thirty reasonably well equipped and professionally trained army divisions and an air force with skilled pilots and more than 1,200 modern aircraft. The soldiers were backed by armored units and plentiful supplies of munitions and oil. "Czech Hedgehogs," massive steel barriers that were connected by giant spools of barbed wire, fortified the border with Germany. Hitler later commented that Germany and Czechoslovakia had been the only states to carry out their war preparations efficiently. The question remained: would those measures be enough?

   With potential enemies on every side, the government gave priority to assembling a network of spies. The commander of the intelligence service, Colonel František Moravec, was a former legionary who believed that his country was living on borrowed time. Espionage was a booming business in 1930s Europe, particularly in such cities as Vienna, Berlin, Geneva, and Prague. Spies operated within a real-life atmosphere of intrigue, enlivened by secret codes, invisible writing, elaborate disguises, and surveillance tricks. Thus Moravec was suitably suspicious when contacted by a man who described himself as a high-ranking German official who was willing to trade information for money. After an interval of hesitation and testing, Moravec arranged to meet the man, who turned out to be exactly what he claimed. Agent 54, as he was called, was a senior military officer who disliked Hitler as much as he liked cash. Beginning in April 1937 and until his arrest five years later, the agent provided a stream of documents and gossip that alerted Czech authorities to Nazi operations that they were nevertheless unable to stop. One such initiative was Berlin's clandestine partnership with Konrad Henlein's quasi-Nazi Sudeten German party.

   Even as he denied any connection to Hitler, Henlein was receiving regular subsidies from Berlin. He was aware that his prestige among the Sudeten Germans hinged on his ability to show support from the chancellor. Thus he was steadfast in his obedience to Hitler's guidance and unrestrained in his flattery. He desired nothing more ardently, he declared in a confidential message, than to see the Sudetenland—in fact, all the historic Czech lands—become part of the Reich. Acting on instructions from Nazi headquarters, he began a propaganda campaign to illustrate the supposed plight of his people. German children from Czechoslovakia were sent across the border to recite tales of suffering at the hands of their alleged persecutors. As agitation intensified, so did British and French pressure on Beneš. The maintenance of peace, they said, was his responsibility.

Beneš was perturbed but not panicked. He thought Hitler an obviously flawed leader whose popularity would wither as the German economy continued to sag. He also felt that most Sudetens were too sophisticated to embrace fascism. Early in his presidency, Beneš toured the region in an effort to reduce the political temperature. He told audiences in their own language that a certain amount of friction between ethnic groups was to be expected but could be resolved through democratic means based on "respect of the human person and complete civic equality." He admitted that Prague had made mistakes in the letting of contracts and in hiring for positions within the federal government. Before closing with a plea for nonviolence, he referred, as he often did, to the nobility of German character as reflected in such great moral teachers as Gotthold Lessing, Friedrich Schiller, and Goethe.

In time Beneš gave in to virtually every demand by Henlein. Even the German ambassador reported, on November 11, 1937, that the president "really wishes to improve the position of the German minority," and, on December 21, that he "has made the internal appeasement of the country the aim of his presidency."

The Czechoslovak leader was always prone to see the glass as half full. When Moravec, his intelligence chief, warned that Hitler and Henlein were determined to spark a war, Beneš told him not to be alarmed. If Hitler, he said, "resorts to force we can rely on our treaty system. Don't forget that our allies, taken together, are still stronger than Germany."

The British journalist and novelist Compton Mackenzie wrote in his biography of Beneš:

> I once saw the president's famous optimism at work. . . . I found
> him in the garden. To me it looked like rain, and I said so. The

president, spectacles in hand, cocked his head at the sky: "I do not think so." So I put my first question. The president held forth. Heavy slow rain-drops, the obvious heralds of a downpour, began to fall. The president frowned. Then he affected to ignore these Henleinist raindrops and continued to talk. Presently he could not ignore them completely and made one of his typical concessions to bring about a compromise. He pulled his chair back under a beech tree. . . . For a minute or two the compromise seemed to be working. The extravagance of the Sudeten rain was being held in check. . . . [Then] an extra heavy drop hit the president's forehead. Another landed on his spectacles. Another fell on my nose. We moved further under the beech tree; but at last the president had to yield. He shook his head over the unreasonableness of the rain and led the way indoors.

Beneš was a prisoner both of his own past diplomatic success and of his rigidly logical mind. My father called him the "mathematician of politics," a man devoted to reason who expected others to be guided by the same star. The world would not wait long to disappoint him.

EARLY IN 1938, Hitler kicked the hornet's nest, neither anticipating nor receiving much of a reaction. He knew by then that Great Britain wouldn't object to a "peaceful readjustment" of Europe's internal borders; Lord Halifax, now the foreign secretary, had admitted as much the previous November. In the führer's speeches, which were listened to around the globe, he claimed that millions of Germans were being forced to live outside the boundaries of their motherland and that other countries—including Great Britain—had never hesitated to defend their own interests. He accused Austria, as he would soon accuse Czechoslovakia, of systematically persecuting its German population. Before long, pro-Nazi demonstrations toppled Austria's governing coalition and opened the way for German soldiers to cross the border, which they did early on the morning of March 12. When it became clear that no resistance would be encountered, the army was instructed to invade "not in a warlike but in a festive manner."

That afternoon, Hitler traveled to Linz, the city of his boyhood, where he visited his parents' graves before proceeding to the capital. Along the way, he was met by cheering crowds. In Vienna, the führer's homecoming was accompanied by widespread public violence against Jews—the most shame-filled moment in Austria's long history. The land where Schubert had been born, Mozart had lived, and Beethoven had composed the *Pastoral* symphony was handed over to barbarians.

The Anschluss—the merging of Austria and the German Reich— was for Hitler the latest in a series of planned provocations. In 1935, he had begun rebuilding his armed forces. In 1936, he reoccupied the Rhineland, thus strengthening his capacity to invade France. In 1938, the conquest of Austria achieved the threefold purpose of uniting Germans, encircling the Czechs, and opening an invasion route to the Balkans. Hitler was on the march, and no one, as yet, dared stand in his path.

On the day of the Austrian invasion, the Luftwaffe sent a small unmarked plane over the Czechoslovak border, dumping leaflets with a greeting: *"Sagen Sie in Prag, Hitler lasst Sie grüssen."** In London that same morning, Halifax told Jan Masaryk not to worry, the Nazis would never do to his country what they had just done to Austria; on that the Germans had given their word. Masaryk observed that "even a boa constrictor needs a few weeks of rest after it has filled its belly." Halifax asked what the Czechoslovaks would do if attacked. Masaryk replied, "We'll shoot."

Notwithstanding the reservations of some senior officers, the German military had already prepared a strategy (Plan Green) for the conquest of Czechoslovakia. Hitler fully intended to press ahead; he did, however, desire a pretext. Czechoslovakia was neither the Rhineland nor a nation of German speakers such as Austria. He told Henlein to demand justice for the Sudeten population and to insist on changes "that are unacceptable to the government." Henlein replied that he understood: "We must always demand so much that we can never be satisfied." Of course, the Wehrmacht's plan was not confined to liberating a discontented

---

* "Tell everyone in Prague that Hitler says hello."

minority; the intent was to subjugate the entire country, seize control of its industry, and "solve the German problem of living space." The timing would depend on Henlein's success in creating a casus belli and on the level of resistance anticipated from the British and French. On May 30, 1938, the führer signed a directive underlining his desire to smash his neighbor by no later than October 1.

The millions who listened to Hitler's speeches and Nazi radio broadcasts were told that the Czechs were conducting "a passionate fight for extermination" against the Sudetens. German-owned businesses were being forced into bankruptcy, children were starving, the level of oppression was incredible. This propaganda was carefully disguised as independent reporting to deceive international audiences. Goebbels subsequently admitted, "It was of utmost importance during the whole period of the crisis that the so-called situation reports . . . should not allow foreign circles [to see] through the tactics of [Berlin]." As the climax neared, sycophantic broadcasters grew hysterical, raving about the "bestial Jewish Hussite Bolshevik monsters" who were preying upon the brave but helpless Germans of Czechoslovakia.

Beneš still had confidence in his alliance with France and felt that if the French were involved in a scrap, the British would join as well. He also had his agreement with the Soviet Union. Zdeněk Fierlinger was the country's ambassador in Moscow. In late April, he reported that Stalin had agreed to act on Czechoslovakia's behalf, provided the French did the same. All well and good, but Fierlinger was stumped when asked precisely what it was the Soviets were prepared to do.

Twenty years later, I wrote my college thesis about Fierlinger, a devious and unsavory man who was skilled at manipulating Beneš but whose principal loyalty was to communism. The son of a teacher, Fierlinger was a mediocre student but gifted at languages; he was also among the many young men who had spent World War I in the Czechoslovak Legion. Toward the conflict's end, he had joined a military delegation to France, where he met and befriended Beneš, his elder by seven years. This connection led Fierlinger into the diplomatic service and to a series of appointments, including, in 1937, minister to the Soviet Union. He was a member of the Social Democratic Party, which occupied a position just to the right of the Communists on the political spectrum. The party was among the more popular despite the fact that some members had joined because they strongly opposed communism and others because they were almost Communists. This tension between the democratic Left and people whom my father referred to as "fellow travelers" was to play a pivotal role in the country's future undoing.

In August, Beneš met with Fierlinger in Sezimovo Ústí, a picturesque country town where they had neighboring summerhouses. The president asked his envoy what the Soviets would do should the Germans attack Czechoslovakia. Both knew that Stalin had spent much of the previous year conducting show trials of civilian and military leaders—a paranoid action that had resulted in the execution of thousands of Communist loyalists and left the Red Army ill prepared for war. Yet Fierlinger insisted that the Czechs could rely on Moscow to render all possible aid. When the president asked what that meant, Fierlinger said it would depend on the course of events. When Beneš probed for a concrete guarantee, the minister had none to give. After several minutes, the usually upbeat president showed signs of depression. To lift his spirits, Fierlinger produced a phonograph album of Russian marching songs, which, he reported in his memoir, "impressed us deeply, because we felt the strength of a great country and her people, ready to defend her freedom and independence to the end." All of which was fine, but it was not the Soviet Union's independence that was in jeopardy.

I REGRET THAT I am unable, at this point in the historical narrative, to offer an eyewitness account of the proceedings; barely one year old, my world was a small one. I can say that while conducting research for this book, I was struck by the sense of helplessness that my parents and many of their countrymen must have felt. This impression was strengthened when, for Christmas 2010, I received a Kindle on which I proceeded to reread *War and Peace*. As fans of Tolstoy will recall, the author makes much of his belief that history is determined far more by the mysterious hand of Providence than by the actions of international leaders. Thus the outcome of the Napoleonic Wars, and the ability of czarist forces to repel the French, depended less on Napoleon and the czar than on the apparently random choices of individuals who together served as the involuntary instruments of some higher purpose. Tolstoy argued that scholars routinely exaggerate the ability of the great and powerful to control events.

There is obviously some truth in that thesis, but the role of leadership cannot be downplayed in the events immediately preceding World War II. If we were to subtract Hitler from the scene, replace the British and French with stronger actors, or bring back T. G. Masaryk for a starring role, the events I am about to describe would not have taken place or would have unfolded far differently, possibly to the extent that World War I would still be called the Great War. As it was, the citizens of the Czechoslovak Republic and many of their brothers and sisters throughout Europe had relatively little say in their own fate, and could only watch as the leading men strutted their hour on the stage.

# 7

# "We Must Go On Being Cowards"

The German invasion of Austria was achieved overnight and to the evident gratification of many who lived in the violated country. There had been no question of intervention by England or France. The danger to Czechoslovakia raised more complicated issues because of the treaties Beneš had negotiated with Paris and Moscow. The Britons were under no legal obligation to Prague, but neither did they wish to see the French ensnared in a losing fight. In the spring of 1938, Neville Chamberlain privately summed up the situation:

> You have only to look at the map to see that nothing that France or we could do could possibly save Czechoslovakia from being overrun by the Germans. . . . The Austrian frontier is practically open; the great Škoda munitions works are within easy bombing distance . . . the railways all pass through German territory, Russia is 100 miles away. Therefore we could not help Czechoslovakia— she would simply be a pretext for going to war with Germany. That we could not think of unless we have a reasonable chance of being able to beat her to her knees in a reasonable time, and of that I see no sign. I have therefore abandoned any idea of giving guarantees to Czechoslovakia or the French in connection with her obligations to that country.

Chamberlain was surely right to doubt that his country, with its undermanned army, could prevent the Nazis from conquering their southern neighbor if they so attempted. He failed to address whether such an assault might be deterred had Hitler cause to fear that a general war would result. The evidence suggests that delay, at least, was possible. Indeed, the führer assured his generals that he would attack Czechoslovakia only if French and British intervention seemed unlikely.

In 1938, diplomats could still believe one thing and say another without having their inconstancy exposed via the leaking of electronic communications. The British, having decided to leave Czechoslovakia to its fate, nevertheless tried to persuade the world that they had yet to make up their minds. Admitting the truth would have embarrassed the French and been interpreted as an invitation for Hitler to invade. In public, therefore, the Foreign Office bobbed and weaved, ruling the use of force neither out nor in.

When, in May, tensions reached a high point, London warned Berlin that if it attacked Czechoslovakia and the French were embroiled as well, "His Majesty's Government could not guarantee that they would not be forced by circumstances to become involved also." At this same time, English officials were telling their counterparts in Paris that they were "not disinterested" in Czechoslovakia's fate. I learned in the course of my own career that British diplomats are trained to write with precision; so when a double negative is employed, the intent, usually, is not to clarify an issue but to surround it with fog. The Germans, unfortunately, were not deceived—Chamberlain's yearning for tranquillity was too obvious. Hitler bragged to acolytes that he had only to mention the word "war" to cause the prime minister to leap out of his skin.

London felt it could best avoid conflict by obtaining from Germany a clear statement of the improvements it wanted to see in Prague's treatment of the Sudeten minority. The British hoped then to be able to press Beneš to accept such a list and thereby leave, in Lord Halifax's phrase, "the German Government with no reasonable cause for complaint." Hitler kept one step ahead by making demands that were so amorphous that they could not be pinned down. He insisted on a halt to persecution of the Sudetens but reserved to himself the right to define the term. Nothing Beneš did could comply fully with his requests, because Hitler didn't care about the rights of the Sudeten minority; he cared about using the alleged sins of Czechoslovakia to take the next step toward the conquest of Europe. The Chamberlain government was slow to realize that the führer was determined to remain indignant. Denied a reasonable ground of complaint, he would quickly invent an unreasonable one.

Compared with the English, the French were as easy to intimidate but less so to fool. In London for a meeting on April 29, their premier, Édouard Daladier, argued that Henlein was out to destroy Czechoslovakia and that Hitler was more ambitious than Napoleon. He added that if Beneš was pushed to make further concessions, the Allies should at least pledge their support in the event Germany was unwilling to take yes for an answer. Daladier, who thought the military situation less dire than did the British, insisted that further capitulation was more likely to produce war than a show of resolve. Chamberlain and Halifax were unconvinced, in part because they thought Daladier's words showed more spine than France actually possessed.

The British also ignored pleas filtering in from Hitler's own countrymen. The shrunken core of antifascists within Germany's military, diplomatic, and industrial elite beseeched England to adopt a sturdier line. They claimed that Hitler was not as powerful as he longed to appear and that the German majority had no wish to follow the "Nazi gangsters" into war. Chamberlain was too timid to take that advice, but he was not entirely blind to the deepening danger. "Is it not positively horrible," he wrote, "to think that the fate of hundreds of millions depends on one man, and he is half mad? I keep racking my brains to try and devise some means of averting a catastrophe."

One option was peace through strength. British rearmament efforts were finally moving ahead, but the country still felt unready for a sustained conflict. The army in 1938 had 180,000 men complemented by a reserve of 130,000 weekend soldiers. The Germans had an army of half a million with an equal number in reserve. The Royal Air Force (RAF) possessed 1,600 planes, the Luftwaffe more than twice that. Only the

British navy was in fighting shape, but it had global responsibilities and could not compensate for the military's shortcomings on land.

The second alternative was diplomacy. The British hoped to avoid war by persuading the potential combatants to take a step back and consider soberly where their best interests might reside. For centuries, British imperialists had been refereeing disputes among fractious groups; why not mediate now among the tribes of Central Europe? Alexander Cadogan, the undersecretary of the British Foreign Office and author of a candid diary, queried:

> What I wonder is, is it even now too late to treat the Germans as human beings? Perhaps they wouldn't respond to such treatment. What I have always in mind these last two years (and urged) is that we should ask them whether they won't let us try our hand at helping to remedy the grievances which they make so much of but which they don't make very clear.

Thus, in the summer of 1938, Chamberlain sent a special emissary to Czechoslovakia with a mandate to mediate. "We cannot but feel," opined Halifax, "that a public man of the British race and steeped in British experience and thought may have it in his power . . . to make a contribution of quite particular value." The public man in question, Walter Runciman, Lord of Doxford, was able enough but no expert on the region. After being systematically lied to by Sudeten representatives, he concluded that a solution could be found only by satisfying Berlin. This was going in circles. For a year, the British had been trying to determine what Berlin really wanted, only to find that however much was on offer, it would be—by agreement between Hitler and Henlein—never enough.

Earlier in the year, Beneš had granted amnesty to Sudeten Germans who had been judged guilty of treason. Henlein had barely acknowledged the gesture and instead demanded full self-government, reparations for past damages, and a pro-German foreign policy. Shedding his earlier restraint, he adopted the Nazi salute, proclaimed the right to promote Nazism openly, and accepted for himself the title of führer. The Henleinists had come to differ from the Nazis only in the color of their shirts (white, not brown) and the design of their banners (scarlet with a white shield, no swastika).

Throughout the long days of that unlovely summer, Beneš sought both to retain his confidence and to parry the arrows and insults that came his way. In response to pressure from the British and French, he strove to pacify the Sudeten Germans, avoided public statements that might provoke Hitler, and granted permission for Runciman to spend August traveling about Czechoslovakia in quest of the magic formula for peace. He even expressed a willingness to participate in an international conference or to accept binding arbitration. To friends, he emphasized his belief that war could still be avoided through a combination of Allied solidarity and the fact that his government had given Hitler no excuse for war. His ultimate life raft was the honor of France.

That country's policy reflected the ambivalence of its authors. Daladier repeatedly promised Beneš that France would live up to treaty obligations that he described as "solemn," "indisputable," and "sacred." At the same time, the French, burdened by labor unrest and high unemployment, had little interest in a quarrel with Germany. Their armed forces had yet to recover from the Great War, which had wiped out a third of France's male population of prime military age. This catastrophe led to a low birth rate and consequently, in the 1930s, to a shortage of new recruits. In addition to size, the French army lacked mobility while the air force suffered from too-few bombers and a reliance on obsolete technology. The country's allies, particularly Poland and the Soviet Union, were at odds with each other and could not be relied on, in a crisis, to close ranks. In the north, Belgium was pursuing a policy of neutrality, effectively barring France from using its territory as a base for military operations. The high command's once bold strategic doctrine had grown defensive, relying on a buildup of supposedly impregnable border fortifications: the Maginot Line. The French hoped that they could protect themselves but had no desire to send the flower of their youth eastward to face German guns on behalf of

Czechoslovakia. Their fears deepened considerably after the Anschluss, when it became clear that to preserve their good name, they might actually have to make such a sacrifice.

It was during this period that the U.S. ambassador to Paris, William Bullitt, reported attending dinners at which French officials began by expressing a determination to avoid war at all costs and ended—several brandies later—vowing to uphold the nation's treaty obligations no matter the price. Striving to rescue France from a decision it did not wish to make, Bullitt urged President Franklin Roosevelt to convene a high-level conference that would bring all the parties together and, he hoped, devise a face-saving exit from the crisis. When the president eventually did propose such a gathering, he was congratulated by everyone involved and otherwise ignored. U.S. leaders simply lacked the leverage to shape events in Europe because the public they represented did not want to get involved. As a result, while the hopes of Beneš rested with France's promise, the hopes of France were invested in Great Britain's ability to push Beneš into appeasing Hitler.

THE FÜHRER, MEANWHILE, was growing impatient. He had boasted to advisers that he would smash Czechoslovakia by October 1, 1938. Three weeks before the deadline, he presided over a late-night meeting at which his senior military staff predicted a swift victory. German propaganda and Henlein's mischief had brought the Sudeten population to the edge of rebellion. The British and French were dithering, the Soviets too far away. Hitler finally had the Czechoslovaks where he wanted them: all alone.

In early September, more than a million Germans swarmed into Nuremberg to celebrate the anniversary of the Nazi Party. On the evening of the twelfth, in a vast meeting hall, an expectant crowd barely listened while an orchestra of modest competence worked its way through the overture to *Die Meistersinger.* As the music reached its crescendo, so did the chanting: *"Sieg heil! Sieg heil! Sieg heil!"* Hitler marched to the podium and motioned for the audience to be still. He spoke then, as he habitually did, in machine-gun-like bursts. "This misery of the Sudeten Germans is indescribable. As human beings they are oppressed

and scandalously treated . . . hunted and harried like helpless wildfowl
for every expression of their national sentiment." "I am in no way willing," he shouted, "that here in the heart of Germany a second Palestine
should be permitted to arise. The poor Arabs are defenseless and deserted. The Germans in Czechoslovakia are neither defenseless nor are
they deserted and people should take notice of that fact."

As Hitler's oration concluded, German thugs in the Sudetenland
began assaulting their neighbors and trashing government offices and
police stations. A British military attaché who was on the scene described the crowds as "not in any way ill-natured, for I walked around
the town for half an hour, except that all the Jewish shops had their
windows shattered." Coming upon a mob beating "a prosperous Jew,"
the official cited his own prudence in turning away. The Czechoslovaks mustered a firmer response. Beneš imposed martial law, sent in
reinforcements, and restored order. "Fight to the end," Henlein had
urged his followers; by dawn, he and his top advisers had fled across the
border to Leipzig.

The Czechoslovaks had responded to Hitler's jab with a vigorous
counterpunch—and were ready for more. In a memo to his civilian superiors, the armed forces chief of staff, General Ludvík Krejčí, argued:

> The morale of the German soldier is being artificially whipped
> up by the cult of the "superman" and intoxicated by the bloodless
> victories during the occupation of the Rhineland and Austria.
> The first failure of this soldier when he approaches our fortifi-
> cations . . . will suffice to break his morale. . . . The artificially
> inflated power of the German armed forces will crumble and
> become a comparatively easy prey for our allies.

The military backed up those words by canceling leave, ordering
a partial mobilization, and sending its best regiments to guard their
country's vulnerable border with Austria.

A MARTIAL SPIRIT could also be detected in London, provided one
looked closely enough. Harold Nicolson, a pro-Churchill member of

Parliament, declared, "We must warn Hitler that if he invades, we shall fight. If he says, 'but surely you won't fight for Czechoslovakia,' we will answer, 'Yes, we shall.'" Nicolson was one of a growing number of hawks who, fed up with Chamberlain, were demanding a more robust policy. The British majority, however, continued to believe that appeasement was the safer and more realistic approach. Such influential publications as the *Economist* and the *Times* remained foursquare behind concessions; their worry was that the government was not doing enough to mollify Berlin.

Hitler's Nuremberg speech rattled the already frayed nerves of the French, who called London to warn that if nothing were done, conflict would soon break out. The British Secret Service agreed, circulating a confidential forecast that, within two weeks, Germany would invade Czechoslovakia. Sir Nevile Henderson, Great Britain's ambassador to Berlin, insisted that the solution to the crisis could be found only in Prague: "None of us can even think of peace again until Beneš has satisfied Henlein. . . . Henlein wants peace and will agree with Beneš if the latter is made to go far enough."

By late in the day on September 13, Chamberlain had concluded that long-range diplomacy was not working. For weeks, he had been contemplating what he called "Plan Z," a direct approach by him to his counterpart in Berlin. He saw no better option than to gamble on his powers of persuasion, which he deemed formidable. A message was sent: would Herr Hitler receive him? The führer replied that he was "entirely at the prime minister's disposal." A meeting was set for two days hence.

When Hitler learned that Chamberlain wished to see him, he anticipated a lecture on the dangers of rash action. He need not have been concerned. The prime minister had no desire for confrontation; he hoped only for peace, a goal he still expected his host to share. All that was needed was a just settlement of the Sudeten question. Chamberlain planned to suggest an internationally supervised plebiscite that would allow the Sudeten Germans to choose for themselves whether or not to remain in Czechoslovakia.

Venturing onto an airplane for the first time, the British leader crossed the Channel early on September 15. From rainy Munich, he went by train to Berchtesgaden and then by car to the führer's residence, where Hitler was waiting to greet him. The two paused in a hallway for tea before climbing the stairs to the same cramped study in which Halifax had been hosted a year earlier. At Chamberlain's suggestion the meeting was restricted to the two leaders, plus an interpreter. Following a brief exchange of courtesies, Hitler launched into his familiar tirade about the cruelties being inflicted on the poor Sudetens. Three and a half million ethnic Germans in Czechoslovakia must be free to join the Reich, the chancellor said, adding that he planned to act on this imperative.

Chamberlain did not attempt to deny Hitler's right to seize the Sudetenland. Instead he sought an assurance that such a concession would guarantee peace. Hitler was evasive, saying that the Hungarians, Poles, Ukrainians, and Slovaks also nursed grievances against Prague that would have to be addressed. Chamberlain pointed out that the means of implementing a territorial transfer might be complicated and proposed a period of peaceful discussion between the Beneš government and the Sudeten Germans. Hitler shook his head, insisting on immediate action. The two agreed that the prime minister should be given a few days to confer in London and Paris before returning to Germany. No mention was made of consulting Prague. A communiqué was issued, stating only that the leaders had met and would do so again.

Returning home, Chamberlain briefed his cabinet, stressing the urgency of the matter and the lack of any viable alternative to ceding the Sudetenland. Though he referred to Hitler as "cruel, overbearing [and] ... ruthless," he also described him as an impressive figure whose word could be relied on with confidence. He believed that Czechoslovakia could survive losing the Sudetenland and that the führer, having won that point, would be reasonable with respect to issues of timing and process. Hitler did not, he had been assured, have any interest in adding racially inferior Czechs to the Reich; thus peace was readily obtainable if Prague would agree to let its Germans go. The cabinet invited the French premier to come to London for a chat.

Jan Masaryk, meanwhile, was trying frantically to discover what had happened during the Berchtesgaden meeting. No one in the Foreign Office would speak to him. On the phone with Beneš, he lamented that the Allies were "talking about us without us."* That was good news in Berlin, which was listening in on every conversation between the ambassador and his boss. The Germans enjoyed the intercepts so much that they decided to share them with the British, including Masaryk's candid, unflattering descriptions of Chamberlain and Halifax. The revelations would detract further from whatever merit the Czechoslovak cause might have had in the eyes of the prime minister.

In London on September 18 the British and French agreed that the Czechoslovaks should give up all areas that were more than 50 percent German. The next afternoon, their ambassadors appeared at the castle in Prague to inform Beneš that he had a choice: accept the loss of one-third of his country or instigate a war that his people would surely lose and for which he would personally be blamed. The president asked for time to consider the question; he was told that a response was needed within the day.

Beneš began consulting with his military advisers, party leaders, and top aides. What options did he have? He knew now what to expect from England and France; what about the Soviets? He asked the Russian ambassador the same questions he had earlier posed to Fierlinger. Late in the day, he received an answer. If the French fought, the Soviets would, too. If the French didn't, the Soviets would bring the matter to the attention of the League of Nations. No help there.

Other Czechs were meeting as well. A group of patriots, among them close friends of my father, had formed a secret council; some were politicians, some journalists, some high up in the armed forces. They were not enemies of Beneš but loyalists who knew him well enough to

---

* One can draw a straight line between Masaryk's complaint and an observation included in President Barack Obama's speech to the people of Prague sixty years later: "Many times in the twentieth century, decisions were made without you at the table; great powers let you down, or determined your destiny without your voice being heard."

doubt that he would act with the requisite strength. That afternoon, they sent him an impassioned appeal:

> It depends solely on you whether we shall capitulate or fight . . . a defeat would not destroy the nation's moral force . . . while capitulation means moral and political disintegration, now and for generations to come, from which we could not recover.

On that day, September 20, Beneš was angry enough at his allies to defer to the judgment of friends. At 7 p.m., he replied in writing to the British and French, complaining that their proposals had been made without input from his government and contrary to the interests of his people. "It is hence understandable," he said, that "the Czechoslovaks would not accept them."

The president's advisers were exhilarated, sure that they had done the right thing while Beneš, who had been up before dawn, retired at 1 a.m. thinking that his country had opted to fight. These feelings were short-lived. Barely an hour later, the French and British ambassadors reappeared at the castle. Their governments, they told the bleary-eyed Beneš, would not accept no for an answer; the arrangement worked out between Chamberlain and Hitler must be allowed to stand. The British emissary warned again of the imminence of war; his French counterpart, weeping, told Beneš that if Germany attacked under the present circumstances, his government would not help, treaty or no treaty.

Amid the deepening gloom, Beneš began yet another series of meetings. Counsel was divided, with the moderate politicians leaning toward acquiescence, and the military, the conservative nationalists, and the Communists initially determined to fight. At noon, the two lugubrious ambassadors returned to the castle, wondering what was taking so much time. Beneš pointed out that the loss of the heavily fortified Sudetenland would leave his country defenseless against a further Nazi attack. When Hitler made his next move, what would England do? The British diplomat said he did not know. What would France, the treaty ally, do? The Frenchman remained silent. Beneš delayed until 5 p.m. Finally, faced by the prospect of leading his people into war alone, he sent word that his government would "with feelings of pain" accept the ultimatum.

As the international community was bullying Beneš in one direction, his outraged countrymen continued to push in the other. On September 22, a general strike was organized, a rally took place in Wenceslas Square, and thousands of citizens—Communists and democrats alike—marched to the castle where they demanded weapons to fight. Beneš sought to restore confidence by changing prime ministers. The new head of government, Jan Syrový, was an army general with a reputation for toughness and the political advantage of possessing, as had the revered warrior Jan Žižka, a black patch over a useless eye.

In London, Alexander Cadogan noted in his diary that some in the press had accused the Britons of betraying the Czechoslovaks. That was "inevitable," he wrote, "and must be faced. *How* much courage is needed to be a coward! . . . We must go on being cowards up to our limit, but *not beyond.*" To that point, the terms and timing of the Sudeten secession had not been worked out. Chamberlain assumed it would happen in a civilized fashion, over a period of weeks, with ample safeguards to protect those living in the region who had no desire to join the Reich. His cabinet had spent hours developing the idea of an international commission to accomplish these goals.

On September 22, while the people of Prague took to the streets, Chamberlain and his ever-present umbrella returned to Germany, this time to the city of Godesberg and a luxurious hotel overlooking the Rhine. He had told his cabinet that he would press for favorable terms—including the commission plan, a smaller area of secession, and a reduction in armaments along the border. Meeting again without staff, the prime minister reported to Hitler that London, Paris, and even Prague were now ready to countenance a change in the status of the Sudetenland. He then outlined the ideas his government had devised for implementing the agreement in an orderly fashion. He thought that although Hitler might quibble over the details, he could not help but be pleased.

Instead the führer threw a fit, informing Chamberlain that his efforts were no longer of any use. Czechoslovakia was an artificial state with a made-up history and no right to exist; moreover, it was becoming a base for Communists. There was only one solution: German occupation of the Sudetenland was to be unconditional and would begin on or before the first of October. There would be no need for international supervision, no thought of compensation, no permission to dismantle infrastructure, and no right to remove military or commercial property; every tank must be left behind and so, too, every chicken.

Hearing the news from Godesberg, Cadogan was appalled.

A week ago when we moved (or were pushed) from "autonomy" to cession, many of us found great difficulty in the idea of ceding people to Nazi Germany. We salved our consciences (or at least I did) by stipulating it must be an "orderly" cession—i.e. under international supervision, with safeguards for exchange of populations, compensation etc. Now Hitler says he must march into the whole area at once (to keep order!) and the safeguards and the plebiscites can be held after! This is throwing away every last safeguard that we had.

Chamberlain returned from Germany shaken but still determined to find the basis for an agreement. He informed his cabinet that Hitler "had a narrow mind and was violently prejudiced on certain subjects, but he would not deliberately deceive a man whom he respected and with whom he had been in negotiation."* The chancellor, he said, was "extremely anxious to secure the friendship of Great Britain . . . [and] it would be a great tragedy if we lost an opportunity of reaching an understanding." The cabinet, however, was now divided and the media increasingly sympathetic to Prague. Jan Masaryk showed up with a letter denouncing the new German demands and invoking the names of Wenceslas, Hus, and his own father. Even Chamberlain was

* It is revealing that Chamberlain thought he had impressed Hitler. In fact, the führer commented to an aide, "All he wants to do is fish. I don't have weekends, and I don't fish."

sufficiently perturbed about Hitler's intentions to inform the Czechoslovak government that if it wished to mobilize its armed forces, England would no longer object.

That message, delivered to Beneš on the evening of September 23,
was joyously received. "It was obvious that he was reading the few sentences on the paper again and again," recalled Beneš's personal secretary. "Then he put the paper on the desk and said, 'Yes' and began to
pace back and forth across the room. . . . I observed that he was as excited as I had ever seen him. Then, he said, 'This means war! The English advise us to mobilize.'"

That night the mobilization order was broadcast. All reservists
under the age of forty were to report for duty. Within hours, men in
uniform were arriving at their assigned posts or heading for the train
station to be dispatched to the border regions. Long fearful of conflict,
the country in its excitement could not wait for the clash to begin. All
Prague was blacked out. The castle took on the appearance of a military
command center, with cots set up in corridors and even Beneš keeping
a uniform and gas mask close at hand. Antiaircraft guns were on full
alert, while friendly planes kept watch from the sky. In Belgrade, my
father prepared to return and take his own place as a lieutenant in the
army. He recalled of that night:

> The national will manifested a resolution far beyond that of its
> leadership. . . . Meetings were organized all over the country to
> demonstrate the determination of the people; resolutions and
> individual messages poured into the Hrad, the seat of the presi-
> dent of the republic, giving encouragement and calling for firm
> resistance.

For a brief moment, a new consensus appeared: Hitler had over-
reached. The Czechoslovaks were ready, even eager, for battle. In Paris,
Daladier was asked what France would do if the Germans crossed
the border. He answered without hesitation that his country would
go to war. On Monday, September 26, the British issued their firmest
communiqué of the crisis, citing all that they had done to achieve an

amicable settlement but vowing to stand with France in the event of a fight.

That evening, Hitler addressed an expectant world once more, this time from the Sports Palace in Berlin. Speaking for an hour, he blamed the Czechoslovaks for failing to agree to a demand made by the British and French; he accused Beneš of seeking the overthrow of Chamberlain and Daladier and of placing all his hopes on Soviet Russia. The matter could be reduced, he said, to a test of wills:

Two men stand arrayed one against the other: there is Herr Beneš, and here stand I. We are two men of a different make up. . . . I have made Herr Beneš an offer which is nothing but the carrying into effect of what he himself has promised. The decision now lies in his hands: peace or war! He will either accept this offer and now at last give to the Germans their freedom, or we will go and fetch this freedom ourselves.

Hitler's nemesis did not hear this melodramatic threat because he had fallen asleep in an armchair at an "undisclosed secure location," where the military, fearing an air raid, had hidden him. When Beneš awoke and was briefed, he felt flattered. For the first time since the crisis had begun, he had reason to believe that France, England, the Soviet Union, and perhaps the United States were all on his side. He thought that Hitler had isolated himself and that now, if war came, Germany would surely lose.

Beneš had not, however, counted on the lengths to which Chamberlain would go in pursuit of peace. While the Czechoslovaks spent September 27 preparing for war, the British sent a special emissary to Germany with a plea for direct Berlin-Prague discussions backed by London in a mediating role. The envoy called on the führer twice— only to be yelled at and dismissed. News of this humiliation had a crushing effect. Although the Germans were not, in fact, yet prepared to strike, Chamberlain and his advisers assumed that an invasion was imminent. They warned Hitler again that if he attacked, Paris would likely respond, meaning that they too would fight. The Royal Navy was mobilized, as was the French army. Civilians began to stream out of Paris, fearful that bombings were about to begin. In London, the cabinet met at all hours, searching for a way out of the box that seemed to have disaster written on every side. "I'm wobbling all over the place," Chamberlain confessed to Halifax, just before stepping in front of a microphone for his 8 p.m. radio broadcast. After a day in which all the news had been bad, Chamberlain's frustration poured forth in words that would define and ultimately desecrate his legacy:

> How horrible, fantastic, incredible it is that we should be digging trenches and trying on gas masks here because of a quarrel in a faraway country between people of whom we know nothing.

He continued in a passage less quoted but more fully indicative of his tortured thinking:

> However much we may sympathize with a small nation confronted by a big powerful neighbor, we cannot in all circumstances undertake to involve the whole British Empire in war simply upon her account. If we have to fight it must be on larger issues than that. I am myself a man of peace to the depths of my soul. Armed conflict between nations is a nightmare to me; but if I were convinced that any nation had made up its mind to dominate the world by fear of its force, I should feel that it must be resisted. Under such a domination life for people who believe in liberty would not be worth living; but war is a fearful thing, and we must be very clear, before we embark on it, that it is really the great issues that are at stake, and that the call to risk everything in their defense, when all the consequences are weighed, is irresistible.

Chamberlain on his best day was not exactly a compelling orator. Here, on one of his worst, he still spoke of fighting bravely against evil. His presentation, however, was confusing and laced with public hand-wringing. He accepted the necessity of war in some situations but conveyed only puzzlement as to whether such a scenario had in fact arisen. He sought to sound analytical but came across as cynical—even afraid. He had dared to share with the public his innermost thoughts but had been too exhausted to speak as he wished to be heard.

His gloom would soon lift. A few hours after his broadcast, the Foreign Office received an intriguing message from the führer that appeared to invite further discussion. The Germans, wrote Hitler, would not move beyond the Sudetenland; a free plebiscite could be carried out; and Germany would join in guaranteeing Czechoslovakia's new borders. One cable with three lies was good enough to hook Chamberlain. The prime minister answered that he thought Hitler could achieve his goals without war; perhaps it would be worthwhile to meet again? After a brief delay, Hitler consented, offering to play host in the capital city of Bavaria.

THE MUNICH CONFERENCE brought together four leaders who had little in common except that none had ever set foot in Czechoslovakia. The deliberations began shortly after 1 p.m. on Thursday, September 29, in the mammoth Führerbau, headquarters of the Nazi party. The delegates arranged themselves in Hitler's spacious office beneath a portrait of Bismarck and in front of a large fireplace. The session was casual, lacking any established agenda, assigned seating, or even official note takers; it was also tedious, because each of the four principals spoke a different language, meaning that every word had to be translated. Hitler and Italian premier Benito Mussolini were perched between the French and British delegations. Hitler began by condemning the Czechs yet again and by insisting that the evacuation (or invasion) start on October 1. Mussolini then put forward a plan that he described as his own but that had been handed to him by the Germans. The document specified what was demanded of Prague. Chamberlain, saying that he could not speak for Beneš, asked that further deliberations be delayed until a Czech representative could be present. The idea was rejected by Hitler as a ploy to buy time.

The Mussolini plan was essentially the same as what Hitler had demanded at Godesberg. The Nazi occupation would commence in

*From left: Chamberlain, Daladier, Hitler, and Mussolini—Munich, 1938*

little more than twenty-four hours. The area ceded would extend well beyond what Great Britain had originally contemplated and would include many small cities and villages where Germans were in the minority; more than 800,000 Czechs would have to move or live under the Reich. A network of zones was drawn on the map to create the illusion of an orderly transfer of authority, but real control would pass immediately to Berlin. The four parties would guarantee Czechoslovakia's new borders, but only against unprovoked aggression; the territorial claims of Poland and Hungary could still be heard.

That afternoon, at Chamberlain's suggestion, two Czechoslovak diplomats, Hubert Masařík and Vojtěch Mastný, arrived in Munich. The prime minister had asked that they be available for consultations, but from the British standpoint, events had rendered their role moot. Instead, the Gestapo escorted Masařík and Mastný to a hotel where the two had unsatisfactory meetings with second-tier officials before being left to languish in their rooms. The conference dragged on past midnight as clerks prepared the texts for signing, a process interrupted briefly by a shortage of ink. The fateful agreement, dated the twenty-ninth, was actually completed in the early hours of September 30. Returning to the hotel, Chamberlain and Daladier handed a copy of the pact to the Czechoslovak envoys, who tried to dispute certain points but were told not to bother; the deed was done. Chamberlain, yawning, referred to himself as "tired, but pleasantly tired," and claimed that the outcome had been the best possible.

Even while the Munich deliberations were under way, Beneš suspected where they were heading. He could no longer hope that the Allies would hold firm. Around midday, he met with the leaders of his military.

The president himself described the scene:

> The representatives of the Czechoslovak army, standing in front of me ... took the floor, one after another. . . . They tried to prove, unanimously and in different forms, this: "Let the big powers decide and agree on anything. . . . The army will not tolerate acceding now to their pressure. . . . We must go to war, whatever the consequences. If we do, the big Western powers will be compelled to follow. The nation is absolutely united; the army is firm and wants to march."

Listening to those words, Beneš was deeply moved but not persuaded. The men in front of him, some in suits, others in uniform, had immersed themselves in the ethics of national honor and had prepared all their lives for just such a moment. He admired their sincerity and the bravery that lay behind their arguments but did not believe in steering by the star of emotion. The facts had become inescapable; he knew this because he had tried so earnestly to find a way around them.

Beneš told the generals that he understood how they felt and why the Czechoslovak people were so determined to fight. But he said he could not consider the sentiments only of the nation and the army. I have to see the whole picture and to weigh the consequences, he said. You are wrong about England and France, he told them. They will not intervene. It would be irresponsible for me to lead our nation into the slaughterhouse of an isolated war, but that does not mean we must despair. "A war—a big European war—will come and there will be great upheavals and revolutions. The allies do not want to fight along with us now [but] . . . they will have to fight hard . . . when we are no longer able to."

In churches and synagogues throughout Czechoslovakia prayers were offered in the name of Wenceslas (by Catholics), Hus (by Protestants), and Moses the deliverer (by Jews). To no avail. German troops crossed into northern Bohemia at 2 p.m. on the first day of October 1938.

*Hitler and troops enter the Sudetenland.*

## 8

# *A Hopeless Task*

The Munich saga was acted out on a global stage by a handful of the world's most powerful people. Its bleak finale formed the first page of numberless other stories centered on the lives of men and women without exalted standing, our family among them. My mother recalled:

> Our personal security was of course touched by it immediately. First Joe, as a reserve officer of the Czechoslovak Army, had to go during the time of the mobilization back to his regiment and I was left by myself with a one-year-old child in Yugoslavia waiting for the war to start. Lucky for me personally, but unlucky for the country to which we both were so devoted, Czechoslovakia was ordered by England and France to succumb to Hitler's demand and so war at that time was not declared.

The events surrounding Munich had a profound and painful impact on the Czechoslovak people, especially of my parents' generation. Feelings of embarrassment for not fighting were mingled with fury at the Allies for their alleged betrayal; both emotions lingered. Writing in 1976, my father assigned the primary blame to France and Great Britain but also lamented that "in her hour of crisis, Czechoslovakia had as her president not a leader, but a negotiator." He acknowledged that much of what Beneš had predicted eventually came to pass but that "the valiant ethos of the nation demanded from its leaders the ethical, not the practicable position. The Munich dictate should have been rejected, no matter what the consequences."

The study of history is surrounded by what-ifs. What would have happened had Beneš chosen—as my father and many others wished—to defy the Munich mandate? Presumably the Czechoslovak armed forces would have fought entirely alone—at least at the beginning. One feels sure that they would have done so with courage for they had the leadership, motivation, equipment, manpower, and training to bloody the enemy badly. Especially in the rain and fog that prevailed in those first weeks of October, this would have been no Blitzkrieg. Fighting from entrenched positions, the defenders would have been difficult to dislodge; but would the superior firepower of the Reich eventually have won out? Almost certainly.

Even if the main German offensive through Bohemia had stalled, the Wehrmacht could have sent troops in from the south (through Austria) and east (through Moravia). Czechoslovakia's antitank weapons and artillery were overmatched, and its core of professional warriors not large enough to hold out indefinitely; the military's own prewar estimate was three weeks. While the conflict raged, the German propaganda machine would have been in high gear, proclaiming the struggle a quest for Sudeten self-determination, a principle already endorsed by the British and French and accepted reluctantly by the Czechoslovaks themselves. The Poles and Hungarians would likely have joined the fight on the opposing side, endeavoring to seize what territory they could from their embattled neighbor. Many, perhaps most, of the Sudetens would have provided the enemy with a fifth column.

In his memoir, Churchill wrote that "Beneš was wrong to yield. Once fighting had begun . . . France would have moved to his aid in a surge of national passion and Britain would have rallied to France almost immediately." With all due respect, the notion that the French would have leaped into the battle seems a pipe dream; they had done nothing in 1936, when Germany had taken the Rhineland; they would do little in 1939, when Hitler invaded Poland; and they had told Beneš outright that they would abandon him if he rejected the Munich

*Hitler in the Sudetenland (October 3, 1938); to his left is Konrad Henlein; on Henlein's left is Wilhelm Keitel, chief of the army high command.*

ultimatum. Yes, there might well have been a flurry of meetings at the League of Nations and numerous unanswered calls for a cease-fire, but before long the Germans would have occupied the country from one end to the other.

In the process of doing so, however, the Reich would have been weakened significantly, especially if the Czechoslovaks had thought to destroy their tanks, planes, and factories rather than permit them to be captured. Such an outcome would have constituted a dramatic gift to Europe on Czechoslovakia's part—an offering for which few would have given the country credit. Tens of thousands of its soldiers and airmen would have died or been taken prisoner, quite possibly including my father. The nation's infrastructure would have been damaged severely. The Nazis, thoroughly incensed by Prague's defiance, would have been savage in victory. If and when the German yoke was thrown off, Czechoslovak storytellers would have had a new generation of tragic yet heroic tales to relate. The country would have endured untold suffering, but its ethos would have emerged unscathed.

Beneš justified his decision to comply with the Munich terms as his best choice from a narrow array of dismal alternatives. A larger European war was inevitable, he insisted, and so was Germany's defeat. By not fighting in 1938, when the odds were so steeply stacked against them, the Czechoslovaks conserved their ability to do so at a more favorable time. This was a judgment echoed by George Kennan, the U.S. political attaché in Prague, who wrote that Beneš had

> preserved for the exacting tasks of the future a magnificent younger generation—disciplined, industrious, and physically fit—which would undoubtedly have been sacrificed if the solution had been the romantic one of hopeless resistance rather than the humiliating but truly heroic one of realism.

Personally, I have as much trouble grasping Kennan's concept of heroic humiliation as I do Cadogan's brief for courageous cowardice. I believe that Beneš should have rejected the Munich terms, but I also find it difficult to condemn him for following the dictates of his own logic instead of the hearts of his countrymen. Abandoned by allies and confronted by enemies on all sides, he was faced with a terrible responsibility. To his credit, he would strive thereafter to make the absolute best of the decision he had felt forced to make.

But what if Beneš had never been put into such an impossible position? What if the British and French had lost patience with Hitler and, instead of pushing Beneš to appease Henlein, had joined forces with Moscow and Prague to take a firm stand? What if they had responded to German military preparations by fully mobilizing their own forces?

Such a strategy would have further motivated Czechoslovak fighters and deepened the misgivings of the German high command. If the Allies had been united, they would have left Hitler with his own set of unappealing options: to back down, endure an open-ended military and diplomatic stalemate, or initiate war in a place and at a time not of his choosing. If the Nazis had decided to attack, the Allies couldn't have prevented them from occupying Prague, but that wasn't Hitler's ultimate objective. Fighting a war against several foes in the fall of 1938 would have subjected the German military to serious pressure on both the western and eastern fronts, while clipping the wings of the Luftwaffe and leaving the country's economy vulnerable to embargo by the Royal Navy.

Western military forces were weaker in 1939 than they would be later, but that was true also of the Germans. The Poles had no love for the Czechs but would still have become allies out of deference to the British and French. Under this scenario, the Nazi plight might fairly have been compared to that of a long-distance runner being forced to negotiate the first mile of a marathon in an all-out sprint. Even if the Nazis had crushed Czechoslovakia, the effort would have prevented—or at least slowed—their march through Europe, and that, in turn, would have opened the door to other possibilities, including a broader rebellion within the German military against Hitler and a shorter, less deadly conflict.

After the war, the imprisoned German general Wilhelm Keitel was asked whether the Reich would have attacked Czechoslovakia in 1938 if the Western powers had stood by Prague. He replied, "Certainly not. We were not strong enough militarily. The object of Munich was to get Russia out of Europe, to gain time, and to complete the German armaments."

Defenders of the British and French leaders have pointed out that the road to Munich was paved before they entered office. The punitive provisions of the Versailles treaty, the defense reductions, the failure to oppose the Nazis over the Rhineland, and the passive tenor of public opinion in the West could not fairly be laid at the feet of Chamberlain and Daladier. I have often told my students that the management of world affairs can be compared to a game of billiards, where every move creates a chain reaction that generates a new set of obstacles and opportunities. A player who begins his turn behind the eight ball should be assessed charitably if he cannot make impossible shots, but ultimately the score will reflect how much was made of the chances given.

In Munich's brief afterglow, Chamberlain addressed a letter to the archbishop of Canterbury. "Some day the Czechs will see," he wrote, "that what we did was to save them for a happier future." We have, he boasted, "at last opened the way to that general appeasement which alone can save the world from chaos." Surely, one definition of inept leadership is to achieve one's goal, take credit for it, then within months have to eat each and every word.

IN LONDON, JAN MASARYK witnessed the exuberant welcome that the British prime minister received upon his return from Munich, including the handshakes and hugs, the claim of "peace with honor," the prediction of "peace in our time," and the joyous shouts of "He's a good fellow!" and "Hip, hip, hurrah!" For weeks, the diplomatic miracle cast a spell over the British imagination. The House of Commons approved the Munich policy by a margin of almost three-to-one; toy shops featured Neville Chamberlain dolls; florists decorated their windows with pictures of the triumphant statesman encircled by roses; and corporations took out full-page congratulatory advertisements. A nation that had been holding its breath felt able to exhale in relief.

As for Masaryk, he had no choice but to resign as minister to Great Britain. Preparatory to doing so, he personally removed his father's portrait from the walls of the Czechoslovak legation and, in keeping with diplomatic custom, paid a courtesy call on 10 Downing Street. Because the prime minister was delayed, Jan was shown in to see the gracious Mrs. Chamberlain. After some desultory small talk, the woman's face lit up. "Oh Mr. Masaryk," she exclaimed. "I must show you the lovely cigarette case that Neville has just received from an admirer." The case was engraved with a map of Europe and adorned with three sapphires—one marked Berchtesgaden, the second Godesberg, and the third Munich.

In Parliament, Churchill was among the few not cheering:

We have suffered a total and unmitigated defeat; you will find that in a period of time which may be measured in years, but may be measured in months, Czechoslovakia will be engulfed in the Nazi regime. We . . . have sustained a defeat without a war, the consequences of which will travel far with us along our road.

Across the Atlantic, the dominant reaction to Munich was fury, less at Germany than at England. Americans were not ready to go to war themselves and had been counting on Europe's leaders to fix the continent's problems before their own involvement became necessary. Thus British representatives, sent to the United States to explain the thinking behind Munich, encountered hostility and ridicule. Dorothy Parker referred snidely to Chamberlain and his frequent air trips as "the first prime minister in history to crawl at 250 miles an hour." Then as now, many people were inclined to express their views through what they wore. In New York, department stores were selling a $1 pin in the shape of a white umbrella—the symbol of Chamberlain in the color of surrender.

BENEŠ WOULD CLAIM that the Soviet Union alone had stood with the Czechoslovak people in their time of crisis; Communist propagandists would make much of that assertion. But is it true? Under the 1935 Czechoslovak-Soviet treaty, the countries pledged to turn to the League of Nations for help if either were threatened. They also promised to aid each other in the event of an armed attack, provided that France also rendered assistance. Soviet leaders had stated on many occasions that they were prepared to meet their obligations, albeit without specifying how. This question was relevant because Russian troops could not reach Czechoslovak territory without passing through either Poland, which refused to grant transit rights, or Romania, which would do so only for airplanes.

One point in the USSR's favor is that, before Munich, its leaders did try to convince Hitler to back down, warning him that the Nazis would face a two-front war if they struck first. The Soviets invited the British and French to coordinate strategies, an invitation that neither acknowledged. As the crisis neared its climax, the Russians claimed to have thirty infantry divisions, reinforced with reservists, near their western border. They also hinted to Poland that, in the event of German aggression, they were prepared to help the Czechoslovaks with or without Warsaw's permission. Ultimately, Moscow was able to keep its word without paying a price. When France ducked its own obligation, the Soviets were off the hook. If France had gone to war on Prague's behalf, the quality and quantity of Russian assistance can only be guessed. Clearly the French had assumed the greater responsibility and their failure to meet it tarnished their good name.

In the end, Munich had three losers: Czechoslovakia, England, and France; it had two winners: Hitler and Stalin. That's a fair one-sentence summary of a historic disaster.

FOUR DAYS AFTER German troops entered the Sudetenland, Beneš abdicated; two weeks later, he left for London. His successor, Dr. Emil Hácha, a sixty-six-year-old former Supreme Court justice, was in poor health and greatly preferred art to politics. Reluctantly, the cautious jurist tried to steer his government of holdovers, second raters, and collaborators in a direction that would pacify the Germans while still preserving national independence. It was a hopeless task.

The Sudetenland is commonly understood to be the northern slice of the country, but under Munich, it was far more than that. As defined by the agreement, the occupied areas extended along the entire western border and also the southern edge most of the way to Slovakia. On the map, the occupied region resembled an open mouth poised to swallow what little remained of T. G. Masaryk's democratic republic.

Adding to the pain, Poland and Hungary pressed their own territorial claims and, with German support, gained lands they had been coveting since World War I. The Czechs, who had been spoiling for battle, were instead asked to acquiesce in the loss of 30 percent of their territory, one-third of their population, 40 percent of their national income, and the majority of their strategic minerals. The powers of their legislature were erased, rendering the various political parties obsolete. Most allies of Beneš were excluded from government jobs, and so too were Jews. The army was cut in half and demobilized. German exiles hiding from the Nazis were exposed and rounded up, while previously captured German spies were released. Antifascists in the Sudetenland were driven out, their property given to followers of Henlein. Slovak nationalists secured autonomy in the form of their own regional administration, their own parliament, and a hyphen—the country's new name was Czecho-Slovakia.

In subsequent months, Slovak separatist parties began to work closely with Henlein and increasingly with Berlin. The many Slovaks who had favored cooperation with the Czechs were pushed aside. If a united republic couldn't stand against the Nazis, why should the Slovaks remain tied to the old capital—especially when the Germans were dangling the prize of national independence?

In Prague the government did its best to avoid the wrath of Berlin, but the Germans devised a sequel to the approach that had worked so well prior to Munich. Nothing the Czechs did was quite enough. Week by week, the list of demands grew: anti-Semitic legislation, economic favors, a share of the country's gold reserves, the dissolution of Communist labor unions, an even more subservient foreign policy. With each new item on the list came the warning that Hitler's patience was again wearing thin.

On October 14, only two weeks after Munich, the Czecho-Slovak Defense Ministry wrote to the Foreign Ministry to request my father's dismissal. The reason given was my mother. Reportedly, she had told some Czech army officers at a lunch in the ambassador's apartment that because of their failure to defend the country, she would rather

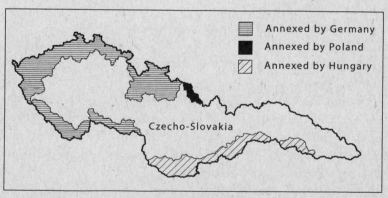

*Czecho-Slovakia after Munich*

marry a street sweeper than one of them. Did she actually say such a thing? I have no idea, but it sounds like her. Did it matter? The issue may have been moot because my father's job would not in any case have long survived the Munich agreement. The profascist leadership in Yugoslavia wanted him out, and so, too, did the Nazis in Berlin. At the bottom of the Defense Ministry letter, there is an addendum in a different size type than the rest: "Dr. Körbel and his wife are Jews."

In late December, my father was recalled from his assignment in Belgrade and given a temporary desk job in Prague. He began immediately to look for a way to move our family to England, where Beneš and other prominent Czechoslovak exiles had begun to gather. Perhaps he could use his contacts in Yugoslavia to obtain credentials as a foreign correspondent in England for a Serb-language newspaper.

Early in 1939, while my father searched for a way out of the country, post-Munich Czecho-Slovakia entered the final weeks of its short life. The Nazis, intent on seizing the whole nation, were once again casting around for a plausible excuse. The year before, the Sudeten cause had masqueraded as self-determination; why not use the same ploy with the Slovaks? The Nazis considered a number of candidates to play the role of a Slovak Henlein, eventually settling on Dr. Jozef Tiso, a conservative party leader, confirmed separatist, and Roman Catholic priest. On March 13, Hitler summoned Tiso to Berlin and gave him an ultimatum: "The question is whether Slovakia wishes independence or not; it is a question not of days but of hours." Tiso had until 1 p.m. on the following afternoon to decide. If the Slovaks did not by then declare independence, Hungary would be invited to swallow them up.

Thanks to its well-placed German spy—Agent 54—the Czech Intelligence Service knew that the Nazis planned to invade, when they intended to do it, and even the invasion's code name (Operation Southeast). The intelligence director, Colonel Moravec, shared the information with Hácha and the rest of the Czech leadership, urging that emergency preparations be made to evacuate military aircraft, blow up munitions factories, destroy secret archives, and transport the country's leaders to Paris or London. Members of the cabinet, convinced that Hitler was content with the status quo, refused to believe that an invasion was imminent. They did, however, decide to seek a meeting to settle the issue. Shortly before dusk on March 14, a few hours after the Slovak parliament voted for independence, Hácha and several of his aides departed by train for Berlin.

MY FATHER'S BROTHER Jan ("Honza") worked for the same construction firm as his father, Arnošt Körbel, and had already set up a subsidiary in England, where it was hoped there might also be room for Arnošt. With Honza employed, it was no problem for his wife and two children to join him. My aunt Margarethe and her husband, Rudolf Deiml, were seeking visas but thus far without success. They had two daughters, Dagmar (Dáša), age eleven, and Milena, just seven. My mother's father, Alfred, had died in 1936. As I would learn much later, Grandmother Růžena lived in Poděbrady, a town about forty miles from Prague. If she made any effort to leave, there is no evidence of it; Czechoslovakia was where she had spent her life and, besides, she had to care for her daughter, my namesake Marie, who suffered from kidney disease.

Of course, no one knew then what was to happen. The war in Europe was still months away. When it did come, it was expected to be over quickly. Nazi prison camps, such as Dachau, housed dissidents regardless of race. For Czech Jews who were not political and who wanted to go, German authorities presented no obstacle; more than 19,000 (or about 16 percent) would leave in 1939. That summer, an office headed by the thirty-three-year-old Adolf Eichmann had been set up in Prague to encourage Jews to emigrate; the challenge was to find governments willing to receive more of those who applied. Every country had a quota of one type or another. The British, who had a League of Nations mandate for the Middle East, had placed an annual limit of 10,000 on the number of European Jews permitted to settle in Palestine. For many older Czechs, the prospect of leaving home was more unsettling than the perceived dangers of staying. As loathsome as the Nazis were, it was unclear how they might profit from persecuting the elderly. "What would they do to me?" asked one mother of her adult son. "I could scrub floors and then the war will be over." Some may also have wished not to take up space on visa lists that, in their view, could better be used by their children and grandchildren.

In the second week of March, my father made brief trips to Paris and London to see whether it might be possible to obtain visas for our family. He was fortunate enough to receive accreditation as a journalist for two Yugoslav newspapers. He returned home the same day that President Hácha left for his conference in Berlin.

EARLY ON MARCH 15, 1939, after waiting for hours while Hitler watched a movie, Hácha met with the führer and his aides. Hitler got right to the point. Because of Bohemian provocations and the unrest in Slovakia, Germany had decided to incorporate what remained of the Czech lands into the Reich. The issue was not up for discussion. The invasion would commence at 6 a.m. Hácha, stunned, refused at first to sign the documents thrust in front of him. Göring threatened that unless he picked up the pen, the Luftwaffe would obliterate Prague within hours. The president consulted by phone with his cabinet, which advised him unhelpfully that active resistance was impossible and explicit acquiescence unconstitutional. Hácha continued to resist, then fainted. At 4 a.m., after Hitler's physician revived him with an injection of dextrose and vitamins, the president, weak in every sense, finally gave in. The statement he signed called on the Czech army to accept the German occupation and declared that he, Hácha, was confidently placing his people's fate in the hands of the führer and the Reich.

Amid the heavy snow that night, one of the few aircraft to take off from Prague's Ruzyně Airport was a Dutch plane sent by the British to rescue Colonel Moravec. He took with him ten senior members of his staff, along with as many secret files and as much cash as they could carry. The next morning, German troops marched on Prague. Because they had not yet received the instruction to surrender, two regiments fought back briefly, thereby earning a permanent place of honor in Czech history. But that was it.

THE NATIONAL MEMORY of any people is a mixture of truth and myth.

For Czechs, 1620 is the year they lost their independence and 1918 the year they regained it. The Ides of March 1939 is when they had their liberty snatched away again. Within days, there were red-bordered posters with an eagle and swastika plastered all around Prague. Storm troopers loitered with fastened bayonets on the streets of Old Town, around Wenceslas Square, in front of cathedrals, and on old Hradčany. The Gestapo set up headquarters. German-language street signs appeared on every corner. According to a March 19 dispatch from the U.S. Embassy:

> There are several thousand . . . political refugees and their families here in hiding and in danger of their lives. Many of the women and children are spending their days and nights in the woods in the vicinity of Prague, notwithstanding that the ground is covered by snow. All relief organizations have been forcibly disbanded . . . the German secret police here are making hundreds and perhaps thousands of arrests in the usual Nazi manner; the Jewish population is terrified; as are . . . those persons closely associated with the former regime.

My parents were among those who had one question uppermost in their minds: how to get out? In the words of my mother:

> To leave Czechoslovakia immediately was technically impossible. There was complete chaos in Prague. Communication was stopped, banks were closed, friends were detained. We learned from competent sources that Joe's name was on the list of people who should be arrested. After leaving Madeleine with my family, Joe and I moved out of our apartment and began sleeping each night with friends, spending the days in Prague streets and in restaurants. It was mostly during the nights that the Gestapo captured people.

After more than a week of living on the run, my parents obtained the necessary paperwork. My mother wrote later that a little bit of petty bribery might have been involved, which would not have been surprising in those days. The Nazis had established an office to process exit visas with an eye to preventing known enemies from leaving, but they were dependent at the outset on Czech inspectors who ignored instructions and allowed hundreds of politically active countrymen to escape.

On March 25, my mother collected me from Grandmother Růžena and, in the afternoon, sat with me in a coffee shop while my father went to the police for the final stamp of approval. When he returned, about five o'clock, we had time to pack two small suitcases before heading to the railroad station. My guess is that Růžena, Arnošt, and Olga were all there to see us off, because in my mother's letter she noted with sadness that that was the last time we saw them alive.

It was ten days after the Nazi invasion. The southeast-bound Simplon Orient Express came through Prague only three times a week; on that day, the waiting platform must have been jammed and the carriages packed. To my parents, the sight of swastikas everywhere would have removed any doubts about their choice. They pushed their way in and handed their tickets to the conductor. The whistle blew, and our long journey began. The sleeping cars consisted of wood-paneled compartments, each with two beds, one above the other, and a tiny washbasin. During the day, the beds could be put up and the space converted into a small sitting room. There being no separate seat for me, I must have been handed back and forth while being encouraged to settle in and sleep. The first border we came to was that of the newly independent Slovakia. Next was the crossing into Hungary, where I imagine every passenger, including every political activist and most particularly every Jew, held his or her breath until travel documents were returned and the train began again to move. After Hungary came Yugoslavia, then on to Greece where we boarded a boat for England. Our destination: freedom.

# PART II

⁂

# April 1939–April 1942

*In our fate, a universal drama is being enacted . . . {as}*
*every resort to brute force is brief compared to the lasting*
*need of man for liberty, peace, and equality.*

—KAREL ČAPEK,
*A Prayer for Tonight,* 1938

———◦———

# 9

## Starting Over

I was too young to remember the tiny rooms in the dank boarding-house where we began our new life, but my mother would not forget those first days in a strange land. We lived among other foreigners, and because England was short of jobs, refugees were barred from seeking employment. Uprooted, we faced a future that was in every way uncertain and had few obvious ways to occupy our time. So as the weather warmed, my mother and I spent many hours in nearby parks while my father sought to reestablish contact with friends.

For my parents, not quite thirty years old, the prospect of starting over amid the alien throngs of London must have been intimidating. The British capital was by far the world's most populous city. Its port was the busiest, its underground rail system the most intricate, its public buildings the most iconic, its financial institutions the center of the economic universe. His Majesty's empire, though aging, still maintained interests on every continent.

It was in London that Beneš had begun his exile, but he soon accepted an offer from the University of Chicago to lecture on democracy. Thus he was in the United States when German troops marched into Prague. He knew from the outset what the invasion signified not only for his country but also for his own credibility. After all, hadn't he warned against trusting the Nazis? Hadn't he said that the Munich agreement centered on a lie? During the First World War, Masaryk had lobbied the world on behalf of Czech and Slovak independence; now

Beneš prepared to lead a comparable struggle to help his country rise from the ashes and, not incidentally, demonstrate the correctness of his own judgment.

In his view, the German attack left Czechoslovakia without a legitimate government, thereby creating a vacuum that only the most recent freely chosen regime—his own—could fill. This demanded that *Professor* Beneš once again play the part of *President* Beneš. Not waiting even a day, he resumed writing letters to world leaders, began issuing instructions to Czechoslovak embassies, and exhorted the media to echo his own keenly expressed blend of indignation and resolve. He made dozens of speeches across the United States, seeking and receiving support from the same immigrant organizations that two decades earlier had aided the cause of independence. He benefited, as well, from the American tendency to rally behind victims of injustice. In New York, Mayor Fiorello La Guardia described Munich as an act of "common butchery" perpetrated by "decadent European democracies and two violent dictatorships."

American audiences may have been deeply sympathetic to Beneš and his plight, but that did not mean they were eager to take up arms. In fact, nothing pleased them more than to be told that it was Europe's job to clean up the mess Europeans had created. On Easter Sunday, President Roosevelt prepared to return to Washington from his retreat in Warm Springs, Georgia. Before boarding the train, he bade farewell to bystanders. "I'll be back in the fall," he promised, "if we don't have a war." This offhand remark set typewriters clacking. Walter Winchell, a popular columnist, wrote, "the future of American youth is on top of American soil—not underneath European dirt." The renowned sage Walter Lippmann urged the administration to use diplomacy to "prevent the hideous consequences of a war." David Lawrence, the founder of *United States News* and *World Report*, echoed the pro-German call for "a second peace conference to undo the wrongs imposed by the Versailles Treaty."

The 1936 Democratic Party platform, on which FDR had campaigned, renounced war as an instrument of policy, pledged neutrality in international disputes, and vowed to resist being drawn into hostilities by "political commitments, international banking or private trading." In 1937, Congress had considered a constitutional amendment that would have required a popular referendum before war could be declared, a potentially crippling measure that was only narrowly defeated. Roosevelt promised over and over again to keep the United States out of war, but conservative commentators noted that Woodrow Wilson had made—and failed to keep—a similar pledge. The president thought privately that a European conflict was inevitable but had not yet decided what America's role should be. To prevent political storms, he typically avoided provocative language, which is why his comment in Georgia generated such a heated reaction. But while the male columnists wrung their hands, one writer seemed ready to raise her fists. "I wonder," wrote Eleanor Roosevelt, "whether we have decided to hide behind neutrality? It is safe, perhaps, but I am not always sure it is right to be safe. . . . Every time a nation which has known freedom loses it, other free nations lose something, too."

On May 28, 1939, Beneš met for three hours with Eleanor's spouse at the couple's residence in Hyde Park, New York. He reported to friends that Roosevelt had greeted him as a fellow president, condemned British and French appeasement, and promised to recognize the old Czechoslovakia in the event of a European war. That description may have embellished the facts. Roosevelt's policies were notoriously hard to pin down. Around the time of Munich, FDR had privately compared British and French actions to those of Judas Iscariot; officially, he cabled his congratulations to Chamberlain. After the invasion of Prague, he warned Hitler against further aggression but without specifying any penalty. Notwithstanding Beneš's description of his meeting with Roosevelt, the State Department had yet to acknowledge him as the legal representative of Czechoslovakia or to support the reestablishment of his shattered land.

Leaving the United States for London, Beneš was in an uneasy frame of mind. How would Czechoslovakia's government in exile fare at the hands of Chamberlain and the architects of appeasement? He soon discovered that as critical as he was of the British leader, so too were many Britons. On July 27, he was the guest of honor at a parliamentary luncheon sponsored by Churchill and his shrewd comrade in arms, Anthony Eden. "I don't know how things will develop," Churchill said, "and I cannot say whether Great Britain will go to war on Czechoslovakia's behalf. I know only that peace ... will not be made without Czechoslovakia." For Beneš, those words must have sounded like a heavenly choir. Many of his countrymen saw him as a failure and Chamberlain considered him a nuisance, but Churchill had written a letter nominating him for the Nobel Peace Prize.[*]

Even before that memorable lunch, the nuts and bolts of a government in exile were being assembled. Jan Masaryk and a small circle of senior émigrés—both civilian and military—were starting work. One morning, a friend of Masaryk's came to see my father and told him, "Here is a key to Jan's flat at 58 Westminster Gardens. He wants you to be his secretary." For my father, this was a career breakthrough at the moment he needed it most. "I was a young diplomatic officer," he later recalled. "Jan was a veteran of diplomacy. We were both without a job. I shall never forget his words of welcome. 'Glad to meet you; heard about you before. Do you need any money?'" Masaryk rented an office near his apartment and recruited a staff that also included Eduard Táborský, a lawyer who had worked in the Czechoslovak Foreign Ministry. Their collective task was to generate articles favorable to their cause in the local media.

At the same time, my father began a lengthy exchange of letters with Hubert Ripka, his elder by fourteen years and a man who shared his commitment to Beneš and the restoration of Czech democracy. Ripka was a broad-shouldered man, well over six feet tall, with short dark hair atop his oval-shaped head, an avuncular manner, and a reputation for being "clever as a bag of monkeys." Ripka had been a diplomatic

---

[*] Churchill was one of many who nominated Beneš; however, the Nobel Committee decided not to designate a winner in 1939—when World War II began—or in any of the next four years.

correspondent for the country's leading newspaper and one of the group around Beneš who had pushed most strenuously for rejection of the Munich agreement. In the fall of 1938, he had moved to Paris, where he had used his many contacts to publicize the country's plight.*

Given the turbulence of the period, it was natural that Ripka had an avid interest in what was happening in London and my father an equal curiosity about the situation in Paris. Their most urgent desire was to help friends who sought visas, often a frustrating quest. A second goal was to discourage the formation of rival power centers; the Czechoslovaks did not need to have more than one government in exile. A third focus was to gain exposure to influential writers, whether in the Slavic countries or the West. One such writer, Shiela Grant Duff, served as an intermediary, carrying letters back and forth between my father and Ripka, whom she had earlier befriended in Prague.

As a female foreign correspondent, Grant Duff was something of a pioneer. She developed a strong following while writing for the popular *London Observer* and was one of the few British reporters to mount a sustained challenge to Chamberlain and appeasement. Her disdain for her colleagues was evidenced in a rhyme she was fond of quoting:

> *You cannot hope to bribe or twist*
> *The honest British journalist;*
> *But seeing what the man will do*
> *Unbribed, there's no occasion to.*

At Ripka's urging, Grant Duff introduced herself to Churchill (she was a cousin of Mrs. Churchill) in order to educate him further about the situation in Czechoslovakia. Her book *Europe and the Czechs* was a heartfelt plea for Britons to take seriously the fate of that small country;

---

* Ripka's letters to and from my father were among the documents made available to me in 2011 by the Czech Institute for the Study of Totalitarian Regimes. The correspondence—parts of which are blacked out—had been maintained in the files of the Foreign Ministry until being removed, sometime in the 1950s, by the Communist secret police.

the paperback appeared on the same day that Chamberlain returned from Munich and sold so well that an updated edition was published just two weeks later.

My father's correspondence with Ripka, which began in May 1939 and continued for two years, had a conspiratorial flavor. Both men were wary about whom to trust. In his first letter, Ripka asked that a message be transmitted to the director of intelligence, Moravec, requesting information about a man in Holland who might or might not have been working on behalf of Jews seeking to flee Hitler. My father in turn described a visit from "a man called Zid . . . [who] elicited in me a peculiar impression." Zid sought a direct audience with Masaryk, flashed money around, and was vague in responding to questions. My father, whose trust had to be earned, thought him a spy.

It was healthy for my father to have a position that demanded his energies, a destination when he left our apartment each morning, and friends with whom he could work and commiserate. For my mother, life was more of a chore as she struggled to occupy her mind and keep me entertained in a city where it was hard for her to communicate and where the overall mood had darkened. As refugees in London that summer, we had plenty of company. Jews and other antifascists arrived from Germany, Austria, Poland, and our own Czechoslovakia. The British had quotas that limited the number of adults, but an exception was made for unaccompanied children under seventeen years of age.

A humanitarian program, the Kindertransport, had begun rescuing Jewish children from Germany and Austria. A similar but separate Czech operation was set into motion by Nicholas Winton, a British stockbroker who had visited Prague at the invitation of a friend, encountered German thugs everywhere, and returned home determined to save whomever he could, especially children. "I wasn't allowed to bring anybody in until I had a family and guarantors that would look after them," Winton recalled, "and it wasn't always easy to get people to make that commitment because some were very, very young." To place more youngsters, he made repeated appeals to the United States, but no help was forthcoming. Of the roughly six thousand children whose names were on Winton's list, only one in ten reached England.

Among those who did was my cousin Dáša Deimlová, then eleven, the daughter of my father's sister. She was aboard the second of the four Winton trains, departing Prague at the end of June. Aside from a small suitcase, she carried only a tiny doll; around her neck was a cardboard sign bearing the number 298. There were six in her compartment, all girls, ranging in age from two to fifteen. Dáša quickly introduced herself to a child with the same first name as her seven-year-old sister, Milena. As the locomotive pulled out of the station, leaving behind parents and friends, the two girls closed their eyes, held hands, and promised each other, "We will not cry." When they reached the German border, the train halted for almost five hours. There had been a mix-up with the paperwork, and proper documents had to be fetched from Prague. The youthful passengers sat, peering anxiously through the windows as Nazis with their fearsome-looking rifles and bayonets marched along the platform. Finally the train resumed its journey westward, passing through Dresden, Frankfurt, and Cologne. The incessant motion upset Dáša's stomach; she accepted an older boy's offer of alcohol, the imbibing of which upset her even more. Not until they reached Holland were the children allowed to stretch their legs, given postcards to send home, and treated by the Red Cross to bananas and cocoa.

*Dáša and Milena Deimlová with the author, one year old*

From there the children took a ferry to Harwich; most then continued on by rail to Liverpool Street Station. Like her companions, Dáša experienced the trauma of sudden separation from parents and homeland. Unlike many, she was old enough to comprehend the reason for leaving and had the comfort of a familiar set of faces at her journey's end. Her sister had been on the list to come, too, but their parents had had a last-minute change of heart, thinking Milena too young. Fifty-seven years later, Dáša told a reporter from the *Washington Post* that the reason Milena had not been on the train was that she had a broken arm. That wasn't true. At the time Dáša did not wish to admit, as she did later to me, that she had never forgiven her parents for their fateful decision; many children younger than Milena had been on the train leaving Prague. The unbearable irony is that my little cousin would have her life cut short not because of her parents' indifference but because of the intensity of their desire to protect her.

My father met Dáša in Harwich and brought her to our apartment. "We took her over in good shape," my father wrote to her parents, Rudolf and Greta. "She was one of the few who was not tired. . . . In a few days we will take her to school. . . . Do not worry, we will take good care of her, and besides I can see that she is a very reasonable little girl." He added:

> Perhaps I will soon learn if it will work . . . it is now harder, because you did not send Milenka. With Canada, Rudo do not have illusions. Kisses—we have not heard from Mother in 2 weeks.

Deciphering those words now, I believe that my father was trying to use whatever connections he had to help Dáša's parents leave Czechoslovakia. He worried that their decision to keep Milena might complicate the matter and was unsure he would succeed.

In the summer of 2009, the exodus of the Winton children was reenacted, using the same locomotive and taking the identical route between Prague and London. Among the passengers were Dáša, then eighty-one, and her old seatmate, Milena Grenfell-Baines. A young girl, dressed in the style of the 1930s (hat, simple coat and frock) was also aboard, representing the travelers from long ago. Around the girl's neck was the number 298, Dáša's number, on the very piece of cardboard that my cousin had worn years before. Waiting to greet them in London was a friend celebrating his one hundredth birthday, Nicholas Winton, the man who—when others had merely shrugged—had acted just in time to save their lives.

FOR NEVILLE CHAMBERLAIN, the Nazi invasion of Czechoslovakia had been a profound embarrassment. The hero of six months earlier had been exposed; Hitler had played him for a fool. The führer's territorial ambitions did indeed extend beyond areas where Germans were already a majority; Poland was likely to be next. Having tried appeasement, the prime minister switched almost literally overnight to deterrence. In late March 1939, he declared that England would rush to the aid of Warsaw in the event of a German attack. It was a tough stance but lacked a military strategy to support it. The British could no more save Poland than they could have interposed their troops between Germany and Czechoslovakia. The hope was to convince Hitler that he could not invade without triggering a larger war. To prepare for that contingency, Chamberlain proposed that twenty-year-olds be drafted, the first peacetime conscription in modern British history. In the villages, people talked about what they would do "If the worst came to the worst." "It was funny," observed the fictional Mrs. Miniver in Jan Struther's contemporary novel of the same name, "how one still shied away from saying, 'If there's a war,' and fell back on euphemisms."

Meanwhile, diplomatic wheels continued to grind. The German Embassy informed the Foreign Office that the Reich would henceforth assume legal authority over persons in England who were of "the Czechoslovak race." The British rejected that but were uncertain who, if not the Germans, could legitimately speak for the occupied nation. To them, Beneš lacked official standing. One question was how to shield people from deportation who, like the members of my family, were traveling on Czechoslovak passports; the solution was to designate us as "stateless persons." An even knottier conundrum was whether to continue inviting the staff of our now-orphaned legation to official parties. After much high-level debate, the Foreign Office settled on a compromise: our diplomats would remain on the guest list through the summer, after which their names, like our country, would be erased.

THE BRITISH GOVERNMENT'S essential tasks were to rearm and to persuade Germany that war would be a mistake. The military buildup began in earnest, but the diplomatic aspect stalled. The logical strategy, endorsed by Churchill among others, was to forge an alliance with the Soviet Union; that would leave Hitler in a position where any conflict would have to be fought simultaneously on the eastern and western fronts. The Russians were eager to conclude such an arrangement, but Chamberlain held back because of his disdain for Stalin and because he worried that the führer would view a London-Moscow alliance as a provocation. He was also mindful of Polish opinion, which was at least as hostile to the Communists as to the Nazis.

This was an opportunity lost. Stalin grew suspicious that the West intended to sit back and pick up the pieces after a war between his country and Germany. He knew that the German factories were still in need of raw materials that only his country could provide. To keep his options open, he fired his foreign minister and replaced him with Vyacheslav Molotov, a tough, unflappable survivor of Kremlin politics. Molotov had no love for the West, nor any trace of sentimentality; he was precisely the sort of man with whom Hitler could do business.

In May, Beneš learned that arms deliveries from the Czech Škoda Works to the Soviet Union were continuing despite the Nazi occupation. He concluded that some sort of secret understanding between Germany and Russia had been reached. This was big news, and he immediately conveyed his suspicions to the Foreign Office. Belatedly, English diplomats made fence-mending trips to Moscow, where they were offered plenty to drink but no deal; the opening for an effective antifascist alliance had closed.

On August 17, the U.S. government sent a cable informing London that Germany and the Soviet Union were about to climb into bed together. The document lacked a priority marking and so was not opened until August 22—the day that Hitler and Stalin announced their shocking agreement to remain neutral in conflicts involving the other. Beneš was at his desk when word of the deal spread. Because it was August, the British government was on vacation: Chamberlain was fishing in Scotland, Halifax was at his estate in Yorkshire, and the senior British military commanders were busy shooting—geese.

The pact between Germany and the USSR appalled the British and most everyone else in the West, even many Communists. Beneš, in his most cold-blooded state, was pleased, aware that Hitler would view the agreement as a license to initiate war—the only means of restoring Czechoslovak freedom. He also understood the logic of the agreement, perhaps even better than Stalin. The Germans had gained a supplier of wheat, petroleum, timber, and minerals, along with a green light to invade Poland from the only country even remotely in a position to stop them. The Soviets had secured access to manufactured goods and the opportunity to seize the Baltics and the eastern half of Poland without worrying, for the time being, that Germany would open fire. Stalin told his colleague and eventual successor, Nikita Khrushchev, "Of course it's all a game to see who can fool whom. I know what Hitler's up to. He thinks he's outsmarted me, but actually it's I who've tricked him." Beneš believed, more accurately, that the Nazis would not wait long to double-cross their partner.

This cynical arrangement meant, almost surely, that Europe would soon be at war. As British leaders swapped their fishing and hunting gear for suits and uniforms, a last round of diplomacy played out. With the Russia card now faceup on the wrong side of the table, the Allies had no choice but to vest their hopes in Mussolini. The Italians—or, in Cadogan's phrase, "the ice-creamers"—had little to gain from a European war. French and British diplomats urged Mussolini to restrain Hitler while they put pressure on Poland to make whatever concessions might be necessary. At the same time, every enterprising businessman who claimed to have a back channel to the German leadership was given a hearing. The Allies called up reserves and mobilized their fleets. Ambassadors exchanged talking points. British aristocrats contacted their Nazi acquaintances. None of it did any good.

Early on September 1, 1939, fifty-six German divisions, supported by

1,500 aircraft, swarmed into western Poland, surrounding and smash-
ing the defending troops while sowing terror in the civilian population. The Poles fought back bravely, but with inadequate numbers and
no reserves. Much of their air force was destroyed before it could take
off, while their horse-borne cavalry was no match for German tanks.
Before succumbing, they launched a desperate counterassault that prolonged the struggle but only until the end of the month. By that time,
the Soviet army had swooped in, vulturelike, to devour Poland's east. A
line separating the German and Russian zones was carved through the
country's midsection. The Second World War had begun.

# 10

# Occupation and Resistance

R eichsprotektor Baron Konstantin von Neurath arrived in Prague by train on April 5, 1939. A daylong celebration ensued in which representatives of local organizations, obliged to take part, did so listlessly and with many a muttered oath. Berlin's intent was to plunder the Czechs without provoking rebellion; thus Neurath hoped to see them adapt quickly and passively to their loss of freedom, minimizing any need for brutality. The silver-haired protektor was by nature more diplomat than firebrand, having been sacked from his previous job as foreign minister because of his distaste for Hitler's war plans. He took care in his new position to show public deference to President Hácha and to maintain the fiction that the Czechs retained a meaningful voice in governing their affairs.

Germany's show of empathy was not confined to that displayed by Neurath. In the first days of the occupation, a Bavarian welfare organization sent a caravan of volunteers to Prague. Their mission was to feed the city's children, who—as Nazi propaganda had it—were starving due to the incompetence of local authorities. In fact, the only youngsters in need of free meals were antifascist Sudeten refugees. When the Bavarians discovered that the hungry were not as numerous as anticipated, they asked some of the more photogenic ones to demonstrate how they said their nightly prayers. The resulting images were sent to Berlin with the caption "Prague children beg for food."

Barely a week into the occupation, the parliament was dissolved and the traditional political parties disbanded. In their place, Hácha created National Solidarity (NS), an organization that included virtually the entire Czech population—except for Jews and Freemasons, who were excluded to please the Germans, and women, who were barred because Masaryk and Beneš were no longer around to insist on fair treatment.* The NS was a pragmatic entity, not an ideological one. It sought to accommodate the German occupation without abandoning the country's indigenous culture and customs.

The Czechs who held official positions during the years of the protectorate would later be vilified as traitors or, as my mother disdainfully referred to them, "collaborants." The labels did not always capture the intentions of the figures involved. In the beginning, Hácha sent messages to Beneš in which he pledged his loyalty: "I am looking forward to the day when I will hand over my office; you know to whom." The old judge had not sought the presidency and always seemed on the verge of resigning. His avowed purpose was to minimize harm, but he failed to recognize that damage can be done to the spirit as readily as to the body. He urged his people to be both good Czechs and good Germans—a possibility in peacetime, perhaps, but not under occupation. Starting as an ally of democracy, Hácha ended as a foe; an exceptionally weak reed, he stood for nothing in a job he never should have accepted.

By contrast, the new prime minister, General Alois Eliáš, turned his face to the wind and refused to buckle. Eliáš too avoided provoking the Nazis, but behind their backs he maintained close ties to the Czech underground, funneled intelligence to London, and did all he could to help the families of those arrested. Many other officials tried to preserve what they could of national identity and independence, hoping that—despite the early German triumphs—the war would be over quickly. However, as the months dragged by, members of the Hácha government found themselves in the untenable position of being hated by Czech loyalists, bullied by German overseers, and respected by no one.

---

* In Czech, the party's name was Národní Souručenství. To many Czechs, the NS badge, worn upside down, stood for "Smrt Němcům," "Death to Germans."

Watching from the U.S. legation was George Kennan. Known throughout his career for penetrating insights and a lack of romanticism, he wrote that "one of humanity's oldest and most recalcitrant human dilemmas" consists of the choice between "a limited collaboration with evil, in the interests of its ultimate mitigation" and "an uncompromising, heroic but suicidal resistance to it." Everyone involved in the drama of post-Munich Czechoslovakia, he observed, would be "tossed, one way or another, on the horns of this dilemma."

BEFORE GOING INTO exile, Beneš had discussed with friends the need to fashion a unified resistance that would articulate a clear political line and operate effectively both domestically and abroad. The president's thirty-nine-year-old personal secretary, Prokop Drtina, was among those who stayed behind in Prague to organize the effort. The dissidents had many friends who were still in government, some in the mayor's office and city council, but also bookkeepers, phone operators, and file clerks who could supply useful information. The network drew primarily upon Beneš's political supporters, the military, former members of the Czechoslovak Legion, the Boy Scouts, the Sokol gymnasts, and such Jewish organizations as the Maccabi athletic club. Early on the resistance helped funnel soldiers and other escapees across the border into Poland and, when that avenue was foreclosed by war, through Hungary.

As in any underground operation, secure communications were vital. In the first weeks, the conspirators had received a cipher (concealed in a toothpaste tube) disclosing an address in Turkey through which reports could be transmitted to Beneš. Drtina used this channel to send regular reports to the president. These, too, required secrecy, and I was intrigued to find in my research that Josef Korbel was credited by one resistance leader with suggesting a clever code involving dictionaries. However, no sooner was the dictionary idea agreed on than it was replaced by a more sophisticated system devised by the army.

Throughout the war, written materials were smuggled by sympathetic railway workers shuttling among depots in Prague, Bratislava, Budapest, and Belgrade. In defiance of all risk, the underground established radio contacts that—although interrupted from time to time—transmitted thousands of messages from the protectorate to England and the Soviet Union. The equipment was serviced by city engineers, who bicycled to the clandestine sites at night. The underground's principal transmitter, code-named Libuše, was a briefcase-size contraption with dials and knobs attached to a barbed-wire antenna, climbing skyward in a stiff array of gnarly knots; the device now resides in the Czech National Museum.

Although the rebels had a central coordinating body (the Committee of Home Resistance), the various groups and cells were by design as independent of one another as possible. Meetings were kept small and confined to a single subject; new recruits were allowed to attend only after thorough investigation. Before returning home, leaders telephoned ahead to be sure the Gestapo was not sitting in their living rooms, waiting. A pot of chrysanthemums removed from a balcony or the altered position of a window shade might also serve as a warning. Important documents were hidden where stacks of paper would not be conspicuous, for example, in the public library, or in a place where few would look, such as inside a department store mannequin.

Couriers were enlisted to carry secrets and also to distribute pamphlets, notices, and antifascist literature. The government's Press Supervision Service controlled every legal newspaper; independent publications, however, still managed to operate, particularly the main resistance paper, *V boj* (Into Battle). One clandestine group produced books of democratic propaganda that, from the covers, looked like conventional detective novels. The telltale sign was on the back, where the publisher was identified in Czech as "G. E. Stapo."

The Nazis were inexperienced in the arts of occupation, but they had an aptitude for repression, infiltration, and terror. Working from lists of suspects, they banged on doors and pulled thousands of people from their beds in the darkest hours of the night. The men and women brought in for questioning needed either a truly convincing story or the ability to withstand excruciating pain. Gestapo headquarters, located near the center of Prague, was in the well-fortified Petschek Palace, used previously as a bank. The safe-deposit rooms, with their windowless walls and heavy doors, were well suited to holding prisoners. Torture was applied without mercy, and the presence of a guillotine rendered verbal threats redundant. Whenever a member of an underground cell was picked up, others went into hiding; the presumption was always that suspects would talk. The Resistance, though, was rarely caught by surprise. Until 1943, a group of German-speaking Czech police, employed by the Gestapo as translators, used their access to report on what prisoners had revealed and to warn dissidents when they were being watched.

Most of the major underground networks were torn apart at least once during a war in which tens of thousands of dissidents were killed. But for all the bloodshed, the Nazis did not come close to breaking the Czechs' spirit or their will to resist. "If German authority in the physical sense is unchallenged," wrote Kennan more than a year and a half after the occupation began, "morally it does not exist. Whatever power the Germans may have over the persons and property of the Czechs, they have little influence over their souls."

*Guillotine used by Nazis in Prague*

From the earliest days, the Czechs engaged in symbolic protests such as boycotting streetcars or, on Hitler's birthday, placing flowers around the statue of Hus. When the Prague Orchestra performed Smetana's *My Country*, the ovation lasted for fifteen minutes. Until the practice was forbidden, some citizens wore homemade badges bearing such inscriptions as "We will not surrender" and "Beneš is not asleep." German officials who were assigned to Prague often found that their telephones did not work, that important documents had been misplaced, or that the fuel tanks in their automobiles had been siphoned dry. On the country's National Day in October 1939, a massive anti-occupation rally so angered Nazi guards that they opened fire, fatally wounding a medical student, Jan Opletal.* At his funeral, friends were bold enough to sing the national anthem and to storm through the city shouting patriotic slogans and tearing down German street signs. When Hitler learned of the disturbances, he demanded reprisals. The Nazis arrested nine student leaders—none of whom had been involved in the demonstrations—stood them against a wall, and shot them dead. Another 1,800 students were detained and held under brutal conditions, many of the boys beaten and girls abused. To punish the local intelligentsia, the führer closed the protectorate's Czech universities and colleges for the duration of the war.

Geography, almost as much as the Germans, limited what the Czech Resistance could do. There were neither ports through which arms could secretly be shipped nor friendly borders across which a secure base of operations could be established. Underground warriors had relatively few guns, a small supply of ammunition, a paucity of places to hide, and little cash. The longer the war lasted, the harder it would be to survive. Also, nearly everyone involved was an amateur. At first, Beneš did not appreciate these constraints. In a broadcast two weeks into the war, he asked the Resistance to deliver a steady diet of heavy blows to the enemy. After the death of Opletal and the subsequent executions, he spoke more somberly, cautioning against superfluous

---

* Fifty years later, a march to commemorate Opletal's sacrifice also veered out of official control, leading directly to the Velvet Revolution.

*Beneš speaking over the BBC*

sacrifice. The president had not changed his view that the Czech underground should make life more difficult for the Nazis. He had come to realize, however, that if it were to administer "heavy blows," it would require external help.

THE FIRST YEAR of occupation was marked by tension between the Czechs' desire for normalcy and their anger that nothing was as it should be. For the majority, life went on as it had—up to a point. Daily routines were not altered, even if shortages had inflated the cost of living and the announcements blaring from street-corner loudspeakers were in what had become a despised foreign language. Food rations were austere but sufficient. Thousands of young men went to the Reich to labor in place of the German youths who had been called to the military. Back home, many Czechs were allowed to retain their government positions. The sense of routine was strongest in the countryside, where a little boy or girl might still enjoy a relatively carefree existence. One such youngster developed a fascination for uniforms. Of course, these were present everywhere—worn by the police, the remnants of the Czech army, and the various Nazi units. Each day, when he could, the youngster went to a store where uniforms and medals were on display in the windows. He stared until an adult grabbed him by the arm and pulled him away. Sitting in his room, he drew pictures of what he had seen, always imagining better and more elaborate outfits. Years later, as the newly elected leader of Czechoslovakia, Václav Havel took delight in authorizing new designs for the uniforms of his presidential guard.

The cinema was less affected by the occupation than many distinctive Czech industries. The Nazis took control of some studios in order to produce German films without having to worry about Allied

*Drawing by the young artist Václav Havel*

bombs, but they also permitted local moviemakers to carry on their business. The best-equipped studio in Europe had been established by young Havel's father and uncle Miloš, a noted producer. The Germans pressed Miloš to make a film that would portray King Wenceslas as the original German collaborator. Havel refused and instead brought to the screen Božena Němcová's *The Grandmother* and other traditional tales. One of the actresses with whom he worked was Lída Baarová, controversial because she had, in the 1930s, an extended love affair with Joseph Goebbels, the Nazi minister of propaganda. Goebbels planned to divorce his wife and marry the glamorous actress until his children's godfather—Adolf Hitler—forbade it. Such a scandal, the chancellor warned, would undermine the Nazi reputation for upholding family values.

The relative normality of life meshed with Germany's long-term plan to transform the Czech lands into an integral part of the Reich. This would be done in stages, by milking the country's resources and gradually altering the population's racial mix. The milking began almost literally, with the inventorying and theft of cows; it continued with the German takeover of Jewish properties and major Czech businesses, including the Škoda industries, the Bat'a shoe factory, the Bohemian Union Bank, and the Sigmund pump works. The bulk of the protectorate's tax revenues now flowed to Berlin, not Prague, and every usable piece of military equipment was confiscated, including 600 tanks, 48,000 machine guns, more than a million rifles, and the entire Czech air fleet.

Hitler envisioned a time, perhaps twenty years hence, when the Czech language would be reduced to a dialect and the people who spoke it to a scattered peasantry with colorful costumes, quaint dances, and no political standing. Nazi officials scoffed at Czechs who complained about the closure of universities, saying that in the future, an elementary school education was all that any of their race would need.

As the months passed, a split developed in the protectorate's leadership. Neurath continued to respect local sensibilities, believing that the population could be kept docile if allowed to maintain its traditions. A harsher line was favored by his deputy, Karl Hermann

"K. H." Frank, a Sudeten who despised Czech culture and wished to Germanize the population without delay. In that he mirrored the frustration felt by many from his region. Sudeten separatists had been overjoyed when Hitler's army had marched in, expecting that they would be placed in positions of authority, but, aside from Frank, few were. Even Henlein, the discount führer of the prewar era, was stuck in his home territory, denied a post in Prague. Worse, when the war began, every registered German male was eligible to be drafted and sent to the eastern front. So while Czech youths were assigned to factories and work details, their German counterparts were camped out in the mud or getting shot. That did not feel like the victory for which the Sudetens had yearned.

Increasingly Czechs were prohibited from any gesture, including making disrespectful comments and booing German sports teams, that hinted at independence. However, repressive measures only stimulated the population's desire to preserve its customs and heritage. One resistance magazine warned, "With great fanfare, the Germans are opening new schools where there used to be none. This is your business, women! It lies in your hands whether our children grow up to be Czechs or Germanized, patriots or traitors." Nationalists had long advocated the use of schools to create a sense of cultural solidarity; now they focused on the imperative of learning at home. Parents were encouraged to teach their children the country's language, stories, songs, and a heroic version of its history. Less helpfully, families were told to ignore the German nutritionists who advised against consuming too much butter; the Czechs saw in this well-intentioned guidance a plot to rob their children of rosy cheeks.

With the Nazis in charge, the only way Czech women could avoid work orders was to become pregnant; this they did with alacrity. Their men, unlike the Germans, were barred from the battlefield and thus more readily available for domestic pursuits. There were also many families that considered the bearing of children a patriotic duty. During the war, Czech couples got married earlier and had children younger. The birthrate increased by 50 percent. Perhaps for that reason, the protectorate was stirred by rumors, thankfully untrue, that German doctors were planning to sterilize local women and inject their babies with poison.

For Jewish families, the tourniquet had started to tighten even before the Nazis arrived. Bohemia and Moravia were not yet subject to the level of persecution that prevailed in Berlin or that would shortly be felt in Slovakia; Jews could still practice their religion and synagogues were undamaged, but discriminatory policies were becoming the norm. Jews were banished from public office and the professions and ordered to sit in the back of streetcars and to avoid many public places, shops, and parks. Restrictions were placed on their access to financial accounts, and valuable possessions were confiscated. Their businesses were seized or purchased at nominal prices, and food rations were even more Spartan than those of their Czech neighbors.

The public's reaction to these measures varied. Many Czechs were indifferent, but others found ways to bend the rules. According to one historian, the German Secret Service was incensed that rather than shunning Jews, "friends went shopping for them. . . . Butchers were selling their best meat only at the time when Jews were allowed to shop. . . . Jews [were] getting help from physicians, lawyers, their former employees, the Czech authorities and sometimes even from gendarmes." Sympathetic Czech judges were quick to rule in favor of applicants seeking to be declared, or to have their children declared, of less than fully Jewish blood—a procedure that often required Jewish women to confess falsely to affairs with Gentile men. When one despairing woman poisoned her two half-Jewish children, her neighbors did not turn away. Instead, four thousand people—including town officials—attended the funeral.

For the exile community in England, letters from relatives back home provided a narrow window on such events. Correspondence to and from the protectorate was irregular. Under wartime conditions, many letters never arrived. Others were sent in care of the Red Cross in Switzerland and delivered only after months of delay. My mother and father received at least a few letters from their parents, but I have not been able to find them and don't know what information they conveyed or when they stopped. Dáša's letters from her mother, Greta, alternated between the practical ("wear warm clothes") and the poignant:

> Milena cried a lot when we came home without you. In the morning when I was combing her hair, she asked me to look to see if she had gray hair from worrying about you. In the evening, she gets into bed and constantly calls with all of her strength: "Dáša," "Dáša," "Dáša," and thinks that you can hear.

As the months passed, the letters from Greta became less frequent. At various points in 1940, Dáša learned that Milena was beginning to ski and had become "a real rogue," studying with diligence but refusing to sit still in class.

In 2011, I asked my cousin what she remembered of her family and of those early years that I had been too young to recall. She told me that her mother had been lovely but also strict, a person who believed that spoiled children would have a hard life. Both metaphorically and literally, she felt that the best way to teach a child to swim was to throw him or her into the water and see what happened. In fact, it was Dáša who had taught Milena to swim, as she would later teach me.

Dáša's father, Rudolf, was a general practitioner who was popular enough among his neighbors that other doctors were jealous. Unlike Greta, he tended to be lenient with children and rarely said a harsh word. Only once did Dáša make him angry:

> Our house was beside a stream over which there was a small footbridge. One day, a friend of mine named Vera was injured by a passing truck. Her father carried his daughter over the footbridge and began calling for my father, who wasn't home. But I was. I opened the door to my father's surgery, which was on the first floor of our house, and began applying disinfectant to my friend's scrapes and bruises, as I had often seen my father do. Just then, he arrived home, took one look at what was going

on, grabbed me by the collar, and spanked me. "What are you doing?" he yelled. "You have no training. Don't you know you could have killed that girl?"

Dáša told me later that she had been raised without religion, attending synagogue but once a year. Nevertheless she was required by her school to participate in a class on scripture and so studied the Hebrew Bible. She got on well with the rabbi who taught the course and, wanting to impress, invited him to visit over the holidays so he could see her Christmas tree. That led to a row between the rabbi and her father, who said, "I'll raise my children as I like."

The Deimls lived in Strakonice, a city of about 20,000. Before their separation, Dáša and Milena had played with neighbors the many games children learned back when entertainment came more from the imagination than from expensive gadgets: marbles, hide-and-seek, stop-and-freeze, musical chairs, blindman's bluff, hopscotch, yo-yos, jump rope, and various card games. No child would have thought to restrict participation according to racial or ethnic background, but the leaders of the protectorate had an agenda to carry out. Milena was forced to transfer to an all-Jewish school where youngsters of every age shared the same classroom; it closed down after a year, and then there was no school for her at all.* Greta wrote that she had had to take the place of Milena's young friends, with whom her daughter was no longer permitted to play.

Despite urgent appeals from Western diplomats (including that of the United States), the opportunity to leave the protectorate legally was dwindling. The time would soon come when the door to the outside world would swing completely shut.

---

* One of Dáša's former teachers was brave enough to go to the Deimls' house and tutor Milena. Greta entrusted her with the family's letters, which were returned to Dáša after the war.

# 11

# The Lamps Go Out

On the morning of September 3, 1939, shortly after Chamberlain informed his countrymen that war had commenced, a French aircraft strayed into British airspace, setting off sirens and causing a brief panic. For the next seven months, aside from the Nazi conquest of the Channel Islands and a few exploratory patrols by the Luftwaffe, that was the extent of military action in Great Britain. The French, fearing retaliation, discouraged England from bombing Germany in support of the Poles; the Nazis were unready to do battle with the West. This was the period—that September until the following spring—that became known as the "phony war" or Great Bore War.

The English made wise use of the interval. Air-defense measures, which had been in train for several years, were now a daily preoccupation. Whole factories were draped with camouflage in the form of brown-and-green netting. Trenches were dug in zigzag fashion through downtown parks, and air-raid shelters were built in backyards, creating, if nothing else, luxurious new homes for dogs and other household pets. The memory of World War I led to the nationwide distribution of gas masks—some, for children, with Mickey Mouse ears attached. Practice sessions were conducted in which adults and teenagers donned the devices and crawled through smoke-filled tunnels made of tin. The masks were supposed to be carried in a cardboard container attached to a string that could be looped around one's shoulder, but inevitably the strings got tangled in handbags, lunchboxes, backpacks, and doors. For a time, everybody carried a mask; by war's end hardly anyone did.

During the late winter and early spring of 1940, the term "jitters" came into fashion. Hitler was forever talking on the radio, but what did he really intend? Restaurant and barbershop conversations buzzed with rumors about the date and landing site of the anticipated German assault. Anxious villagers raised their binoculars to scan the skies. Churches fell silent, as bell ringing was reserved for spreading an alarm in the event of invasion. Meanwhile, hands were kept busy stitching quilts and rolling bandages. To children, some sacrifices were greater than others. The Stop Me and Buy One ice-cream salesmen vanished from the streets, their refrigeration units to be held in readiness for the transport of blood.

When World War I began, it was said that "the lamps are going out all over Europe." In World War II, the sense of encompassing darkness was reinforced by measures taken to deprive enemy bombers of potential targets and illuminated landmarks. In London and the surrounding countryside, windows were covered at night with thick black curtains that proved stifling. Air-raid wardens checked on every house to guard against any telltale glow. For the same reason, streetlights were turned off and car headlights masked. Although intended to save lives, the blackout in that first year had the opposite effect. For every Briton killed by the Germans, more than a hundred died in automobile or pedestrian accidents.

Worries about an invasion had preceded the declaration of war. In addition to civil defense measures, plans were made to send children and caregivers from London and vulnerable coastal regions to inland villages and remote towns. Lists were drawn up and the countryside canvassed to identify spare bedrooms, a task made easier by the departure of young men and women to military service. The children with their manila name tags departed from Paddington Station, each carrying a small bundle of tinned milk, tinned corned beef, tinned chocolate, and a packet of Woolworth's cookies. After consuming those delicious items, the youngsters collected the metal containers for recycling. Despite the meticulous planning, the evacuation project soon lost momentum. Families were understandably distressed at being separated, especially when the enemy had yet to show up. Many of the children who were supposed to leave London never did; many who did soon returned.

THE DECLARATION OF war accelerated the flow of Czech and Slovak refugees to the British capital. Soon the office established by Jan Masaryk began to take on the aspects of a full-fledged shadow government. My father was given a new job, to organize and manage radio programs for the audience back home; these began on September 8, 1939, just a week after the Wehrmacht had crashed its way into Poland. "The hour of retribution is here," proclaimed Masaryk:

> The limits of the patience of the Western democracies have been reached and the struggle to exterminate Nazism has begun. Our program is a free Czechoslovakia in a free Europe and for the attainment of this we are ready to sacrifice all.

Radio in the 1930s had an impact comparable to that, in more recent years, of television and the Internet. For the first time, a world leader could project his voice thousands of miles into the kitchens and living rooms of strangers, creating a seemingly boundless opportunity to share information—or propaganda. For citizens at war, particularly those kept close to home by rationing and blackouts, the radio became a center of existence.

The facilities where my father began work were on George Street, near Marble Arch. The BBC made its airwaves available to him at scheduled intervals, as it was doing for the Poles, the Serbs, and the representatives of other occupied lands; the British either originated or sponsored programs in sixteen foreign languages.

In the war's first year, it was possible for friends in Prague to advertise the BBC broadcasts openly, albeit indirectly. In Czech parlance, the pet name for Jan is "Honza," also the name of a hero in Bohemian puppet stories who regularly outsmarted Austrian authorities. When signs appeared in windows suggesting that *The Tale of Honza* could be heard that evening, people knew to tune in. The Nazis eventually

*Wartime broadcast range of the BBC*

caught on, but by then the programs had become, for Czech loyalists, an exhilarating addiction. "We huddled around the radio each night as if gathering for prayer to hear fifteen cherished and illegal minutes of news from the BBC," recalled one wartime resistance leader. "I can hear the theme music now, through the static of fifty-odd years."

To receive the signal, Czechs had to install in their radios a home-made device—called a "little Churchill"—involving a bedspring and a toilet paper roll. The authorities required that every radio bear a sticker warning that the penalty for tuning in to a foreign station was death. Schoolchildren were urged to inform on anyone who attempted to defy this prohibition, including classmates and neighbors, even parents. Accordingly, regular listeners were careful to hide the device and to retune their radio after signing off.

In addition to his other duties, my father had responsibility for negotiating with the Foreign Office concerning program contents. Each script was prepared in English as well as in Czech and was reviewed by British supervisors for adherence to security requirements. Nothing could be said that might disclose information useful to the enemy. The

Czechoslovak leaders, too, had to approve, which brought my father into frequent contact with Beneš and other senior officials.

The broadcasts were recorded on aluminum disks that did not survive because they were immediately melted down and made into new ones. However, the BBC archives in Caversham Park do have the minutes of staff meetings—in which my father participated—and also broadcast transmission logs that list the program subjects and include comments about what did and did not go well.

Each broadcast was slotted for exactly fifteen minutes, including introductions. This meant that the scripts had to be the right length and the announcers had to read at the correct pace. The writers did their best to tailor the texts, but some people talked more rapidly than others. This sometimes led to furious signaling, indicating that some material should be dropped or that the reader should slow down. In one case, an announcer was given a list of news briefs, the ninth marked "Must be read last," followed by items ten and eleven. Timepieces that were out of sync could generate headaches as well; if the BBC's clock disagreed with that of the studio, the program might begin in the middle or with twenty seconds of dead air.

After months of experiments, the producers settled on a three-part menu: political instruction, the "main talk," and "news of the day." Thus the most vital portions—exhortations to the public and coded signals to the resistance—could be read without fear of running over. The main talks, often written by my father, were limited to six minutes.

Producers frequently used music and sound effects such as recorded gunfire and the whoosh of planes. The technique added a touch of drama but also the risk of mechanical malfunction. One program was interrupted by an unwanted Beethoven scherzo; others were marred by discordant sounds—grinding engines, quacking ducks—that crept in when the tapes ran on too long. Rustling papers and coughing fits (most announcers smoked) added to the cacophony.

Before long, the programs were running thrice daily: at 7 a.m., 6:30 p.m., and around midnight. My father wrote scripts continually, while also editing and reviewing the work of others. He did not take the microphone according to any set schedule but could be heard several times a week, usually in late-night commentaries. These generally covered either patriotic topics (such as T. G. Masaryk's birthday) or those tied to current events—the latest speech by Hitler or FDR. Because of his fluency in Serbo-Croatian and knowledge of the local culture, he also broadcast to Yugoslavia. My father worked hard because of his passion for democracy but also because we needed money; the more he wrote, the more he was paid, which was still not much. From start to finish, the government in exile was a bare-bones operation.

These late-night broadcasts were designed to provide what we, in our era, might call "rapid response." Agents monitoring radio programming in Prague and Bratislava were asked to report each evening on the latest Nazi propaganda. Whenever possible, London's rebuttal was conveyed within hours. This required quicker than usual writing, translating, and vetting, a process that worked sometimes smoothly, other times not. The British censor, for example, got on people's nerves. As the program on November 23, 1942, began, he moved backward toward his customary chair and, per a subsequent report, "sat down on air." The ignominious pratfall was greeted by the Czechoslovaks with "restrained hysteria."

This incident, though juvenile, illustrated the tension that existed between the two cultures. The Britons felt they were within their rights to dictate what was said from their facilities. The exiles, from Beneš on down, remained angry about Munich and impatient with Chamberlain's foreign policy. They had no choice, however, but to accept the role of junior partner. When my father was contacted by the sympathetic H. G. Wells about producing a week of anti-Munich broadcasts, he had to decline out of deference to the sensitivities of his hosts.

Another aspect of my father's job was to decide who among the many Czechoslovak politicians in exile would have access to airtime. Beneš and Masaryk received priority, but many less senior officials yearned for the chance to hear themselves talk. The matter was so sensitive that a thirty-member committee was appointed to provide advice. Decisions were highly political because some speakers were more divisive than others, a balance was needed between democrats and Communists, and Beneš wished to ensure that Slovaks felt included. Accordingly, my father hired Vladimir "Vlado" Clementis, a Slovak Communist, to assist in the broadcasts. I remember Clementis well, for reasons that will be described later, but also because he had a prominent bulge in his forehead caused by, I was informed, a steel plate. Why? No one seemed to know. It should be noted that the desire for balance in selecting speakers did not extend to gender. Although my father raised the possibility of including women, the idea was dropped for fear that female voices would not be taken seriously.

WHEN THE WAR began, Beneš was sure that the West would win. The British were not. On September 6, 1939, Cadogan confided to his diary, "We shall fight to the last and may win—but I confess I don't see how!" A month later, former prime minister David Lloyd George urged his compatriots to make peace with Hitler. "People call me defeatist," he said in an interview, "but what I say to them is: 'tell me how we can win.'"

That autumn, the Red Army and Wehrmacht completed their conquest of Poland. Tens of thousands of soldiers, civilian officials, intellectuals, Jews, aristocrats, and priests were murdered and dumped into mass graves. Soviet and German officers met in the middle of the country, where they erected border markers and stationed sentries who faced one another daily without exchanging a word. The new boundary left Polish families divided and without recourse on either side. One refugee compared the choice between living under German or Soviet rule to that of standing bareheaded in the pouring rain or beneath a running gutter.

At the end of November, the Soviets sought to secure their northern flank by invading Finland, hoping to replicate their partner's Blitzkrieg tactics and conquer their peace-loving neighbor in two weeks. Thousands of tanks charged across the border, only to be slowed by forests and swamplands. The gutsy (and angry) Finns, with their white camouflage uniforms and skill at cross-country skiing, were able to harass the invaders and inflict heavy casualties. Lacking an effective antitank gun, they invented a means of attack—consisting of a bottle of flammable liquid and a match—that they named after the Soviet Union's foreign minister: the Molotov cocktail. The invasion dragged on for four months before the overstretched aggressor and the outnumbered defender agreed to an armistice. Finland survived but with the loss of a tenth of its territory and 30 percent of its economic assets.

As 1939 finally drew to a close, the foul British weather matched the people's mood; rain soaked the countryside in December, followed by the coldest January in more than forty years. The Thames froze solid. Heavy snows delayed coal deliveries and made transportation of any sort a trial. Chamberlain had recruited Churchill for the war cabinet, an encouraging move but one that had the effect of quieting criticism of the government by the hawks in Parliament.

Each day, the few hours of light were spent in preparation: rubbing the rawness off military recruits, stockpiling equipment, filling sandbags, and building new and more daunting barriers to attack. Across the English Channel, the French were content to sit behind their fortifications, making no move against Germany. The Luftwaffe flew reconnaissance missions; the RAF dropped pamphlets. There were skirmishes at sea, but all was generally quiet on the western front. As the weather grew warmer and the daffodils bloomed, spirits began to lift; perhaps the worst was already over. In a cheery talk to the House of Commons, Chamberlain announced that the British armed forces had made great strides. He was now, he said in April, "ten times as confident of victory," claiming that the Germans had failed to strike in time. "One thing is certain," he declared, Herr Hitler has "missed the bus."

Few pronouncements had a shorter life. Within days of Chamberlain's boasts, the Nazis had seized the capital, principal ports, and airfields of Norway. The British, caught off guard, tried to counter with an expeditionary force that put ashore in various spots along that country's coast. However, their troops were poorly equipped to fight in the snow, hampered by a disorganized chain of command, and forced to confront German units that were well dug in and generously supported from the air. None of this was reported by the War Ministry, which instead portrayed the defensive operation as a stunning success. These false accounts were intended to boost morale and did, in fact, cause hopes to soar; but the brief taste of victory made the truth even harder to face.

On May 2, 1940, Chamberlain returned to the House of Commons with a less sprightly step and news that the British forces, having failed to dislodge the enemy, were being withdrawn. The opposition indignantly demanded a review of the war's handling, leading to a debate five days later. Standing before his colleagues, the prime minister downplayed the recent fiasco and issued an appeal for unity, suggesting—as beleaguered war leaders habitually do—that criticism would only provide aid to the enemy. This was not the message that the country hungered to hear. Instead of an admission of mistakes and a call to arms, Chamberlain offered a litany of excuses and a suggestion that everyone remain calm. Parliament being Parliament, barely a sentence escaped his lips without being interrupted or jeered.

The debate droned on for hours but climaxed in early evening when a legislator from Chamberlain's own party closed with the words Oliver Cromwell had said to Parliament three hundred years previously: "You have sat here too long for any good you have been doing. Depart, I say, and let us have done with you. In the name of God, go!" After that bruising rebuff, Chamberlain's vote of confidence was so narrow he felt it necessary to try forming a new government with broader representation. The opposition agreed but with one condition: Chamberlain must indeed go. On May 10, the prime minister reluctantly informed King George VI that he was resigning his position. The king wondered hopefully whether his replacement might be Lord Halifax. No, came the reply, not Halifax, the other fellow.

WINSTON CHURCHILL WAS stately, plump, and sixty-five years old. He had held virtually every important official position except those of prime minister and foreign secretary. In so doing, he had attracted acclaim and derision in roughly equal measure. The two-time prime minister Stanley Baldwin once observed:

> When Winston was born lots of fairies swooped down on his cradle with gifts—imagination, eloquence, industry, and ability. Then came a fairy who said, "No person has a right to so many gifts," picked him up and gave him such a shake and twist that with all these gifts he was denied judgment and wisdom.

In 1915, as first lord of the admiralty, Churchill had presided over the disastrous British attack on the Gallipoli peninsula in the Dardanelles. In the twenties, as chancellor of the Exchequer, he had overseen damaging reductions in the British defense budget. In the 1930s, he had railed against Gandhi and staunchly opposed loosening the imperial rein in India. Churchill could always be counted on to defend freedom with matchless tenacity—provided those exercising it spoke with the right accent and had the proper skin color. Yet for all his faults, the new prime minister would quickly validate the views of those who believe that when history most requires it, Fate lends a hand.

On the day Churchill claimed his new office, Germany attacked the Netherlands, Luxembourg, and Belgium in preparation for an assault on France, the great prize of continental Europe. One of the lessons of World War I had been that the aggressor is eventually driven back, so perhaps this, finally, would be Hitler's Icarus moment. The French had confidence in the Maginot Line, as did the British, many of whose leaders had been treated to a tour. No one anticipated that the Nazis would quickly penetrate the French defenses, not even German general Erwin Rommel, who wrote:

> The flat countryside spread out around us under the cold light of the moon. We were through the Maginot Line! It was hardly conceivable. Twenty-two years before we had stood [against] . . . this self-same enemy and had won victory after victory and yet finally lost the war. And now we had broken through . . . and were driving deep into enemy territory.

The Germans circumvented the heaviest barriers and concentrated their tanks on the weakest points. Armored units, aided by the terrorizing effect of Stuka dive-bombers, shattered the enemy forces in both north and south, leaving the French in disarray. Cadogan wrote of May 15 that it was "an awful day. . . . Don't know where this will end. News still v. bad. . . . Now the 'Total War' begins!"

The same afternoon, Churchill dispatched the first of many

passionate letters to Franklin Roosevelt seeking the loan of ships, aircraft, ammunition, and steel. "I trust you realize, Mr. President, that the voice and force of the United States may count for nothing if they are withheld too long." The letter arrived in Washington at the same time as a warning from Ambassador Joseph Kennedy that America could be left "holding the bag for a war in which the allies expect to be beaten." If we have to take up arms, continued the envoy, "we would do better fighting in our own back yard."

Churchill went to France for a firsthand view of the hostilities; he returned disgusted and shocked. Tens of thousands of British troops had crossed the Channel to aid the French. The Royal Air Force was running hundreds of missions a day. Many Czech, Slovak, and Polish soldiers had joined the fight. Yet Boulogne was taken, then Calais. One air officer, Antoine de Saint-Exupéry (the author of *The Little Prince*), compared the Allied effort to tossing glassfuls of water at a forest fire. The French government appealed to Churchill for more planes, but the Britons, who lost more than 950 aircraft in the campaign, refused to risk further losses.

Jan Stránský belonged to one of the Czechoslovak Republic's leading democratic families; he would later join the government in exile. In July 1939, he escaped from Prague to Poland in a coal truck, then found space on a ship to Marseilles and signed up with the French Foreign Legion. For months, he and his fellow Czechs slept in dirty, rat-filled barracks and spent days marching around in ragged outfits with sun-bleached caps. When the Nazis threatened, the volunteers were finally given uniforms and transported north to the front. There they encountered

> drunken French soldiers deserting right and left; disorganization and utter chaos; lack of a unified command, lack of food; cartridges which did not fit rifles, positions supposed to be ours but long since taken by the Germans . . . a retreat and more trenches to dig, another retreat and a rout.

Stránský's crew commandeered a truck that they drove night and day "on those terrible encumbered and bombarded roads . . . sometimes machine-gunned, sometimes stopped by the gendarmerie who took us for German parachutists and often having to fight our way through."

As the Germans closed in on Paris, the only question was whether the British Expeditionary Force and its allies could be saved for future battles. On May 27, Cadogan was near despair: "position of B.E.F. quite awful, and I see no hope for more than a tiny fraction of them." The last remaining open port was Dunkirk. Another who was there reported:

From the margin of the sea, at fairly wide intervals, three long thin black lines protruded into the water, conveying the effect of low wooden breakwaters. These were lines of men, standing in pairs behind one another far out into the water, waiting in queues till boats arrived to transport them, a score or so at a time, to the steamers and warships that were filling up with the last survivors. The queues stood there, fixed and almost as regular as if by rule. No bunching, no pushing, nothing like the mix-up to be seen at . . . a football match.

The epic evacuation provided a welcome lift in that otherwise disastrous spring. If the expeditionary force had been demolished, the Commonwealth would have had to face the Nazis alone and with its army in tatters. The Royal Navy, the coast guard, and a flotilla of volunteers pulled first 100,000, then 200,000, then more than 330,000 British and Allied troops from the beach. Just as the English were preparing for the worst, they engineered a miracle.

No orator anywhere at any time could have outperformed Churchill in those weeks. In May, he had nothing to offer except "blood, toil, tears, and sweat." On June 4, he pledged to "fight on the seas . . . oceans . . . air . . . beaches . . . landing grounds . . . fields . . . streets [and] . . . hills." On June 18, he declared:

The Battle of Britain is about to begin. . . . Hitler knows that he will have to break us in this island or lose the war. If we can stand up to him, all Europe may be free and the life of the world may move forward into broad, sunlit uplands. But if we fail, then the whole world, including the United States, including all that we have known and cared for, will sink into the abyss of a new Dark Age made more sinister, and perhaps more protracted, by the light of perverted science. Let us, therefore, brace ourselves to do our duties, and so bear ourselves that, if the British Empire and its Commonwealth last for a thousand years, men will still say, "This was their finest hour."*

As stirring as Churchill's rhetoric was, oratory alone could not transform the abysmal spectacle presented by the world in midyear 1940. On June 10, Italy had joined the war on Germany's side. The Nazis overran Paris four days later, and within seventy-two hours, the French gave up. The ceremony of surrender was held in a clearing in the Forest of Compiègne. Hitler, arriving first, paused before a monument commemorating the 1918 defeat of the "criminal . . . German Empire." The führer turned and entered the same railway car (carefully preserved by the triumphant French) in which German officers had admitted to defeat twenty-two years previously. Following in his wake, the French officers took their seats and listened with blank faces as the terms of capitulation were read. After an exchange of salutes, the fifteen-minute ceremony was finished—and so, for the time being, was France.

Between April and June, Germany had seized 400,000 square miles of Europe, taken control of air and naval bases from the North

* The House of Commons did not, at that time, have a recording system. Some historians have suggested that the BBC hired an actor, Norman Shelley, to impersonate the prime minister reading his most famous speeches. In fact, when Churchill lacked the time to record a speech on tape, the broadcasters simply summarized and quoted from his remarks. For archival purposes, many recordings of Churchill were produced following the war. However, Shelley does deserve an honored place in history: he was the radio voice of Winnie-the-Pooh.

Sea to Marseilles, secured access to a bounty of ore and oil, and wiped out the only significant opposing army on the continent. There was scarcely anyone left—only Greece, Great Britain, and the ragtag bands of exiles that had washed up on English shores. Beneš, who rarely exhibited anger, pointed out that fully one-third of the German tanks rolling into France had been built in the Škoda Works, now turning out munitions for the enemy. He was bitter, as well, that to celebrate the Dunkirk rescue, the BBC had played the national anthems of all the countries in the Allied coalition—except that of Czechoslovakia.

# 12

# The Irresistible Force

My first memories are of an apartment where my parents slept on a Murphy bed, the kind that comes out of the wall. We had a green telephone and a cabinet radio several centimeters taller than I was. The radio fascinated me because it was our sole source of entertainment and because, hearing my father's voice on the BBC, I thought that he was *in* the radio and tried to get him out.

The apartment was a welcome step up from the cramped, dingy boardinghouse. We lived on the third floor of a redbrick building called Princes House at 52 Kensington Park Road in the neighborhood of Notting Hill Gate. The building was four years old; I was three. We had a tiny kitchen, a small bath, a little hallway, and two main rooms with central heating and milk delivery to the door. Dáša and I shared one room; my parents slept in the other, which had wooden floors and a trio of windows overlooking Portobello Road, a busy thoroughfare. In front, across the road, was Ladbroke Square Gardens, a lovely park where I was taken when the weather allowed. The building was U-shaped in order to accommodate a large horse chestnut tree; there were shrubs, too, along with pots of marigolds and violets.

The residents of Princes House made up a miniature League of Nations—including British, Polish, Spanish, German, Canadian, and other Czech families, among them that of Prokop Drtina, the friend of my father who had been Beneš's private secretary at the time of the Munich Conference. Drtina had stayed on for almost a year in Prague

*The author in front of Princes House, 1940*

to help organize the underground. He had escaped in February 1940 and resumed his role as an adviser to Beneš, working closely with the head of intelligence, Moravec. I must have liked Drtina, and he me, for in his memoir, he characterized little Madlenka Korbelová as "charming . . . a pleasure and an entertainment."

Near the same time, Dáša wrote to her parents, "Madlenka is very cute. . . . She prays to God every night. One evening she thought she could pray with her little feet [instead of her hands]." When Dáša was not helping to care for me, she was in boarding school, trying to adjust to an alien culture, a different language, a new way of writing, shoes that gave her blisters, a mandatory brown skirt that she thought drab, and the temperamental British weather. "When it rains," she complained to her parents, "I am always in a bad mood, because they take us into a kind of big room and there we occupy ourselves." She met a Czech girl whom she liked but from whom she was promptly separated because the teachers would not allow them to speak their native tongue.

Life in exile had its irritations for the older generation as well. My mother remembered:

> We were living in a foreign country but surrounded only by Czech people, without making friends with the English except for a very few. . . . English people have a different temperament than those coming from Central Europe. It was pleasant to be there as only temporary guests. That was what we wanted and they wanted as well.

Even with Churchill now in the prime minister's chair, the legacy of Chamberlain and appeasement was not forgotten. My father told a story about that period. He had been on a bus and tripped over an Englishman's foot. Instead of apologizing, he said, "I am not sorry, that is for Munich." Then there was the immigrants' ironic prayer: "Please, O God, give the British all the strength they will need to withstand the beating they deserve."

Because his mother was American, Jan Masaryk had spent more of his early life in the United States than in his native land. During his years across the Atlantic he had acquired a reputation as something of a playboy, a man who enjoyed a good time and loved music. Few had a better ear. Not only could he speak English, he could do so in a variety of accents equally suitable to a barroom argument or a Buckingham Palace dinner. He could make a joke out of almost anything and complained that he spent much of his time as ambassador correcting Britons who insisted on referring to his country as Czechoslovenia or Czechoslavia. But although he was at ease in the United Kingdom, many of his countrymen were not. We Czechoslovaks felt looked down upon for the clothes we wore and for our cuisine—such as the traditional Christmas Eve dinner of potato salad and fried carp.* Of course, we were hardly thrilled with British food. London loaves were bland and white instead of earthy and dark. The omnipresent beverage was not coffee but tea, which was ruined in any case by an excess of sugar and milk—and how could any nation whose elite ate cucumber and watercress sandwiches hope to defeat Nazi Germany?

Evidence that Jan Masaryk might have been in England for too many years arrived at Princes House one afternoon along with the foreign minister. As always, he aimed to please, and when he rang the bell of number 35, we found that he had brought with him a huge platter of red Jell-O mold with slices of various fruits suspended within. Our family gathered round the table. As the eldest child, it was Dáša's responsibility to express delight and gratitude. The problem was that she had never seen anything so revolting. With all the adults, including the famous Jan, looking on, she had no choice but to dig in. The fact that she recalled the scene so vividly seven decades later should be a warning to guests about bearing edible gifts.

Language, too, was a barrier. Czechoslovaks sound the r differently from Britons and, like Germans, pronounce w as if it were a v. Any Slav, seeing a word in English, would likely emphasize the wrong syllable; as for London slang—incomprehensible. My mother cringed at an oft-told story about one of our soldiers who visited an aquarium soon after learning how to order fish and chips. He gazed happily at the swimming creatures and pointed. "See," he said, "chips." Another soldier answered the phone in a house where he was staying and instead of identifying himself as a guest replied to the inquiring voice that he was a "ghost." Even Beneš pronounced "theories" like "tories," spoke of using public debate to "make my luggage" instead of "make

* Recipe from an 1825 Czech cookbook: Scale and split the carp, cut into pieces, rinse and sprinkle with salt, and leave it salted for half an hour; then wipe each portion with a clean cloth, roll in flour, then in beaten eggs and bread crumbs; then fry in hot butter until golden.

my case," and expressed his determination to "take the bull by the corns."

IN 1938, BENEŠ had been required to choose between a fearsome war and an ignoble peace. In 1940, he was an exiled leader with a diplomatic agenda; for him, a more comfortable role. In the next two years, he was the irresistible force crashing against the immovable object of the British Foreign Office. Cadogan's diary includes a number of references to meetings with him accompanied by such comments as "pretty awful" and "Beneš, for an hour and a quarter!"

Diplomatically, the Czechoslovak leader knew exactly what he wanted to achieve and whom he needed to convince. His country, he insisted, had not really ceased to exist. The legal situation was the same as it had been before the now-discredited Munich pact. That meant that he was still president, the government in exile should be accorded full recognition, and the Allies were obliged to restore the nation's prewar boundaries. If the legal arguments did not persuade, the moral rationale should. Czechoslovakia deserved support because the country had been badgered into submission by its supposed friends and because its traditions, in the spirit of the great T. G. Masaryk, stood for everything that Hitler was fighting to tear down.

All that and more Beneš was prepared to explain at length to whoever would listen, but he did so from a position of weakness. Half his homeland was under Nazi occupation; a second portion had declared itself independent; and other slices were controlled by Hungary and the Soviet Union. In the week after Munich, he had formally resigned his presidency; his successor, Hácha, sat in the castle office. In London, there were contingents of Czech, Slovak, and anti-Hitler Sudeten Germans who refused to acknowledge Beneš as their rightful leader. Whether or not Foreign Office lawyers were sympathetic to the Czechoslovak plight, there was no legal precedent for what he was asking.

Further, the president was hampered by his personality, which took flight for one cause only. He was incapable of the kind of sophisticated banter or feigned interest in others that helps to grease the wheels of diplomacy. He did not make jokes, coin witty aphorisms, or indulge in irony. He was sensitive about his height and tried to avoid being photographed while standing next to taller people. Edvard Beneš was not a natural leader.

Yet he succeeded. He pushed ahead relentlessly but knew when to ease the pace. He moved from one step to the next instead of trying to leap the whole stairway in a single bound. He learned to restrain his frustration with British policy and took care in public statements to avoid giving offense. By sharing intelligence information and his own expertise, he made himself as useful as possible to the Allied cause. Even without the power to arrest, banish, or discipline foes, he gradually consolidated a position of undisputed leadership among Czechs and Slovaks.

Beneš also chose competent aides, such as my father's friend Hubert Ripka, now in London, to manage the day-to-day affairs of government. Ripka was a skilled administrator, the man to whom people inside the government turned to get things done. The broadcasting operation was under his jurisdiction, so my father went to him with problems and requests. Of these there were many because, as in any group under pressure, not everyone got along with everybody else. Drtina, for example, thought Jan Masaryk received too much attention and did too little of the hard work; Ripka coveted Masaryk's job; the spy chief, Moravec, was envied for the secrets he knew and would not share with anyone except the president. Yet despite the petty resentments, their loyalty to the nation and its leader was never in doubt.

All in all, Beneš performed remarkably. Notwithstanding his limitations, he was a genuine symbol of Western democracy. Helping him meant rejecting Munich, and with Hitler now unmasked, Munich had become a synonym for feckless leadership and cowardice. Besides, the British were in urgent need of people who would fight.

During France's precipitous fall, Beneš had appealed to Czech soldiers and airmen to escape by any means possible and find their way, if they could, to England. For many, the journey was both perilous and

*Meeting at the Foreign Ministry of the government in exile; (from left) Jiří Špaček, Josef Korbel, Hubert Ripka, and Jan Masaryk*

circuitous. Within a short time, however, some 4,000 soldiers were stationed at a temporary camp in Cholmondeley Park next to the castle in Cheshire of the same name. Most were still garbed in the uniforms of the French Foreign Legion; the more fortunate wore the British battle dress with a narrow patch over the shoulders that read "Czechoslovakia." Some of the units were made up of young soldiers and exhibited an air of confidence and vigor; others consisted of older reserves who didn't belong in a professional military but had nowhere else to go. Age and appearance, however, were less serious barriers to cohesion than ideology and prejudice. Among the ranks were Czechs, Slovaks, Sudeten Germans, Jews, anti-Semites, and radical nationalists—also Communists, who, taking their cue from the Hitler-Stalin pact, were reluctant to fight on the same side as British "imperialists." Ignoring a personal appeal from Beneš, more than four hundred of the leftists joined an international brigade pledged to carry on the battle but without the taint of capitalism.

*Beneš confers with Czechoslovak exile officers.*

Through a program of rigorous training, the early difficulties were resolved, and within a year or two, the Czechoslovak army was as able as any of its size. By contrast, there was never a question about the professionalism of the country's airmen, of whom about nine hundred (including eighty-eight experienced pilots) made it to Great Britain. In July, fighter and bomber squadrons were formed, while additional pilots were assigned to Polish and other RAF units.*

Years later, a friend of my family recalled visiting the Czechoslovak club on Clifton Road near Cholmondeley. "That is where the officers went for coffee and to schmooze," wrote Renata Kauders, "and sometimes to pick up girls. It was a lively place. Surrounded by the noise and movement, you felt as if you were in the middle of the war." Among the

---

*One of the Czech exiles who served with the French military and then the RAF was Karel Mahler, the eldest child of Marta Körbelová, my grandfather Arnošt's older sister. Following the war, Karel moved to Brazil, where his son, Pedro, grandchildren, and great-grandchildren live today.

officers with whom Renata remembered sharing drinks were two army
men, Jan Kubiš and Jozef Gabčík. Both were in their late twenties,
handsome, and full of coiled energy; Kubiš was tall, slender, and as reserved as the muscular Gabčík was outgoing; each smiled easily. Renata
did not know (and neither did the men) the role that they would soon
play in the history of the war.

BY MIDSUMMER 1940, most of Europe was either neutral or under occupation. Hitler felt he could be magnanimous. At home, he gave medals
to his favorite generals and announced that public dancing would
again be allowed on Wednesdays and Saturdays. Abroad, he resumed
his quest for just the right kind of peace. The author of *Mein Kampf*
was a man of many hatreds, but Great Britain was not among them.
He had long believed that the English and Germans had important interests in common. Both nations were economically ambitious, racially
"advantaged," anti-Bolshevik, and concerned about "the unbounded
French drive for hegemony." There was no reason, in Hitler's view, why
the two empires could not coexist.

In June, Hitler told Göring that "the war is over, Hermann. I'm
going to reach an agreement with England." The führer had already
begun sending messages to London that, although arrogant, also held
out the promise of a permanent settlement. He must have thought he
was being generous because he felt invincible. The British had failed
to defend Czechoslovakia, shown themselves impotent in Poland, and
been routed in Norway and France. The mighty Soviets were preoc-
cupied with their own conquests. Surely Chamberlain and his advisers
would choose the safety of renewed appeasement over the risks of invasion and bombardment. Potential peace terms were not hard to formulate: Germany would control Europe and parts of Africa; the British
would be allowed to survive and retain their overseas holdings; the
Americans would be expected to mind their own business.

By mid-July, however, the Germans were becoming frustrated. Back-
channel contacts showed that many Britons were desperate for peace at
any cost, but Churchill had no interest in such a bargain. Not only did
he talk defiantly, he seemed to welcome the prospect of fighting. Hitler

found that illogical. The prime minister, he concluded, was not quite civilized. On the sixteenth, Hitler told his military, "Since, despite its desperate military situation, Great Britain shows no sign of good will, I have decided that a plan of invasion will be prepared for and, if necessary, carried out."

Three days later, in a speech to the Reichstag, he made one last public appeal:

> I feel it is my duty to ask once more, in good faith, for reason and wise counsel on the part of Great Britain. . . . My position allows me to make this request, since I do not speak as a defeated man begging favors but as the victor speaking in the name of reason. I can really see no cause why this war should continue. Mr. Churchill ought, for once, to believe me when I say that a great empire will be destroyed—an empire which it was never my intention to destroy or harm. It gives me pain when I realize that I am the man who has been picked out by destiny to deliver the final blow.*

German propagandists assured their listeners that the mighty and fearless Reich craved peace above all else. Increasingly that summer, this seductive message assailed English ears. Many families were frustrated by the lack of war news on the security-conscious BBC and so tuned each night to a broadcast that began "Germany calling, Germany calling, Germany calling!" Then would come the voice of William Joyce, alias Lord Haw-Haw, an ultraconservative Irish-American politician who, though a British citizen, had sought refuge in Berlin just before the war. When Churchill and Halifax responded dismissively to Hitler's Reichstag address, Joyce shed crocodile tears: "It is a pity! It is a

---

* The British, schooled in the classics, may have heard in Hitler's words an echo. As Herodotus told the story, the king of Lydia, a man named Croesus, was noted for his wealth and power. Wishing to attack Persia, he sought guidance from the Great Oracle at Delphi. "If Croesus goes to war," the Oracle predicted, "he will destroy a great empire." Infused with confidence, the king did indeed go to war, and, of course, the empire he destroyed was his own.

thousand pities! This is the tragedy that the führer went out of his way to avoid. But if those who rule England . . . care less for their country than the führer has cared, force . . . must arbitrate."

Joyce claimed to speak for the common man and drew a contrast between himself and England's haughty upper class. Although he had been born in Brooklyn, he professed to love "the sceptered isle" so much that he had to betray it in order to reawaken its true spirit. A skilled writer, he composed his own scripts and also prepared propaganda for the New British Broadcasting Station, based in Germany but purporting to represent the views of English working families. More subtle than most Nazi offerings, the programs portrayed Churchill as a warmonger eager to repress the Irish, Welsh, and Scots and to promote the interests of London banks. The station also did its best to engender fear. One summer morning announcers spread an alarm that during the night "German parachutists wearing civilian clothes were dropped in the vicinity of Birmingham, Manchester and Glasgow. They carry capsules to produce fog and so avoid capture. Some are equipped with an electromagnetic death ray." Many listeners, having been warned for so long about the terrors of war, were quick to believe the worst.

For England, the land battles had not begun, but ports and shipping lanes were under attack and so were cargo vessels. This made maritime transportation dangerous and raw materials precious. Ration books were distributed to every head of household, but many common goods were rarely available or only in poor quality. Real butter was scarce, and margarine came in brittle patties that broke immediately when placed in contact with a cracker. Shop owners put signs in their windows with the wry message, based on a Broadway show tune, "Yes, we have no bananas." Candy stores, to remain in business, had to diversify by selling such items as flashlights and electrical tape. Members of the public were asked to eat less cheese so that more would be available for coal miners and vegetarians. The new, cheaper matches were reluctant to ignite, and toilet tissue was sold in individual rolls.

For some, the hardships of war became real only when the minister of food, Lord Woolton, announced that rationing would be extended to tea. One woman from a prosperous neighborhood found a note

*Edvard Beneš, Hana Benešová, and Queen Elizabeth*

signed by her maid and pinned to the door of her flat: "Madam, there is no honey, no sultanas, currants or raisins, no mixed fruits, no saccharine at present, no spaghetti, no sage, no herrings, kippers or sprats (smoked or plain), no kindling wood, no fat or dripping, no tins of celery, tomato or salmon. I have bought three pounds of parsnips."

A flood of volunteers signed up for the home guard, organized to defend against the anticipated German invasion. Members of Parliament dutifully took their turn on the nightly fire watch, cabinet members prepared for battle, and even Queen Elizabeth practiced her shooting. Overall, the number of recruits far outpaced the storehouse of weapons. Anthony Eden, a future foreign secretary and prime minister, joined a rifle corps that had no rifles and was forced to train in the aisles of a department store. Lacking hand grenades, some platoons made do by concealing razor blades in potatoes.

Working feverishly, the navy laid minefields around the coasts. The army, with the help of civilian volunteers, covered the beaches with barbed wire, dug tank traps, and set up pillboxes and concrete walls. Coastal areas were ruled off limits to visitors. Road signs were removed to confuse the enemy. Families in the countryside were instructed to stay put in the event of an invasion so that roads would not be clogged, as had happened in France. Each day when radio transmissions ended, loyal subjects rose from their living room chairs and sang "God Save the King."

As the British girded for what was to come, the Czechoslovak government in exile achieved its first diplomatic milestone. On July 23, the United Kingdom formally recognized Beneš as head of a provisional government representing the Czech and Slovak peoples. Beneš was so pleased that he came to our apartment house in person to break the news to Drtina and my father. With Masaryk visiting the United States, Drtina would have the honor of reading the announcement over the BBC. The English decision was a victory but a limited one. The "provisional" label was demeaning, and the Allied war aims referred only to the liberation of the Czechs from German domination. There was no mention of restoring the country's prewar borders or even its independence as a nation. Nevertheless, five days later, the Czech anthem was played by the BBC for the first time and the king and queen invited Dr. and Mrs. Beneš to lunch.

# 13

# Fire in the Sky

August 13, 1940: From Reichsmarshal Göring to all units of air
fleets two, three and five: Operation Eagle. Within a short period
you will wipe the British air force from the sky. Heil Hitler.

The Luftwaffe had gained experience supporting fascist forces in
the Spanish Civil War and had effectively complemented German
ground invasions in Poland, the Low Countries, and France. Embold-
ened by those adventures, the Nazis approached the Battle of Britain
with celestial confidence. Their strategy was to batter their foe—which
they saw as isolated and trembling—by destroying its planes, antiaircraft guns, and munitions plants and to do so quickly enough to open
the way for invasion by mid-September 1940. Göring assured Hitler
that, English weather permitting, it could be done.

As soon as the Germans overran France, they began to build air
bases along the northern coast, just a short trip across the Channel to
England. On July 10, seventy-five Nazi bombers, escorted by forty-five
fighters, attacked a convoy of ships in Dover harbor. In public Hitler
was still talking peace, but his military was waging war.

IN 1935, AN engineering team headed by a researcher with the quin-
tessential British name of Professor Robert Watson-Watt discovered
that radio waves were deflected by passing aircraft. To capitalize on

the revelation, the Air Ministry established a chain of twenty coastal stations, each including a pair of massive towers, one to send signals, the other to receive. The engineers referred to their new technology as "radio direction finding," known now as radar. With this tool, the Britons could detect aircraft at a distance of a hundred miles, and, because their own planes were equipped with transmitting devices, technicians could differentiate between friend and foe.

When the Battle of Britain began, the radio towers were among the first targets the Luftwaffe sought out. By mid-August, Nazi pilots thought they had eliminated all or most. They were wrong. The towers had been built with backup generators and could withstand enormous damage; that mattered because although radar gave only several minutes' warning, it was enough—when response decisions were quick— for pilots to scramble in time to intercept incoming planes.

To protect their bombers, the Nazis deployed escorts of Messerschmitt fighters, in single- and double-engine versions; both were highly capable, but each had a flaw. The fuel gauge on the single-engine plane would begin blinking after a mere hour in the air, thus forcing the pilots to turn back even as a bombing run was under way. The twin-engine version was not as fast or maneuverable as the British fighters— the quick-climbing Spitfire and the sturdy Hurricane. The British Air Ministry endeavored to assist its fighters through the use of blimp-size barrage balloons that forced enemy pilots to fly so high that they could not properly identify targets on the ground. Londoners soon learned to equate "when the balloon goes up" with an impending attack. In addition, some air stations were equipped with five-hundred-foot-long rocket-launched steel cables that, once extended, descended only slowly, supported by a parachute. If sent skyward at the proper moment, the devices created a lethal, hard-to-avoid obstacle for fast-approaching aircraft.

Despite these preparations, the outlook for the British was grim. The bombers of the day had an advantage over even the best air-defense systems, especially if the attackers arrived in sufficient numbers. British manufacturers were turning out 450 fighters a month, but the Luftwaffe began the battle with a substantial edge. Because the outcome would

hinge on the rate of attrition, the English had to inflict a disproportionate amount of damage. Unfortunately, the RAF was many pilots
short of its required minimum strength; squadrons that were supposed
to have twenty-six pilots had to get by with sixteen. To make up the
difference, training courses were shortened to less than two weeks and
novice pilots were forced to practice in World War I biplanes.

HISTORIES OF THE U.S. Civil War typically begin with an account
of Washingtonians carrying picnic baskets to the Virginia battlefields
to observe the fighting. For a time, there were comparable scenes in
Great Britain as families that were safely inland gathered on high spots
to watch the Spitfires and Hurricanes duel with the Messerschmitts,
cheering each success and oohing and aahing at the spectacle. It was a
breathtaking show. The enemy planes began as specks on the horizon,
rapidly growing larger; within seconds the *kah-chunk, kah-chink* of engines could be heard merging into a hum and soon a roar. The planes
approached in lines before breaking into groups, then individual arcs
as they were challenged. The bombers ducked and swooped, their pilots
eager to discharge their deadly cargo and head home. The fighters on
both sides buzzed frantically, seeking the cover of clouds or the advantageous glare of sunlight at their rear. As the combatants darted, puffs
of smoke from their guns were etched against the blue or black sky,
and on the ground, blasts of fire marked the spots where five-hundred-
pound bombs punched holes in the earth. Villagers and farmers were
startled when, from time to time, an injured plane landed in their fields
or a parachutist descended in a web of tangled nylon. It was not uncommon for an RAF pilot to be surrounded or even shot at by suspicious
townsfolk before making his identity clear.

For those working in or near one of the preferred German targets—a dock, an airfield, a gun battery, a munitions plant—the view
of the fighting was less enthralling. There, the siren's banshee wail carried an especially ominous meaning, followed as it was by a hail of bullets kicking the ground, ugly cylinders falling, and explosions that, for
many, would be the last thing felt or heard.

After a successful strike, the German pilots returned to their bases to report on the damage done. Yet even before the dust had begun to dissipate, British ground crews were at work tending to the wounded, clearing debris, activating generators, getting power stations back up to speed. Men and women, including switchboard operators and other civilian personnel, carefully inspected airstrips, marking unexploded ordnance with red flags; then came the disposal units and after that the shovel brigades mixing concrete to fill craters left by the bombs. More quickly than the enemy thought possible, guns were made ready to fire once more and airstrips patched well enough for fighters to land and take off. Docks were rebuilt. Crippled planes and injured pilots returned to the air. One bombed-out factory functioned for months without a roof, its tools sheltered from the elements by a huge tarpaulin. The unending stress exhausted the maintenance and repair technicians, many of whom remained on round-the-clock duty for months without a break, dozing on cots or floors or, in clement weather, grass.

For all the blood and strain, the Britons could be satisfied that Germans, too, were experiencing an acute sense of tribulation. The attackers felt that after a month of heavy bombing, victory should have been theirs. Yet the enemy radars had not been destroyed, the supply of RAF planes and pilots seemed bottomless, and the path to a risk-free invasion was still blocked. The Germans retained a superiority in numbers, but their losses in planes and personnel were far higher than had been anticipated. The Stuka dive-bombers that had commanded the air in France were too slow to evade the fire of machine guns mounted on British fighters. A Luftwaffe pilot, describing the balloons and anti-aircraft guns as "pretty hot," complained to the journalist William Shirer that he and his comrades had expected to find London a city afire but instead had been impressed by how much was still untouched. Fliers were in the air every night, and scores were dying; yet the British wouldn't yield. The unpleasant surprises would continue: by mid-August, the Germans believed that the enemy's fighter fleet had been reduced to 450; in fact, the RAF had almost twice that number. Hitler's anticipated cakewalk had turned into an uphill climb.

The Luftwaffe's target list did not include London, but on August 24 a pair of pilots, looking for a petroleum depot further up the

Thames, misread their charts and dropped several bombs on the city's East End; the explosives damaged the Church of Saint Giles and blew up a statue of John Milton. Churchill was unaware that the strike had been a mistake or that the offending pilots had been reprimanded and reassigned. Considering the attack an intentional provocation, he ordered reprisal raids on industrial facilities in Berlin. Owing to clouds and enemy guns, the British pilots also missed their targets, hitting instead a residential area, killing ten civilians and injuring twenty-nine. Now it was the turn of German opinion to be outraged; Göring had promised that their capital would be kept safe. Hitler was both furious and opportunistic, seeing a chance to smash London under the cover of self-defense. On September 4, he went before a raucous audience in Berlin. "Since they attack our cities," he thundered, "we shall extirpate theirs!" Amid hoots and applause, he added, "The British have been asking, 'Why doesn't he come? Why doesn't he come?' We reply: 'Calm yourselves. Calm yourselves. He is coming!' "

Three days later, on the afternoon of September 7, 1940, the Blitz of London began. The German shift in strategy was prompted by more than a desire for retribution. As autumn approached, and with it the expectation of high winds and stormy seas, the opening for invasion narrowed. Hitler needed to land a decisive blow, and for that he had to find a way to force Churchill to put more of his fighters into one place. What better way than to fill the heavens above Saint Paul's and Buckingham Palace with German bombers? It wasn't too late—if the attackers were relentless enough—to destroy the adversary's capacity to resist.

With Göring watching expectantly from the French coast, an armada of more than three hundred bombers and six hundred fighters ascended to 17,000 feet and soared toward England in two massive waves. British radar operators could scarcely believe what they were seeing. Every squadron in the area received an order to scramble, their pilots anticipating, as usual, the need to defend airfields and gun batteries. Instead, the bombers and their escorts veered abruptly toward London. The city's airborne protectors were outnumbered by more than ten to one.

The first bombs hit the Royal Arsenal at Woolwich, the surrounding rail stations and factories, whole rows of houses, and the Surrey docks. In the words of one witness:

> Suddenly we were gaping upwards. The brilliant sky was crisscrossed from horizon to horizon by innumerable vapor trails. . . . Then, with a dull roar, which made the ground around London shake as one stood upon it, the first sticks of bombs hit the docks. Leisurely, enormous mushrooms of black and brown smoke shot with crimson climbed into the sunlit sky. There they hung and slowly expanded, for there was no wind, and the great fires below fed more smoke into them as the hours passed.

So it went through that tumultuous Saturday night. The all clear sounded at 6:30; two hours later, the air raids resumed. The rattling of gunfire was followed by ear-popping explosions; bombs slammed into houses along Pond and Victoria Streets, Westminster, and the East End; 1,000 fires were ignited; 430 people were killed and more than 1,600 injured. On the night following, another round of raids caused heavy damage to rail stations and track lines; by dawn, hundreds more lay dead.

From September 7 until the end of October, over fifty-seven consecutive days, an average of two hundred bombers assaulted London. The globe's leading city had become a battlefield. Great buildings were no more. Streets were impassable because of the countless shards of shattered glass. Each morning, rescue workers scurried about the charred cinders, methodically bandaging survivors, practicing triage on the badly wounded, and trying to piece together the remains of neighbors for burial. These efforts were made more perilous by delayed-action bombs, which had to be defused or carted carefully away. There was no reliable place of refuge. The shelters, whether in home gardens or public parks, provided protection only against collateral blast and debris. Families who retreated to basements were often crushed or suffocated by the collapse of buildings above. In the first six weeks, 16,000 houses were destroyed and another 60,000 seriously damaged; more than 300,000 people were displaced.

*Bombing of Surrey docks, September 7, 1940*

Londoners, however, proved an adaptable species. Knowing that they might be stranded for days between trips home, office workers arrived at their desks with toothbrushes, pillows, blankets, and extra clothes. As evening approached, the parade of mattresses began into cellars, shelters, and the Underground. Weather data were classified, so people made their own forecasts—fair skies meant a favorable day for Hitler; clear nights at certain times of the month provided a bomber's moon. The social divisions that defined British culture momentarily melted away as people from all walks of life wished one another well. Defiant shopkeepers displayed signs: "Shattered, not shuttered" or "Knocked, not locked." Banks and the postal service promised "business as usual"; enterprising streetwalkers did the same.

Some disruptions, however, seemed only prudent. Senior members of the Foreign Office were under strict orders to retreat to an air-raid shelter near Berkeley Square as soon as the siren, known as "Weeping Willie," began to scream. On the morning of September 13, a prestigious group of middle-aged men sat huddled together, impatiently awaiting the all clear. Hours passed without any work being done. Suddenly there came a loud rapping on the door. Outside was a teenage Czech girl who wished to deliver a letter from President Beneš to the Foreign Office. Her mission complete, she turned and walked unhurriedly back across the exposed streets of London. The mandatory shelter order was soon rescinded.

FROM NEWSBOYS TO members of Parliament, the British pondered the ominous implications of the attacks. Was this the final shoe falling before the invasion? Churchill warned the cabinet that a German force was gathering; large concentrations of enemy barges had been sighted along the French coast.

On September 22, the prime minister received a call from an unusually excited Franklin Roosevelt. The United States had received word that Germany planned a surprise military landing on British soil. When? That very day. As soon as he rang off, Churchill was on the phone again to Anthony Eden, who was in the southeast of England

*Londoners camp out in the Tube, October 21, 1940.*

within walking distance of the Dover cliffs. Eden made a rapid reconnoiter and reported back that the seas were rough and the fog impenetrable. An invading force, he informed his boss, would either become lost or arrive in an advanced state of seasickness. The next morning, Roosevelt phoned back. "I'm so sorry," he said. "Our codes got mixed. The invasion was of Indo-China, not England, and by Japan, not Germany."

More than once that month, bombs fell on or near Buckingham Palace, causing significant damage to the historic building but no harm to the residents. The attacks helped to cement a love affair that had sprung up between the British people and the royal couple. King George's brave attempt to overcome his stuttering was well known and deeply admired—as were visits by the king and queen to areas that had been bombed. Occasional remarks made over the wireless by the young Princess Elizabeth struck a chord as well. Historians have noted that the Nazis would have been smarter to confine their attacks to London's grittier neighborhoods, thus aggravating the divide between rich and poor. Instead the opposite took place. A popular song from the period went "The King is still in London, in London, in London; and he would be in London Town if London Bridge was falling down."

I WAS BY then a full-fledged toddler. My family's routine, when the warning sounded, was to hurry down the cramped gray concrete stairwell of Princes House to the cellar, which was divided into several small rooms and one larger. There were about two dozen of us at any one time, occasionally more when buildings nearby had to be evacuated. We sipped tea or coffee prepared by the air-raid wardens and shared snacks of bread and biscuits. We slept—when we could—on camp beds or mattresses in the biggest room. Although the building was structurally sound, the basement had large hot-water and gas pipes suspended from the low ceiling; they warmed the rooms, but had a bomb fallen nearby, we'd have been scalded or asphyxiated even if uncrushed. As a child, I did not think of such possibilities, instead relishing the excitement. In the morning when the all clear sounded, we burst into the street or climbed to the roof to survey the damage.

Lacking any strategic value, Notting Hill Gate was hardly a prime target for the Luftwaffe, but bombs still struck more than a dozen locations and killed fifty people. One of our neighbors was Orlow Tollett, an original tenant at Princes House who would still be there at the turn of the century. In 2011, at the age of 103, Mrs. Tollett kindly consented to be interviewed for this book. She remembered that there was a degree of separation between the refugees and the Britons who lived in our building, but that it was "generally a very pleasant kind of group with a friendly warmth between the two sides; the people were very supportive of one another. They used to play great bridge games and share out their supplies." In the full bloom of youth and still single, Orlow rarely went down to our basement; she thought that, if the worst happened, it would be healthier near the top of the rubble than underneath. One evening at the height of the Blitz she tempted Fate and went with a friend to the Freemasons Arms, a small Portobello Road pub, for a game of darts and whatever liquid concoctions might be available. She recalled:

> The pub had a direct hit that night. I fell under the counter. I was squashed there and couldn't get out. Then the fire brigade came and they were ever so kind and pulled me through. By the time they did, I hadn't much on; they took me to the convent.

Orlow recalled that her mother was less upset about the terrible danger than about the lack of clothing.

Another time, a bomb landed nearby but did not detonate, so all the buildings in the area were evacuated and an emergency team arrived. After a careful investigation, the crew members told us not to worry; the explosive was a dud. Inside the casing they had found a note written by Czech factory workers. "Don't be afraid," it said. "The bombs we make will never explode."

One morning in mid-September, the Germans came early. My father and Mr. Drtina decided to ignore the sirens and remain in our apartment working on a radio script. It would be quieter up there, they thought, than in the crowded shelter. This was a fair assumption but— as it developed—an overly optimistic one. Drtina remembered:

The whizz of a flying bomb was so loud that we both threw our-selves down and Dr Körbel quickly jumped under the table. The airborne assault was deafening and our house swayed so much that it reminded me of a ship on the high seas. I would never have believed a huge iron and concrete building could vibrate that dramatically and still not fall to pieces. When we felt ourselves out of danger, we could not resist a laugh of relief.

Bombs continued to fall; for the intrepid pair, enough was enough. Together, they descended the dust-filled staircase to join the rest of us.

THE CZECHOSLOVAK 310 Fighter Squadron was formed at Duxford on July 10 and became operational five weeks later. Based in central England, the squadron chose as its emblem a sword and lion, and as its motto "We fight to rebuild." The commander was Major Alexan-der Hess, a veteran of World War I who, on the last day of August, had crippled an enemy bomber, forcing it to land in a field. His blood up, Hess put his Hurricane into a dive with every intention of shoot-ing the three-man crew, then hesitated to fire his gun when he saw the men waving up at him. Taking another pass, he steeled himself, re-solved that there should be no survivors. Again the Hurricane dove. This time the downed airmen had found something white to hold aloft, and Hess, cursing, restrained his trigger finger once more. Re-porting on the incident, the commander complained, "I have become too bloody British!"

Stanislav Fejfar, a ruggedly handsome graduate of the Czechoslovak Military Academy, shot down his first enemy plane on the ninth of September. As he related the story:

We were flying at 27,000 feet and it was very cold. As we came through some clouds, we could see Luftwaffe bombers escorted by many fighters. We were given the order to attack but had to be aware of the German fighters since they spotted us and were above us. I found a ME-110 to attack and promised myself that this German swine would not sleep in his bed that night. I

maneuvered behind him and fired all my machine guns. He tried
to escape by climbing steeply and turning but I managed to deliver three more bursts and he began to smoke, then went down.

Fejfar was a native of Štikov, a small town in the northern reaches
of the republic, near the border with Poland. His father had died in
the Great War, fighting on behalf of Austria-Hungary. The twenty-
nine-year-old pilot was a cheerful sort, loved to fly, and would continue
doing so until May 17, 1942, when his Spitfire was hit during a day-
time raid over France. His remains were recovered by the Germans and
buried in Calais. Fejfar's mother, never accepting his fate, died in 1960.
Her last words were a plea to leave the front door unlocked because
"Stanislav did not take a key."

When not aloft, the Czech and Slovak airmen occupied their time
reading newspapers and books, listening to the gramophone, and playing games of cards and chess. Sleep came at odd hours and in irregular positions on metal benches, cots, and chairs. The pilots were never
without their yellow life preservers, called Mae Wests in honor of the
vests' inflatable attributes. The planes were always ready for takeoff,
and everyone had an ear alert for when the scramble call might come.

Of the Czechoslovak pilots not assigned to the squadron, the most
accomplished was Sergeant Josef František. Like many of his colleagues, František had fled to Poland at the time of the Nazi invasion
and campaigned there in an obsolete Pulawski fighter. After Poland
fell, he escaped from an internment camp in Romania and made his
way via Syria to France. There he flew brilliantly with the French air
force. Following the Dunkirk evacuation, he was assigned to the Polish
squadron training in England. František—who had a boyish face, thick
black eyebrows, and a piercing stare—was known for his bravery and
for what the British referred to as "bottle." That September, he shot
down seventeen German planes, more than any other Allied airman of
any nationality. On October 8, his plane disappeared from the view of
his fellow pilots and was later found, smashed up, in Surrey. František's
body, its neck broken, was discovered in a nearby hedge.

———————

EVEN BEFORE THE end of September, Hitler had concluded that the primary goals of the bombing campaign were beyond his reach. The RAF had not been destroyed; an invasion was impractical; the enemy's will to fight had, if anything, increased. Nonetheless, he ordered that bombing continue. In October, more than seven thousand tons of explosives were dropped on London. Liverpool, Manchester, Birmingham, and other cities were also hit. In November, devastating raids were directed against Coventry. In December, it was London's turn once again.

Christmas 1940 is remembered by all who spent that holiday in or around the British capital. Although the nightly bombing had stopped, the raids that did occur forestalled any sense of celebration. In our minds, if not in Hitler's, the prospect of a German invasion still loomed. There were no lights on in our home in Notting Hill Gate, but seasonal decorations still went up; we had our tree. At year's end, I am sure that my parents' thoughts were there with me, but also with their loved ones in Poděbrady and Prague.

Despite the sadness and worry, there was also a note of satisfaction. Hitler wasn't losing at that juncture, but neither had he rolled over the British as he had the French. A joke circulated among my father's friends in which the führer telephones Mussolini to chide him for the shoddy battle record of his troops. "You promised that you would be in Athens by now and in Cairo the following month," he complains. "Those dates have long since past, and still you sit in Rome." Mussolini is quiet for a bit, then replies, "Sir, I am having trouble hearing you. The connection must be weak." Hitler, with raised voice, repeats his criticisms. "My apologies, sir, but I still can't hear you," says the Italian dictator. "You sound so far away. May I inquire from where you are calling? Is it London?"

# 14

# The Alliance Comes Together

Early in the second week of 1941, Franklin Roosevelt's confidential emissary Harry Hopkins flew to London to consult with Churchill. The idiosyncratic Hopkins was less a diplomat than an all-purpose problem solver who acted as FDR's eyes, ears, and—because of the president's limited mobility—often his legs. In May of the previous year, Hopkins had moved into the White House where he would continue to live until Roosevelt's death. More than any other official, Hopkins could be counted on to speak for the U.S. commander in chief.

During the visit, Hopkins spent hours with the prime minister, reviewing Great Britain's keenly felt need for replacement ships and aircraft. The meetings went well. In November, Roosevelt had won re-election, based in part on his pledge to keep the United States out of war. Though he was not ready to renounce that promise, he was determined to help England. In mid-December, he unveiled his innovative lend-lease program, under which—in return for minor military basing considerations—a portion of U.S. defense production would be loaned to the British and other allies for the duration of the conflict. When pressed by reporters about the cost, he observed that "a man would not say to a neighbor whose house was on fire, 'Neighbor, my garden hose cost me fifteen dollars; you have to pay me fifteen dollars for it.' He would lend the neighbor his hose and get it back later."

Although most Americans remained wary of direct involvement

in the European conflict, they were gradually coming to share FDR's sentiments. From New England to California, they had followed the Blitz closely and admired England's resolve. Correspondents such as James Reston, Edward R. Murrow, and John Gunther dipped into a palette of colorful adjectives to paint a flattering picture of Great Britain under siege.

Some of their pieces were melodramatic:

They're sustained in part by folklore, the tradition, and the history of Britain; but they're an undemonstrative lot. They don't consider themselves to be heroes. . . . These black-faced men with bloodshot eyes who were fighting fires; the girl who cradled the steering wheel of a heavy ambulance in her arms; [and] the policeman who stands guard over the unexploded bomb down at St. Paul's tonight. . . . There is humor in these people, even when disaster and hell come down from heaven.

Some were reflective:

There is a tremendous vitality behind the quiet thoughts of the people of this country. . . . All the popular time killers of modern civilization have been crippled by the war. It is hard to get to the movies. There are no dances, football, or dog racing. The people have time on their hands now. They are reading more and like all sad men they are thinking hard. A new England is being born in the subways and shelters of this brave island.

And some were just stories:

"Please pass the marmalade," said the little old lady. I was having a breakfast in a small hotel in an English south coast town a very few hours ago. At this moment the air raid sirens began to wail, and the man at my elbow looked up at the clock. "A little ahead of himself this morning," he remarked. . . . I drank my coffee and tried not to gulp it. No one around me budged. "Would you please pass the marmalade?" said the little old lady again, more firmly this time, as the warbling siren died away.

Tributes from U.S. journalists were reinforced by the BBC's own American broadcasts, which ran for six hours each afternoon. The programming featured firsthand accounts of the Blitz and dramatizations accompanied by sound effects. The scripts were designed to nudge the United States toward war but without advertising that intention. Instead, commentators found suggestive links between the Magna Carta and the U.S. Constitution, between Parliament and Congress, and between the struggle for freedom in Europe and its survival in America. Participating celebrities included *Gone with the Wind* star Leslie Howard, who told heart-tugging anecdotes with just the right accent, and the left-leaning novelist J. B. Priestley, who spoke not of war's glamour but of working-class grime and guts:

It is industrial England that is fighting this war ... those scores of gloomy towns half buried in thick smoke with their long dreary streets of little houses all alike and the rather short ... folk usually with bad teeth, who aren't much to look at, but who happen to be among the most highly skilled and trustworthy workmen in the world.

All the while, British agents were spreading rumors in the American media about Hitler's alleged plans to dominate the Western Hemisphere and outlaw organized religion. The combination of respect for England and contempt for the Nazis pushed an ambivalent population toward the brink of commitment; a survey at midyear revealed that although 70 percent of U.S. voters still opposed entering the war, an equal percentage favored defeating Germany at all costs, even if it meant jumping into the fray.

The improving transatlantic relationship was enhanced further by the arrival in London of a new U.S. ambassador, John G. Winant. During the Blitz, Ambassador Kennedy had retreated each night to the suburbs, discouraged expatriate Americans from participating in the home guard, and been openly pessimistic about British prospects. The new emissary took a flat in central London, cheered the volunteers on, and expressed full confidence in England's eventual success. Delight in him was evident from the beginning; Winant's train was met at Victoria Station by the king, the first time in history that a monarch had so welcomed a foreign diplomat.

As his meetings in Great Britain neared their climax, Harry Hopkins dined with Churchill in Glasgow. Before taking his leave, he looked directly at the prime minister and said, "I suppose you wish to know what I am going to say to President Roosevelt on my return. Well, I'm going to quote you one verse from that Book of Books: 'Whither thou goest I will go; and where thou lodgest, I will lodge; thy people shall be my people, and thy God my God. Even to the end.'" This recital of Ruth's pledge to her mother-in-law is still referred to whenever British and U.S. leaders express public affection for each other. At the time, it caused Churchill to weep. In subsequent months, hundreds of millions of dollars' worth of tanks, trucks, torpedo boats, food, and firearms of all description were transferred from the arsenal of democracy to English hands.

LIKE AHAB PURSUING his whale, Edvard Beneš was intent on hunting the Munich pact across every diplomatic ocean "until it spouts black blood and rolls fin out." Beneš felt that his decision not to fight had been the right one but knew that many of his followers disagreed. He was deeply hurt by their criticisms and by the widespread assumption that T. G. Masaryk would have exhibited more spine. Had Beneš shown weakness at the moment of crisis? He did not think so, but if he were to save his reputation—and his country—he could not waste a minute brooding; Munich must be harpooned.

His first step had been to gain recognition of the provisional government in exile. The second would be to persuade England to drop the modifier. As Jan Masaryk reminded any who would listen, the Czechoslovaks who had died fighting the Nazis were not "provisionally" dead. To my parents and their friends, the unfairness with which we were being treated was plain. The Crown had appointed ambassadors to the other exile groups; we had to settle for a liaison officer. At diplomatic gatherings, our representatives, although back on the guest list, were given the worst seats at every table and the rear positions in any line. The Polish and Serb contingents in London had no humiliating adjectives before their names. Beneš knew that he would never be able to erase Munich if his government was not viewed as fully legitimate. For him to be taken seriously, he must first be treated equally, especially since his principal goal—to see our country restored to its prewar borders—was not a priority for any other international leader.

In addition to his quest for full recognition, Beneš had three major concerns: to settle differences among Czechs and Slovaks without damage to the cause; to maintain contact with supporters back home; and to demonstrate his country's commitment to the Allied success. Propaganda played a key part in those efforts. In January 1941, a Czechoslovak Institute began operating in London with the purpose of promoting the country's culture and educating the English about the "people of whom we know nothing." To generate enthusiasm, the government printed martial posters ("Czechoslovaks! The hour of your liberation is coming!") and distributed "V for Victory" patches ("A Free Czechoslovakia in a Free Europe; Czechoslovakia Fights for Victory!").

Because the headquarters of the government in exile had suffered damage during the Blitz, new and expanded offices were set up at 8 Grosvenor Place in London's center, while Beneš moved his residence to Aston Abbotts, a village of four hundred on the outskirts of Buckinghamshire. There he lived with Hana in an ivy-covered two-story house complete with a croquet set on the lawn, a study piled high with books and maps, and—on his desk—a framed copy of "If," Rudyard Kipling's ode to courage under fire. Approaching his fifty-seventh birthday, Beneš was showing his age. His hair had turned silver, and his habitually grave face was marked by ever-deepening bags beneath his eyes. As always, he worked constantly, conducting business at Aston Abbotts on weekends, on Mondays, and in the evenings. On other days, he made the ninety-minute commute to London in a chauffeured Daimler. Like many Europeans, he communicated with his hands as well as his mouth. When not reading, he used his eyeglasses as a prop, waving them about, holding them pensively, then raising them again to drive a point home.

Madam Benešová usually kept to the background, but her passions, too, ran deep. During World War I, her husband had offered her a divorce to shield her from political persecution. She had refused and donated most of her personal wealth to the independence campaign. Unable to escape Austria-Hungary, she had been arrested for revolutionary activities and imprisoned for eleven months, during which she had endured severe interrogation. Hana Benešová was of stocky build with a pleasant round face and pinned-up brown hair; she wore hats, as most people did at the time, along with sensible coats and often a pearl necklace and earrings. She was the honorary president of the Czechoslovak Red Cross, established a nursery for exile children in London, and helped procure basic living supplies for impoverished soldiers. Although she spoke only rarely in public, she did take an occasional turn at the BBC microphone, advocating democracy, patriotism, and community service. Like most people within the exile community, she was counting the days until she could return home.

WINSTON CHURCHILL'S COUNTRY estate, Chequers, was but a few miles from Aston Abbotts. On February 26, 1941, the prime minister's luncheon guest was Dr. Beneš, who then and at subsequent meetings delighted in his company. The president described Churchill to a friend as "at last an Englishman who understands the fundamentals of this war and what it means to Europe." To Beneš, the Second World War was partly a continuation of the first—a struggle between a militaristic Germany and the West, but with Russia better positioned than before to tip the balance. He was convinced that, despite the Hitler-Stalin pact, Germany would soon invade the Soviet Union and that Moscow and London would end up fighting on the same side. This view was strengthened by reports from his intelligence service that he dutifully passed on to Churchill.

During the meal, the president invited the prime minister to visit Czechoslovak troops. The invitation was accepted, and on April 19,

Churchill journeyed by car to the military encampment, which had moved south from Cholmondeley to a base near Leamington Spa. There he inspected the soldiers, who stood ramrod straight in their best uniforms and wore helmets that looked like inverted soup bowls. After lunch Beneš pressed into the hands of Anthony Eden, now foreign secretary, a memo making the case for unrestricted recognition of Czechoslovakia. As Churchill prepared to leave, the soldiers broke into a chorus of "Rule, Britannia!" in heavily accented English. Slavs are generally rousing singers, and Churchill quickly maneuvered his portly frame out of the car to join in. The following day, he sent a note to Eden: "I see no reason why we should not give the Czechs the same recognition as we have given the Poles." Eden replied, "I agree."

Was this another step forward for Czechoslovak democracy? Not yet. Before taking official action, Eden felt obliged to submit the memo to legal experts, who found the paper's anti-Munich tone offensive and were unimpressed by its core argument. Beneš had resigned, and another man had taken his place; by what logic could the Foreign

*Churchill and Beneš inspect Czechoslovak troops.*

Office conclude that he was still president? Beneš was a lawyer himself and should have understood this, yet he continued to pester; he was without question an irritating man. This perception was shared by the most influential U.S. diplomats. Hopkins had refused to meet with Beneš during his London visit, and Kennan held the bizarre view that the puppet president, Hácha, was the better leader. In any case, Kennan did not believe that Czechoslovakia either would or should be reestablished regardless of who won the war. Another impediment arose when, in April, Germany lashed out in the direction of Yugoslavia and Greece. The incursion created a danger to British interests that would engage the attention of Churchill and Eden for the next two months. While London delayed, Beneš sat at his desk, perhaps pondering the fifth line of Kipling's poem: "If you can wait and not be tired by waiting . . ."

SATURDAY NIGHT, MAY 10, 1941, the Luftwaffe dropped more than seven hundred tons of bombs on London, starting two thousand fires and damaging such symbols of the empire as the British Museum, the Tower of London, the House of Commons, and Westminster Abbey. More than 1,400 Londoners were killed. For the English, it was the cruelest bombing of the war.

Soon after, around the time of my fourth birthday, my parents decided that we had had enough; it would be safer to move outside the city. Fortunately we had somewhere to go. My father's brother, Honza, lived with his wife, Ola, and children, Alena and George, in a stately sixteenth-century house adorned by wisteria and yellow roses in Berkhamsted, northwest of London. Years earlier, my uncle had begun working with Grandfather Arnošt in the field of construction supply and prefabricated houses. In 1937 or 1938, he had established an outpost in England for the multinational firm that employed them. In the spring of 1939, his family followed. Alena, three years older than I, would be told later that her family had left Prague because of my father's involvement in politics. She does not remember any arguments, but I can recall loud disagreements between Uncle Honza and my father. Perhaps the cause was their differing temperaments, or maybe it was just a case of sibling rivalry. In any event, from my bedroom above the kitchen, I often heard the two men quarreling late at night, even if I didn't know why.

On weekends, Czech friends came to visit, bringing with them—due to the wartime shortages—a contribution of food. It was with the encouragement of such company that, one afternoon at the end of that May 1941, my parents were baptized into the Roman Catholic faith in a ceremony at Sacred Heart Church. I was baptized as well, though I have no memory of the ceremony.

When, in 1997, I learned that my family heritage was Jewish, I assumed initially that my parents had converted to Catholicism in order to escape the Holocaust. This was, of course, inaccurate. We had already been living in England. In any case, conversion had meant nothing to the Nazis. So why did my parents make this choice? Certainly, they had not been attempting to deceive their friends and acquaintances to whom their Jewish ancestry was no secret. Surprised, and with no parents to ask, I could only speculate from the distance of more than half

*In Berkhamsted: (front, from left) George Korbel, Alena Korbelová, the author;*
*(back) Ola Korbelová, Dáša Deimlová, Mandula and Josef Korbel*

a century. I have nevertheless thought hard, trying to understand their decision.

To begin, I doubt that theology played any role. My father had been raised in an almost entirely secular household; according to my cousin Alena, Grandfather Arnöst forbade the family from attending synagogue. To my knowledge, neither of my parents was much influenced by the Jewish thinkers who flourished in the early twentieth century, among them Martin Buber, whose *Three Addresses in Prague* (1909– 1911) helped lay the groundwork for Czech Zionism. At the time of their marriage, my parents had recorded themselves as being without religious confession.

This does not mean that their feelings were exactly the same. On the surface, at least, my mother was more emotional and less cerebral than my father. Like many Czechs and Slovaks, she was a spiritualist who believed that there were mysteries for which science had no answer, and that the boundary between life and death was not as unbridgeable as commonly supposed. The fear and stress of the war years, made more painful by separation from loved ones, could only have deepened her quest for reassurance. Among my parents' closest friends at this time were Jaroslav and Milada Stránský, both observant Catholics.* Jaroslav, a diminutive former professor and newspaper editor, was an official in the government in exile and a frequent contributor to the Czech-language radio commentaries. His family, also Jewish in heritage, had converted in the 1890s. Milada had grown up in a devout household and was eager to save souls. The encouragement of the Stránskýs might well have made the idea of conversion more inviting, especially to my mother.

A second, and I think even more telling, factor might have been my parents' desire to underline our family's identity as Czechoslovak democrats. Our home country was overwhelmingly Christian, and many Czechs and Slovaks unfairly associated Jewish culture with the enemies of their national aspirations. These prejudices, which had their

* Jaroslav was the father of Jan Stránský, whose adventures during the fall of France are described in chapter 10.

roots in the days of the Austro-Hungarian Empire, had softened during Tomáš Masaryk's republic, but the majority of Czechoslovak Jews still spoke either German or Hungarian. The yearning to be—and to be seen as—fully Czechoslovak probably explains why, during the war, my family dropped the umlaut from our last name, although the absence of that symbol on British typewriters may have contributed.* The name "Korbel," accented on the second syllable, sounded more Czech and less German than "Körbel."

Finally, and I suspect most crucially, I believe that my parents joined the Church because of the child they had had and the children they planned to have. My aunt Ola and cousins Alena and George had obtained baptismal papers while still in Prague, so their example might have had an influence. I expect that my parents thought life would be easier for us if we were raised as Christians instead of Jews. The reasons for such a conclusion, in the Europe of 1941, need little explanation.

People ask me now whether I regret my parents' decision; I don't know how to respond. It's hard for me to imagine a life different from the one I have known or to compare what might have been with what was and is. I am a firm admirer of the Jewish tradition but could not—beginning at the age of fifty-nine—feel myself fully a part of it. Celebrating both Christmas and Hanukkah with my grandchildren, I have reasons for gratitude that my origins are richer and more complex than I had thought; but still, I wish that my parents would have explained to me, when I was old enough to understand, what they had done. I would like to have had a chance to discuss every aspect of their deliberations. Exactly when did they decide and for what reasons?

Although wary of addressing such a hypothetical question, I feel it is important to add my belief—given all I know about their values—that my parents would not have made the choice they did had they waited four more years. The world in 1945 differed from that of 1941, as it has ever since. Nazi persecution of Jews was well under way at the time

---

* The missing umlaut explains why, on some British documents, my father's name is spelled "Koerbel."

of our baptism, but the grim unfolding of the Holocaust was still in its earliest stages. Czech Jews had not been forced into concentration camps, nor were they yet required to wear the yellow star. My parents would have viewed their decision to convert as difficult but made with the next generation of their family uppermost in mind. By war's end, the desire to be associated with Czech as opposed to German culture would have been even more powerful, but acting to substitute a Christian identity for a Jewish one would have been—in the absence of a genuine religious calling—hard to conceive. When viewed through the lens of the Holocaust, the moral connotations of such a choice had been altered irrevocably. Perhaps that is why my parents never found a good time to discuss the decision with me and seemed to avoid doing so with others. Before the slaughter of six million Jews, they might have found the words; after it, they could not.

JOSEF STALIN WASN'T ordinarily given to wishful thinking. He tended to expect the worst of others, which is why he engineered the murder of so many colleagues. Odd, then, that in the spring of 1941, he should choose to don rose-colored glasses. The previous year, he had been shocked by the speed with which Germany had rolled over France. He had hoped for an evenly matched contest that would have left both sides bloodied, drained of resources, and ripe for revolutionary change. Instead Hitler had felt confident enough to take on Great Britain, then plunge into the Balkans; he also had troops in North Africa and seemed intent on capturing Egypt and Crete. Those battles were still being fought.

The previous November, Molotov, the Soviet foreign minister, had met with Hitler in Berlin. The führer had assured him that the English were done for and that their efforts at retaliation were ineffectual. No sooner had he made this boast than the two leaders were hustled—due to Allied bombing—into an air-raid shelter. Stalin did not think that Hitler would move against the Soviet Union until he was certain that he had won in Europe. Surely the Nazis were wise enough to avoid having to fight a two-front war? As a precaution, the Soviets did nothing to awaken Hitler's ire. In the first four months of 1941, they had sold to Germany a quarter of a million tons of oil and 750,000 tons of grain. Stalin, like Chamberlain before him, was reassured by the knowledge of what he would do if he were in Hitler's shoes; like Chamberlain, he was wrong.

Beneš, however, was right. He did not expect Hitler to act in a manner others thought logical but instead to fulfill his own imagined destiny. The Nazi dream depended on expansion to the east, which made a clash with Stalin inevitable. If the führer waited, thought Beneš, he would give the Soviet military a dangerous amount of time to prepare. Moreover, to retain the element of surprise, Hitler had to move before enemy analysts thought he was ready. Throughout that spring, the Czechoslovak president insisted that the Nazis intended to invade the USSR soon and without warning. On June 22, that forecast came true. Within a week, German tanks and troops advanced more than two hundred miles into the Soviet Union, killing and taking prisoner a massive number of Russians. The Red Army, caught unawares, retreated in confusion. Stalin, who was more than a trifle paranoid, worried that he would be ousted or even possibly shot by his own aides. The outcome of the assault seemed certain. Military experts were unanimous: the Germans would overrun Moscow within two months.

In London, these events led to a drastic and immediate reassessment. The Soviet Union, reviled both for its Bolshevism and for its infamous pact with Hitler, had become overnight the sworn enemy of civilization's most dangerous adversary. The Soviets needed help; the West trembled at the thought of Hitler enthroned from Paris to Vladivostok. Broadcasting to his nation on the night of the invasion, Churchill repressed his deeply felt anticommunism and declared, "Any man or state who fights against Nazism will have our aid. Any man or state who marches with Hitler is our foe. . . . It follows therefore that we shall give whatever help we can to Russia and the Russian people."

The Nazis' destruction was appalling, but it also meant that the Soviets would be shopping for friends—and the affections of Czechoslovakia were available. Beneš told the Russian ambassador in London that his country would do all it could to assist, provided only that

Moscow grant full recognition to the government in exile. Representing a nation under siege, Soviet diplomats had little interest in fine points of law; the answer was yes.

Beneš immediately informed the Foreign Office that he was being courted. "I am worried," he said in his most sincere tone, "that Russia will claim the complete allegiance of my people and that England, as at Munich, will be left behind." For emphasis, he pointed out that the Soviets had promised to set up a Czech and Slovak military legion on their soil and to begin Czech-language broadcasts from Moscow. This maneuvering, transparent as it was, had precisely the desired impact. With the support of Churchill and Eden, legal qualms were finally swept aside, the "provisional" was dropped, and on July 18, the government in exile was officially recognized by both the Soviet Union and Great Britain.*

The Czechoslovaks were now on the same plane as other exiled leaders in London—but membership in that group had grown. Not only were there Poles and Yugoslavs, but also French, Belgians, Greeks, and a fair sampling of crowned heads: King Haakon VII of Norway, Queen Wilhelmina of the Netherlands, Emperor Haile Selassie of Ethiopia, and Albania's King Zog. All were allies but also, in a sense, rivals. Each had distinct interests, and all wanted attention and help from the British.

The Czechoslovak cause was aided in this competition by the brave service of its soldiers and airmen. Beneš was concerned, however, that the Czech puppet government in Prague would prove spineless and embarrass him by endorsing the German invasion, possibly even sending troops to fight the Soviets. He dispatched a firm message to Hácha and Eliáš demanding that they not show any sign of support for the Nazis; he also told Hácha that his government had just about exhausted its value and that he and his top advisers should be ready at any moment to resign. No response was received.

---

* The Roosevelt administration responded to the British and Soviet decisions by appointing a full ambassador to what it still referred to as a "provisional" regime, again suggesting that the Hácha government was legitimate and had more domestic support than Beneš. Not until October 1942 did the United States drop the "provisional" and begin to address Beneš as "president of the Czechoslovak Republic."

The time had also come to reevaluate the role of the Czech resistance. Twenty-two months had elapsed between the signing of the Hitler-Stalin pact and the German betrayal of the Soviet Union. For Communists in the protectorate, this had been a period of enforced silence and confusion. Under orders from Moscow, they had refrained from actions that might upset the authorities in Berlin. On June 22, that particular signal switched from red to green. Suddenly the Communists—or, as they preferred to be called, the "heroic vanguard of mankind"—were free once again to vent their rage at the "hordes of fascist beasts." Through messengers and radio broadcasts, the Russian leadership appealed to Czechs to move from passive to more active resistance. Members of the underground were urged literally to throw sand into the gears of the Nazi war machine by tossing gravel into the places in munitions plants where lubricating oil was normally applied. The impact of this call to action, though incremental, was still tangible. Week by week, there were more factory breakdowns, accidents, cases of sabotage, railway fires, and displays of anti-Nazi graffiti.

At a remote site along the Scottish coast, the British Special Operations Executive began training a select group of exiled airmen and radio operators to participate in clandestine operations. Efforts were also made to upgrade the Allied communications network; messages could now be sent and received by transmitters from as far west as Portugal, as far south as Cairo, and as far east as the Soviet Union. The potential for coordinated resistance actions had expanded.

The BBC programs supervised by my father assumed a more dynamic role as well. In September, he and his team helped to formulate and publicize two campaigns. The first was a "go slow" initiative that encouraged Czech workers to dawdle. "You who labor in factories operated by the Germans, do not be in such a hurry," Jan Masaryk advised. "If you go and fetch tools, do not run because you may get out of breath; and if all of you will work just a little more slowly you will hasten victory. Therefore quickly and slowly, each doing his own job in full cooperation with Beneš and London."

The second initiative was a summons to boycott newspapers that were pro-German, a category that—because of censorship—now included virtually every publication available on the streets of Prague. Both campaigns were launched early in the month and sustained by daily appeals. The sabotage and slowdowns reduced Czech industrial production by an estimated 30 percent. The press boycott cut newspaper purchases by more than half. At a staff meeting on September 24, my father reported that the strategy had been "an outstanding success."

But if there was one thing the Germans would not tolerate, it was a Czechoslovak success.

# 15

# The Crown of Wenceslas

President Hácha held the keys so that his guests might see; one for each of the seven locks to the royal chamber and crown jewels of Bohemia, representing the heritage of a thousand years of Czech history. Before him, laid out on a center table, were the king's scepter, orb, and cloak, the coronation cross and sword, and Wenceslas's gleaming gold crown. Slowly Hácha made a half turn and surrendered them. Reinhard Heydrich, the acting reichsprotektor of Bohemia and Moravia, grasped the keys firmly, and he, too, held them aloft. Here, on November 19, 1941, in the Wenceslas Chapel of Saint Vitus Cathedral, the Czechs' supposed leader formally entrusted his nation's most cherished heirlooms to Germany. Heydrich smiled amiably as he inspected the treasures, gripping the sword's handle and lightly brushing the fleurs-de-lis atop the spears of the jeweled crown. With a friendly gesture, he returned three of the keys to Hácha, cautioning him, "View this equally as trust and obligation." The ceremony's meaning was as plain as the body language of the participants. Heydrich, over six feet tall and finely turned out in his military uniform, exuded strength and order; Hácha, nearly a foot shorter, stood stoop-shouldered and expressionless. In German eyes, the rightful relationship between the two peoples had been set a millennium earlier when Wenceslas had first made peace with Saxony and begun paying an annual tribute. The Germans were destined to rule, the Czechs to serve.

*Hácha presents keys to Heydrich; on the left is Heydrich's deputy, Karl Hermann Frank.*

EIGHT WEEKS EARLIER, in his bunker in East Prussia, Hitler had met with K. H. Frank, the Sudeten leader, who favored taking a harsher line against the Czechs. At issue was the growing alarm caused by the Resistance. With the invasion of the Soviet Union beginning to stall, the German army could not afford a slowdown in the production and shipment of war materiel. Frank, ever mindful of his career, steered all blame in the direction of Protektor Neurath, who, he said, was coddling the local populace and failing to demand respect. Perhaps there might be a stronger man for the job? Hitler agreed but, instead of turning to Frank, sought help from one of the busiest figures in the Reich. Reinhard Heydrich, deputy to Hitler's security chief, Heinrich Himmler, supervised all police operations in Germany. Since August 1940, he had been head of the International Police Commission, or INTERPOL. He was also a leader of the German sports federation. On September 27, he flew to Prague to begin additional duties as acting protektor; Neurath was directed to take time off "for his health."

Blue-eyed, with fair hair and chiseled features, the thirty-seven-year-old Heydrich was the ideal national socialist: dedicated, organized, ambitious, and without pity. He had inherited his extreme nationalism and anti-Semitism from his father, a composer of little distinction who was rumored, falsely, to be Jewish. Before the young Heydrich found his calling as Himmler's protégé, he had been expelled from the navy for spending a night with one woman shortly after proposing to another. A Nazi since June 1931, Heydrich had made his reputation by identifying and dealing with internal enemies. In recognition of his diligence, he was among the first to be awarded the SS's coveted Death's Head ring. When the war began, he continued to act where others might have hesitated. Under Himmler's direction, he organized the mobile death squads that massacred Jews, churchmen, aristocrats, and intellectuals during the invasion of Poland. In the autumn of 1941, he ensured that similar barbarities were perpetrated against the Russians.

For most Czechs, the first two and a half years of the Nazi occupation had been both annoying and mortifying. Jews faced severe discrimination; universities were closed; curfews were still in effect; and streets, shops, government offices, and factory boardrooms swarmed with German soldiers, bureaucrats, spies, and profiteers. Yet the typical Czech felt more angry than frightened. Those who kept their mouths shut and heads down could go on about their lives. Even the majority of people arrested were soon released; executions were infrequent. Czechs could show pride in their identity, provided they did so without disrespecting Germans. Over time, the relatively relaxed atmosphere had an effect. The underground gained confidence. The BBC broadcasts evolved from a minor annoyance into a real threat. Czechs began wondering how hard they could push. To this last question in particular, Heydrich was intent on providing an answer.

THE ACTING PROTEKTOR's plan for asserting dominance was based on the principle of mixing carrots and sticks or, as Heydrich preferred, "whips and sugar." From his first day in Prague, he instilled fear by imposing martial law and by ordering the apprehension, questioning, and torture of thousands of Czechs. Among those arrested was Prime Minister Eliáš, whose ties to the Resistance had been known for some time but who had been shielded from punishment by Neurath. Now

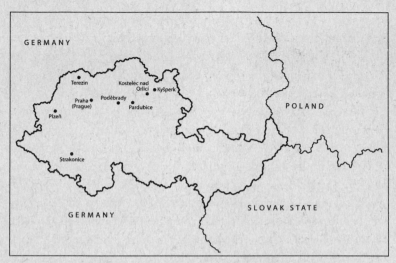

*Protectorate of Bohemia and Moravia, 1939–1945*

Eliáš became the only prime minister in a German-occupied land to be tried and sentenced to death. Each afternoon, police cars transported prisoners to the shooting grounds. In the morning, neatly typed placards were posted on streetlamps with the names and birth dates of the victims; relatives were required to reimburse the Gestapo for the cost of the executions and placards.

On October 2, a week after his arrival, Heydrich gathered his staff. He demanded that the recent surge in resistance and sabotage be met with "unflinching hardness." Every Czech must understand his duty: to work ceaselessly in support of the German war effort. That could be accomplished only through brutality, because the local population equated kindness with weakness. "Not a single German should forgive anything to the Czech.... There should not be one German who would say: 'But this Czech is a decent person!' "

The goal, he said, was not to push the wretches to the point of exhaustion but to reap the full fruits of their labor. This meant that for the duration of the conflict, workers had to be "given their grub." However, the long-term goal must remain. "This region must once more become German ... the Czechs have ... no right to be here." Heydrich unveiled his plan to order medical exams for the protectorate's entire population, beginning with its children, to determine which portion could be saved for possible Aryanization and which eliminated. The most worrisome, he warned, were Czechs "with hostile intentions but of good racial extraction. These are the most dangerous."

The ceremony in which he took possession of the Bohemian crown jewels provided a climax to the first phase of his strategy: German supremacy had now been established symbolically as well as politically and economically. In phase two, Heydrich lifted martial law and emphasized the dividends of collaboration. Workers in defense plants received higher wages, free shoes, cigarettes, more food, and extra days off. Czechs who informed on their countrymen were rewarded, police and other officials who cooperated with the Nazis promoted. By blending cruelty with a tempting promise of special favors, Heydrich did much to weaken the Resistance. For the first time, the system of clandestine wireless links between the underground and the government in exile was disrupted. More than four hundred people were executed; thousands more went into hiding. The feeling grew that effective rebellion was not possible. Hácha, pliable now to the point of treason, publicly denounced both Beneš and the BBC's troublemaking broadcasts. For the Nazis, the results were gratifying: the munitions plants were again operating at full speed; there were fewer incidents of sabotage; and Czech schoolchildren collected mittens, scarves, sweaters, and skis to send to Germans fighting in Russia. Even Hitler was pleased. "If we give these gourmands double rations," he gloated, "the Czechs could be made into fanatical supporters of the Reich."

Heydrich had sought the assignment in Prague because he saw it as a stepping-stone to a position in Paris or to an even loftier job in Berlin. Martin Bormann, the führer's private secretary, noted how much the young man resembled Hitler in his creative energy. Heydrich, he said, "always remained a happy, strong optimist. How much human weakness, inadequacy and evil did he see! Nonetheless, he always remained an unworried, aggressive national socialist, whose faith in the mastery of tasks simply could not be shaken!"

The acting protektor was indeed tireless, for on top of all his other responsibilities, he had volunteered for yet one more: to find an answer to the Jewish question.

BUREAUCRACIES SPAWN SPECIALISTS; the Nazi system produced experts in persecuting Jews. Members of that cadre assembled first in Germany and Austria, then in other occupied lands. They developed their own jargon, replete with euphemisms for genocide and murder, and found their home in the Gestapo, where they were accountable only to Himmler and ultimately the führer. Heydrich was one of their guides. He explained that all Jews, whether religious or secular, bankers or bricklayers, were part of a multigenerational conspiracy to dominate the world and annihilate Aryan values. "The Jew," wrote Heydrich in 1935, is "the mortal enemy of Nordic-led and racially healthy folks"; even the many Jewish soldiers who had fought for Germany in the Great War had done so to deceive real patriots and divert attention from their self-aggrandizing goals.

During the 1930s, the combination of Hitler's verbal bigotry and Nazi intimidation had prompted hundreds of thousands of Jews to flee Germany. The exodus was encouraged by the Reich, which sought in the process to separate the emigrants from their property and money. This policy was refined in Austria after the 1938 Anschluss by a Heydrich acolyte, thirty-two-year-old Adolf Eichmann, whose office streamlined the handling of paperwork and financed emigration by taxing wealthy Jews. His "Vienna Model" pushed 110,000 people out of Austria in five months. In the summer of 1939, Eichmann established a comparable office in Prague, where he declared, "I too am a Zionist; I want every Jew to leave for Palestine."

Before the war and in its first two years, the Nazis considered emigration a means both for raising funds and of removing an unwelcome population. Hitler even appealed to Western leaders to back their statements of concern for the well-being of Jews by opening their borders, a challenge that the West shamefully failed to take up. In 1940, Nazi planners devised a scheme to send one million Jews annually to the French colony of Madagascar. That brainstorm, supposedly blessed by Hitler, came to naught when the British survived the Blitz. The

German fleet was not large enough to fight His Majesty's Navy while simultaneously conducting a ferry service to Africa.

Operation Barbarossa, the German invasion of Russia, appeared to open the door to another option. Assuming a quick victory, the entirety of the USSR's frozen East would become available. The Nazis soon found, however, that success there would not come as rapidly as hoped; meanwhile, the Wehrmacht had first claim on rail facilities for the movement of troops. The bureaucrats concerned with the Jewish question had to improvise.

From the fall of 1941 until January of the following year, Heydrich presided at a series of meetings intended to settle what he referred to as the *Endlösung*, or "final solution," of the Jewish question in Europe. Emigration had provided a start but was clearly insufficient given Germany's recent conquests. Labor camps and prisons could accommodate but a fraction of the three and a quarter million Jews who had come under Nazi jurisdiction. A comprehensive strategy was needed that would take into account war requirements, the führer's urgent wish to expel Jews, and the Foreign Ministry's desire to avoid unnecessary damage to the country's reputation. Heydrich decided on a step-by-step approach: evacuees would be sent first from the Reich and protectorate to ghettos; then, for purposes of labor and "special handling," to points east.

The systematic deportation of Jews from the Czech lands began in October, when the first of five transports to Poland left Prague. The passengers included many of the city's leading professionals and businesspeople. Upon their arrival in the Łódź ghetto, they were assigned to work details. After months of exploitation, survivors were sent thirty miles to the village of Chelmno, where the first Nazi death factory was operating. The installation of mobile and later fixed gas chambers was referred to by the Nazis as Operation Reinhard in honor of Heydrich's leading hand.

IN NOVEMBER, THE Germans also began the relocation of Jews to Theresienstadt or, as the Czechs called it, Terezín. This was the star-shaped army fortress that Emperor Joseph II had named for his mother

150 years earlier. The town was located near the Czech-German border, forty miles north of the capital. Prague's Jewish leaders had been shocked by the earlier transports. The sight of their neighbors lined up and packed into trains under the eyes of the Gestapo had caused them to accept with relief the idea of a Jewish ghetto inside the protectorate; this was before anyone outside the Nazis had imagined death camps or gas chambers. If German occupation meant that Jews would be forced to live separately but nearby, so be it; matters could be worse. Heydrich and Eichmann promised not only that Czech Jews would be "self-governing" but that they would be allowed to remain at Terezín and not required to move again. That was a lie. Heydrich had already informed Eichmann and other associates that Terezín would be but a "temporary assembly camp." In time, if all went according to plan, the Jews would be gone and the area repopulated by Germans.

I TOURED TEREZÍN while secretary of state and again in the spring of 2011. The facility is in two parts. The Small Fortress, in its history both guardhouse and penitentiary, is as one might expect: cold, grim, and hard as the concrete in its floors and the iron in its bars. Visitors are told of its history, first as home to a core of gunners under Joseph II, then in the Great War as a jail for more than 2,500 political prisoners, most notably Gavrilo Princip, the assassin of Archduke Ferdinand. During the years of independence, a corps of Czechoslovak artillerists was stationed there, after which came the Germans, who, beginning in June 1940, used it as a place to detain, torture, and often execute alleged enemies of the Reich. These included leaders of the patriotic Sokol gymnastics organization, participants in student protests, perpetrators of sabotage, and others who had sheltered fugitives or in any way displeased the Nazis. There is such a terrible cruelty about the fortress that one can stand in a cell and readily imagine the cramped space stacked with so many captives that they could not lie down.

By contrast, the so-called ghetto of Terezín looks nothing like what one might imagine of a concentration camp. There are no thick walls surrounding it, no dark dungeons, no rusted shackles. Terezín today is a town again, though sparsely populated. The buildings where the

*Terezín*

inmates were once crowded together are handsome enough; the grass is thick and green, the sense of pain and despair less easy to conjure up. This is perhaps appropriate for a place that masqueraded as a spa. Fittingly, the displays put together by the Terezín Memorial project emphasize different aspects of what the experience had been like: the rail depot, the boys' dorm, the barracks, the administrative headquarters, the crematory. The exhibits are there, too, of the old suitcases and clothing, the arrival and departure cards, the artwork and music, the newspapers, and most hauntingly, photographs of the children.

THE FIRST CZECH Jews to come to Terezín did so in November 1941; they were skilled laborers whose job was to prepare the old fortress town for its new function. By year's end, trains were arriving on a more or less weekly basis. Although the ghetto was originally intended for protectorate Jews only, the Nazis found it convenient to send German, Austrian, and later Dutch and Danish Jews. Because Terezín was described publicly as a self-administered retirement community, Eichmann felt confident that "Theresienstadt will allow us to preserve our appearance abroad."

It didn't take long for the Nazis to break their promise that the facility would serve as a permanent residence. Records indicate that my grandfather's younger sister, Irma (Körbel) Paterová, was the first in my family to be sent there, arriving on December 10 along with her husband, Oscar, and their twenty-eight-year-old daughter, Herta. Five weeks later, the three were among those transported by unheated cattle car to the German-controlled Latvian city of Riga—the site of hideous pogroms the previous year. There the passengers were unloaded and taken to a nearby forest, where they were shot.

IN THE SUMMER of 1941, the Czech underground had intensified its pressure on the Nazi occupiers. In September, the Germans struck back. Now it was Beneš's turn. Heydrich's campaign of terror demanded a dramatic response, something memorable to convince the Allies that Czechoslovaks were not to be pushed around. Behind closed doors, the president suggested "a spectacular action against the Nazis—an assassination carried out in complete secrecy by our trained paratroop commandos."

For months, an elite unit of Czech and Slovak officers had been undertaking missions in their homeland. Supervised by the British, parachutists were taught to operate radios, shoot, fight, read maps, live off the land, resist interrogation, handle explosives, and jump out of airplanes without injury. They carried materials useful to the Resistance such as ammunition, batteries, cash, forged papers, and information about codes. To reduce the likelihood of arrest, they wore Czech-made clothing and were supplied with locally manufactured toiletries, cigarettes, even matches. Before departure, they wrote their wills and were given tablets—sealed in paper—of cyanide. The missions of the men,

*Jozef Gabčík*

who were organized into groups of two or three, sometimes involved sabotage but more typically the repair and replacement of radio transmitters. Operation Anthropoid would be more ambitious.

It was that rarity in time of war or peace—a secret well kept. Only a tight circle of British and Czech officials were in on the planning, my father not among them. Political sensitivities were considerable, the likelihood of success low, and the prospect that the parachutists might survive almost nil. Even on routine operations, those sent were often apprehended within days or weeks. For this mission of unprecedented daring, everything would depend on the quality of the men.

The British recommended a combination of firearms and bombs. Jozef Gabčík would use a Czech-made Sten machine gun, lightweight and, when broken down, easily concealed. Jan Kubiš was given a supply of grenades, designed originally for crippling tanks in North Africa. In mid-December, the two were transported to London and installed in a safe house, awaiting suitable weather and the availability of a plane.

*Jan Kubiš*

While in the capital, the officers met Beneš, who thanked them for their bravery and stressed the importance of what they had been asked to do. The president's personal aide, Eduard Táborský, remembers how young they appeared. "One of them seemed to me more a boy than a soldier, let alone a parachutist, ready for anything and setting off right in the midst of that hell." A fellow trainee recalled that "they were both ordinary chaps. . . . Kubiš was a quiet fellow; he would never have hurt a fly. Gabčík, on the other hand, was fiery and enthusiastic. . . . As soldiers, they felt that orders were orders—no argument. The news from our country, telling us about the tortures and the killing of our people, had worked them up to a high pitch."

To fly from London to the Czech countryside and back on a single night without being detected required the many hours of darkness available only in winter. To identify a landing spot accurately demanded an amount of moonlight available only ten days a month—and a relative absence of clouds. Not until December 28 did those conditions coincide. The plane's course took it over France, then Germany; for twenty tense minutes it was trailed by enemy fighters that either lost sight of it or ran low on fuel. In the early-morning hours, the aircraft decelerated and dipped to a point several hundred feet above the snow-covered countryside south of the city of Plzeň. At 2:24 a.m., the hatch was flung open, and moments later camouflaged parachutes descended from the sky.

# PART III

❧

## *May 1942–April 1945*

*What good to mankind is the beauty of science?*
*What good is the beauty of pretty girls?*
*What good is a world when there are no rights?*
*What good is the sun when there is no day?*
*What good is God? Is he only to punish?*
*Or to make life better for mankind?*
*Or are we beasts, vainly to suffer*
*And rot beneath the yoke of our feelings?*

*What good is life when the living suffer?*
*Why is my world surrounded by walls?*
*Know son, this is here for a reason:*
*To make you fight and conquer all!*

—HANUŠ HACHENBURG (1929–1944)
*Terezín*

# 16

# Day of the Assassins

Marie Moravcová (Moravec) was in her forties, tall and amply built. She had brown hair, round cheeks, lively eyes, and a carefree laugh that when the Nazis invaded all but disappeared. She lived in a two-bedroom flat in the Žižkov neighborhood, a working-class area on the outskirts of Prague, named for the Hussite warrior Jan Žižka and known for its many pubs. Marie shared the small apartment with her husband, Alois, a pensioned railway worker, and their son, Ata, age twenty-one. She was a good-hearted woman who volunteered with the antituberculosis league and served as secretary of the Sisters of the Red Cross. The organization was influential in Prague, and its members were naturally alarmed when friends had to go underground for fear of Nazi arrest. Such families dared not register for ration coupons and so were in danger not only of jail but of starvation. The Red Cross volunteers adapted by holding unpublicized meetings and learning the tradecraft involved in smuggling food. Madam Moravcová was not normally a political person, but she had acquaintances all through the city and assured the chapter president, "If you need anything at all, I am entirely at your disposal."

The time was February 1942. Heydrich had been in Prague for five months. The Czech underground was still functioning, but people's nerves were taut. Each arrest led to speculation: Who could withstand the torture and who could not? How much did the arrested person know? Which of us will be next? One day, the Sisters received an urgent message: could they arrange to shelter some young men? The first to step forward was Marie Moravcová.

There were three of them at first, all in their late twenties, known to her as Little Ota, Big Ota, and Zdenda. Marie found places for the men to stay, then moved them around, supplying clothes, razor blades, cigarettes, and food. She introduced them to her apartment building's caretaker, František Spinka, a coin collector by hobby who lived on the ground floor and who consented, when the visitors came back at night and whispered the correct password, to unlock the door. Spinka agreed, as well, to look after Zdenda's large black sheepdog. The strangers, it seemed, would be keeping irregular hours.

What were they up to? The men spent much of their time exploring the routes that separated the capital from Panenské Břežany, the town where Heydrich had secured a palatial château for his family. Trying not to attract attention, Zdenda and his comrades trudged along the roads and surveyed the surrounding bushes and clumps of trees. They paid particular heed to places where the road from Panenské Břežany curved, deciding finally on a hilly stretch where cars heading for Prague had to slow before turning sharply to the right and crossing a bridge into the city. This was in a residential neighborhood consisting of narrow streets and small houses, without a police station near. Through contacts among Czech staff at Hradčany Castle, the men learned of Heydrich's daily routine. They knew that his car would carry him through the winding intersection each morning and evening, sometimes with a security escort, usually not.

When asked by others in the Resistance what they were up to, the men laughed and said that they had come to "count the ducks on the Vltava." Each had a briefcase, carefully concealed. Little Ota acquired a girlfriend, a young acquaintance of Madam Moravcová. Big Ota promised to marry the nineteen-year-old daughter of the family in whose apartment he was staying. At times, the men asked Marie, or "Auntie," as they called her, for something a little different: a length of rope, a place to hide a transmitter, a bicycle whose serial number had been filed off.

Little Ota's real identity was Gabčík; Big Ota's was Kubiš; Zdenda was Sergeant Josef Valčík, a radio operator whose team had been inserted separately on the same night as the other two. In early April, they were joined by Lieutenant Adolf Opálka, the senior officer of a group that had arrived by parachute in late March. He had been accompanied by a man called Vrbas but whose birth name, destined for infamy, was Karel Čurda.

The parachutists were able to communicate with London via a transmitter set up in the village of Pardubice and monitored by other members of their team. Couriers traveled on foot or by bicycle and employed a full set of special knocks, passwords, and ciphers. Any message, once decoded, was rewritten with a similar meaning but in different words so that it could not be used to break the code even if intercepted. Sometimes instructions came from Beneš himself.

One morning in late April, Madam Moravcová asked her son to accompany Zdenda on a trip to the countryside to retrieve a radio beacon hidden by another team of parachutists, recently arrived, who had run into trouble upon landing. Before they could complete their mission, the two were discovered by a Czech policeman, who warned them to vacate the area because it was heavily patrolled by Germans. Ata, upset by the close call, was further shaken by a warning from Zdenda: "You see that wooden crate, Ata? The Huns might knock it about so hard it would begin to talk; but if that happens to you, you mustn't say anything, not a word, you understand?"

Although the purpose of Operation Anthropoid was supposed to be a mystery, various figures in the underground came to suspect what was being planned; furious arguments broke out between the parachutists—who had been given an order—and local leaders who feared that the mission, whether or not it succeeded, would doom their own future efforts. The Resistance sent a message to London urging that the operation be canceled or at least transferred to a less provocative target. On May 15, speaking over the BBC, Beneš appeared to deliver his answer:

> In this situation, a proof of strength even in our own country—rebellion, open action, acts of sabotage and demonstrations, may become desirable or necessary. On the international plane, action

of this kind would contribute to the preservation of the nation itself, even if it had to be paid for by a great many sacrifices.

The instruction seemed clear: the countdown was on. In London, the word went out: now that Beneš has pronounced himself—don't stir things up. On May 21, my father received an unsigned memo on a blank sheet of paper: "BBC broadcasts call too much attention to sabotage. . . . Sabotage still goes on but less said about it the better." In Prague, the team had to act quickly; word from the castle was that the target would leave soon for a new assignment in France.

ON THE EVENING of May 26, 1942, Heydrich inaugurated the Prague Music Festival, featuring a concert of chamber works composed by his father; the proud son wrote the program notes. It was a memorable moment.

The following morning, a Wednesday, the acting protektor was driven from his estate toward his office in Prague. Despite warnings from Berlin, he traveled without a police escort, believing that no Czech would be so foolhardy as to attack him. His open-top Mercedes tourer nonetheless maintained a high speed until forced to slow as it approached the hairpin curve. Opálka and Valčík, acting as lookouts, signaled the arrival. When the car entered the turn, a figure arose from the side of the road, shook off his raincoat, and pointed a machine gun at the vehicle. Nothing happened: Gabčík's weapon had jammed. Instead of ordering his driver to accelerate, Heydrich drew his pistol, stood in his seat, and motioned for the vehicle to brake. At that instant, Kubiš stepped from the shadows on the far side of the pavement and tossed one of his powerful antitank grenades toward the right rear tire. A loud explosion broke the morning quiet. The bomb had fallen a few inches short, but the force of its detonation propelled shards of metal, glass, and seat stuffing from the Mercedes into the passenger's guts.

Heydrich fell back into the car, clutching his stomach with one hand, waving his gun with the other. Kubiš, despite bomb splinters in his own chest and forehead, swung himself onto his bicycle and ped-aled furiously toward the nearby neighborhood of Liben. The chauffeur

*Heydrich's damaged car*

jumped from the car and, having failed to intercept Kubiš, took off after Gabčík, who had swapped his machine gun for a revolver and was sprinting—tie flying in the breeze—up the hill in the direction from which the car had come.

The two men ran, trading shots, until Gabčík, out of breath, slipped down a back street and into a butcher shop that, as poor luck would have it, was owned by a fascist. The startled butcher dashed to the sidewalk, where he motioned frantically to the chauffeur, who took cover behind a post and began firing into the shop. Gabčík's pistol barked in reply, and the driver grunted, grabbing his leg. Seeing his chance, the young man darted back into the street and fled, this time chased by the butcher, whom he soon outran.

Valčík and Opálka got away undetected. That evening and in the harrowing days that followed, Madam Moravcová and resistance leaders moved stealthily to hide the assailants, bind Kubiš's wounds, and fashion a plan for what to do next.

Heydrich, meanwhile, had been rushed to Bulovka Hospital in a commandeered two-cylinder Tatra van, stretched out in the back among crates of floor wax and tins of furniture polish. To the eyes of the van's frightened driver, the wounded man looked "in a bad way, yellow as a lemon and hardly able to stand." A Czech doctor swabbed Heydrich's three-inch-deep wound, but almost immediately German physicians took control. They determined that the patient required an operation and proceeded that afternoon to reinflate his left lung, extract the tip of a fractured rib, suture the torn diaphragm, and remove the spleen, which contained a mix of grenade fragments and upholstery fibers.

Himmler was quick to visit his fallen protégé in the hospital and to send his personal physician to monitor the patient's care. For a time, Heydrich's condition appeared to stabilize, but then he developed a fever. On June 3, he lapsed into a coma before dying the next day at 4:30 a.m. The apparent cause was blood poisoning. His body was taken to Prague Castle to lie in state, and on June 9, the funeral was held in Berlin. Hitler spoke, and later honored the corpse by attaching its name to an SS unit operating on the eastern front.

"THE SHOTS WHICH sounded in Prague on the 27th of May," declared my father in a broadcast three days after the attack, "weren't an isolated event . . . they showed the tension which started on the 15th of March 1939. . . . No nation can accept the fate of slaves or give up the right to exist. The proud Czech people cannot do that."

The killing of the Butcher of Prague, as he was called in the West, was front-page news on both sides of the Atlantic. There were, however, no claims of responsibility. The Germans had not identified any suspects and were without firm leads. In London, Beneš said nothing; Jan Masaryk, in New York, was not so tight-lipped. Asked by NBC whether Heydrich might have been struck down by a Gestapo rival, Masaryk could not resist the most transparent of hints. "From certain indications upon which I would not like to enlarge today," he said, "I am definitely of the opinion that it was the Czech people who did this beautiful job. I would even go so far as to say that there were people living at home or people who came from some free country, perhaps

England, to perform this duty to humanity." If that were not clear enough, Masaryk added, "You know . . . there is a gadget called a parachute."

The assassination led to the final break between the London exiles and the protectorate's puppet government. Hácha attended a memorial service for Heydrich in Prague, urged the public to cooperate in the investigation, and joined in offering a reward for help in identifying the killers. Further, he blamed Beneš for every hardship being experienced by the Czech population, even identifying the president as the nation's number one enemy. That was too much for my father. In his broadcast on May 30, he explained that the government in exile had never accused Hácha of being a "traitor or a quisling because we were aware of the conditions under which the so-called protectorate was formed." But he said that the cabinet should have resigned rather than serve one day next to Heydrich. "They would have done better if they had gone at the right time, saving themselves from this heavy responsibility and dishonor."

For the exile government in England, this was a period of intense anxiety. Beneš and his intelligence staff were gratified by the success of the mission but in the dark about what had happened to the assassins. As their mounting animus toward Hácha reflected, it was essential to win the competition for public opinion back home. Each day the BBC broadcasts reminded the Czechs and the world of Heydrich's crimes. "The leaders of today's Germany and the whole German nation are responsible," said my father. He added, quoting Stalin, "We must hate our enemy wholeheartedly and from the depth of our soul if we are to beat him."

On the evening of June 5, my father was in the middle of reading a news bulletin about the death of Heydrich when the British censor terminated the audio in midsentence. Apparently, the text had not been fully vetted. There is no record of my father's reaction, but I did come across one piece of evidence. In a letter of complaint to the BBC, the censor asked, "Can Korbel be told to keep his shouting to himself?" Clearly, emotions were running high.

———————

WHEN FIRST NOTIFIED of the shooting, Hitler ordered the execution of all Czech political prisoners and the random arrest of 10,000 more. Warned by aides against such massive reprisals, he reconsidered, settling on a more tactical but no less barbaric response.

Lidice was a mining village about twelve miles northwest of Prague, not far from the Masaryk family's country home. The Gestapo had received a tip that the townspeople had given aid to parachutists, possibly even the killers of Heydrich. The report was untrue, but on the evening of June 9, a few hours after Heydrich's funeral, SS troops surrounded the village. They searched each house, confiscated valuables, and ordered the residents to assemble. At dawn, the men, 173 in all, were separated from their families and shot.

The women and children were trucked to a school gymnasium in the next town, where they were questioned and inspected. A number of the youngest, those with fair hair and a Nordic face, were given to German families to be raised as their own. The remaining children (about eighty of them) went to Poland, first to Łódź, then to Chelmno, where on July 2 they were murdered in the gas chambers. The women were sent to concentration camps. Citizens of Lidice who were away at the time of the massacre, or working a night shift, were tracked down and killed. Every building was burned or dynamited and the cemetery plowed under. The name of the town was excised from maps. Even a river running through it was diverted.

Pictures of Lidice taken before the massacre feature a church steeple and the sharply slanted roofs common among residences in the Bohemian countryside. The houses are of good size and arrayed in an irregular pattern atop gently sloping land on three sides of the church. A line of poplar trees stands guard against the northern wind. Photographs made after the killings show only a large area of grassland, scarred by a squarish shape where all vegetation had been scraped off. There are no broken timbers visible, no burned embers, no stone foundations or other signs of settlement. There is nothing. The line of poplars was left behind, but each was beheaded—chopped off a few feet from the ground. The Nazis filmed it all. Returning to

*The burning of Lidice*

Prague, one SS man confided to a Czech security official, "We didn't find any traitors, but the day was glorious."

TO THE NAZIS, the attack on Heydrich was a challenge to their dominance that not even the liquidation of Lidice could extinguish. From Berlin, the command went forth that the killers must be held to account. Thousands of homes, stores, and warehouses were searched. Hundreds of potential suspects were picked up and interrogated. The exhibits in the inquiry—Gabčík's briefcase and Kubiš's bicycle—were put on public display. Anyone with information leading to the identity of the killers could count on the gratitude of the führer and a generous reward. Meanwhile, the wrong remark overheard in a bar or a passing comment on the street could mean death. "Approval of the assassination" was deemed a capital offense, for which 477 Czechs would be executed.

Tereza Kašperová, the mother of a seven-year-old, recalled that "right through the whole city of Prague, the Gestapo and the SS searched flats and houses, shouting and roaring, looking everywhere for the men who were responsible for the attack." They searched her house, too, but neglected to look behind the large blue-and-yellow cushion that had been wedged between a sofa and the wall, thus failing to note that behind the cushion was a cupboard and that inside the cupboard was Lieutenant Opálka.

Despite the frenzied searching, no parachutists were captured between the day of the attack and the destruction of Lidice. Seven were being sheltered in Prague, including the four participants in the assassination (Kubiš, Gabčík, Opálka, and Valčík). The Resistance decided that it would be wiser to bring the men together in one place than to leave them dispersed and at risk in safe houses with security patrols combing the city. Jan Sonnevend, the local leader of the Christian Orthodox Church, observed that the Nazis had not been searching religious buildings with any degree of rigor. He suggested as a hiding place the crypt beneath Karel Boromejsky, a sanctuary dedicated to Cyril and Methodius, the two saints who had brought Christianity to the Czech lands a thousand years before.

When Marie Moravcová was sure the parachutists were safely hidden, she and her family left Prague for several days. Part of her journey took her to Pardubice, the site of the transmitter; there she asked for and was given a cyanide capsule. Returning to Prague, she began again to bustle about, taking parcels of food, coffee, tobacco, and kerosene to intermediaries, who saw that they reached the church. Sometimes the wife of the caretaker handled the packages; they took different routes and found various spots in which to rendezvous.

Marie told no one where the men were but reported to intimates that they were in high spirits, even though she knew it to be untrue. The men slept in spaces cut from the wall that had been used to store the coffins of monks. Even in June, the crypt was cold. A couple of small stoves were all the fugitives had for cooking and heat. A bigger problem was morale. The men had access to newspapers and knew that hundreds of Czechs were being killed and thousands more detained and harassed as a result of the assassination. Gabčík and Kubiš discussed ways to take full responsibility, then commit suicide.

Resistance leaders urged them to forget such thoughts and concentrate on escape. A scheme was devised whereby four of the seven parachutists would be taken to a nearby town in a police car. The others would be concealed in coffins and transported to a second town. The whole group would then be sent to a clandestine airstrip in the mountains from which a plane would carry them to London. The rescue operation was scheduled for Friday, June 19.

That Monday, Madam Moravcová set off again with a parcel. When she returned, she told the caretaker of her apartment house that she would be preparing something special for Wednesday—one of the parachutists had a birthday.

AMONG THE DISTINCTIVE figures sculpted into the Saint Vitus cathedral overlooking Prague is that of the devil tearing Judas Iscariot's soul from his mouth.

Karel Čurda had been in the protectorate for two months. His parachute team had had the job of planting radio beacons to aid in the Allied bombing of the Škoda Works—a mission that did not work out. He had then gone to Prague, where he had met some of the other parachutists but played no role in the assassination. After the attack, he fled to his family home in southern Bohemia, where he took refuge in a barn. As the hours and days crawled by, he began to review his options. He knew about Lidice and the Nazi threats to kill yet more innocent Czechs. He had barely evaded capture himself and was, by his presence, putting his entire family at risk. He had also learned of the rich reward that was on offer for information leading to the assassins. On June 16, he came to a decision, then set out for Prague and the headquarters of the state police. He was ready to betray his country and friends. But how much damage could he inflict? He was unsure exactly who among his fellow parachutists had taken part in the plot against Heydrich. He had no idea where the conspirators were hiding. He knew only one name that might matter, that of a middle-aged woman who had briefly arranged for him to stay in Prague, a woman from the Žižkov neighborhood known as "Auntie," Marie Moravcová.

BEFORE DAWN ON the seventeenth, the German police superintendent, a man named Fleischer, stormed past the caretaker's wife and up the stairs. He pressed the bell outside the Moravecs' flat. The door opened and the police rushed in, expecting to find the assassins. "Where are they?" he demanded of Marie, who was standing near the wall with her husband and son. "I don't know anyone," she replied and asked to use the toilet. The police commander refused but was soon called from the room as the search continued.

When he returned, he demanded to know where the woman had gone. Cursing, he tore open the bathroom door and found Marie glassy-eyed and unable to speak. Within minutes, the poison had done its job; she was dead. Her husband and son, still in their pajamas, were hauled off to the cellar of Peček Palace.

Alois, the husband, would not talk and may not even have known where the parachutists were hiding. The Gestapo tortured young Ata throughout the day. He held out for hours, refusing to say anything, but as his strength failed him, the interrogators plied him with brandy. Next they rolled in a fish tank and—with a sick flourish—removed the covering. Suspended inside the tank, as Ata could see, was his mother's severed head. Broken, he told all that he knew; not where the parachutists were but that Marie had advised him, if trouble arose, to go to the catacombs of the Boromejsky church.

It was the middle of the night. In the darkness, the Gestapo established a thick cordon and posted guards on roofs and at every manhole and sewage outlet. More than seven hundred armed men had been called to duty; their instructions: take the assassins alive.

At 4:15 a.m., the Gestapo entered the church, seized the keys, and moved in a swarm from behind the altar into the sanctuary. They were in the nave when shots rang out from above, hitting one of the Nazis in the arm. Kubiš, Opálka, and another of the parachutists had been caught outside the crypt on the balcony that surrounded the worship space. Because they had pillars to hide behind and only a single winding stairway to guard, capturing them would be not be easy. For almost two hours the parachutists and their predators fought in noisy desperation, ducking into and out of concealment, firing shots, trying to avoid the ricochets. As wounded Germans were pulled out, fresh marksmen were sent in, this time equipped with machine guns and grenades. Finally the firing stopped and the blood-drenched bodies of the wanted men were brought to the street, one dead, two dying. The traitor Čurda was motioned forward. He identified the corpse as that of Opálka, the man with whom he had dropped into the country less than three months before.

The Gestapo soon reasoned that the remaining fugitives, however many there might be, had taken refuge in the crypt beneath the church. At first they could find no entrance except for a small ventilation window about eight feet up, facing the street. They ordered a Czech fireman to break the glass; then they tossed in tear gas grenades, covered the opening with a mattress, and stepped away. Immediately, the mattress was shoved outward and the grenades thrown back amid a hail of gunfire; the parachutists had a ladder pressed against the other side of the window. The Germans set up a floodlight with the idea of blinding the hunted men; instead the bulb was shot out before it could be turned on. The next plan was to flood the crypt; fire hoses were inserted, but they, like the mattress and grenades, were promptly pushed away. The Germans then tried to break through the surrounding wall; it refused to crumble. As the sun rose higher, the Nazi commanders argued among themselves. K. H. Frank, who had been Heydrich's deputy, arrived on the scene. Reputations and careers were at stake.

Eventually the Germans found the secret opening within the church that the parachutists had used to climb down into the crypt. A priest, in handcuffs, was ordered to shout in Czech that the men must surrender and that, once in custody, they would be treated humanely. The reply was more shooting. A heavily armed German "volunteer" was lowered into the narrow hole on a rope, wounded within seconds, and quickly hauled out. Then the sanctuary rug was drawn back and a hollow spot detected beneath the floor. Using dynamite, the Germans blasted away a slab, exposing another set of stairs. A killing squad was sent in and soon repulsed. As the Nazis regrouped once again, four shots sounded from below. The assassins had used their last bullets.

THE DEATH OF the parachutists was a prelude to more killing. With the traitor's help, the radio teams were again broken up. The small village of Lezaky, where the shortwave transmitter nicknamed Libuše was located, met the same fate as Lidice. The families of the parachutists and the neighbors and priests who had sheltered them, as well as Alois and Ata Moravec, were executed.

WAS THE ASSASSINATION of Heydrich wise or foolish, a bold strike for justice or an impetuous blunder by a leader trying too hard to make an impression? Beneš himself may not have been too sure, for he never took credit for the attack. The Germans' vengeance claimed thousands of Czech lives and made it impossible for opponents of the regime to do much more than hide and hope to survive. They had, however, already been under enormous pressure, and the operation's daring raised Allied spirits, which had been at a low point. Heydrich was the first— and last—senior Nazi official to be successfully targeted by clandestine agents.

The Czechoslovak intelligence chief, František Moravec (no relation to Marie), was among those who considered the assassination a success. The action garnered worldwide attention, elevated the country's status among exile groups in London, and deprived the Nazis of one of their most capable leaders. The English diplomat Bruce Lockhart, a friend of Jan Masaryk and ordinarily Czechoslovakia's most steadfast ally within the Foreign Office, held a contrary view: the incident had magnified the pressures faced by the Resistance, he asserted, while adding nothing to the Allied cause.

The plot against Heydrich illustrated the complex choices faced by leaders and citizens alike. Beneš had to weigh, on the one side, the political benefits of landing a dramatic blow and, on the other, the inevitable repercussions—the Nazis had the ability and the will to retaliate harshly. Inside the protectorate, many Czechs faced a more personal dilemma. Active members of the Resistance had already decided to sacrifice their lives if necessary but many others were in the position of having to make instantaneous judgments: to inform or to keep quiet; to bear witness or look away. The caretaker and his wife had never signed up for the underground, but when Marie turned to them for help, each replied, "Here I am," at grave danger to themselves. The same is true of friends and relatives who found room for the fugitives in their basements, garages, or attics "for just a few days." It is little wonder that the priests at the Boromejsky church quarreled among themselves about the proper course to take.

Still other Czechs were morally at risk because of their professions, including the first doctor to examine the wounded Heydrich, the interpreter present at the interrogation of Ata Moravec, the firemen ordered to direct their hoses into the church basement, and the police called to secure the site. These men weren't being instructed to kill anybody but to make life easier for those who would. Put into that position today, how would we respond? What is accomplished by refusing to obey? Are there not other doctors, interpreters, firemen, and police who would do what they were told if we did not? Wouldn't we be sacrificing our lives for nothing?

Čurda was a villain, but what about the Czechs—and there were hundreds—who came forward with information concerning what they had seen on the days surrounding the assassination? Were they greedy, or were they honestly trying to save lives by removing the immediate cause of Nazi brutality? What should we think of the weak-kneed president, Hácha, who condemned the assault on Heydrich and begged his countrymen to cooperate in the investigation? What about the local officials who did all the Germans asked them to do but, like the good soldier Švejk, with as little competence and efficiency as possible?

The above questions bring to mind a Czech variant of the excuse "I was just following orders." It translates as "I was not the conductor of the orchestra, only a musician." My own reaction is to feel disdain for the outright traitors and unrestrained admiration for the heroes who chose bravely. As for the many who kept their eyes averted and mouths shut, doing all they could to avoid involvement, I feel neither respect nor any sense of superiority. Placed in the same circumstances, would I have shown the courage of a Madam Moravcová? As much as I would like to think so, I can make no such claim.

The assassination yielded a mixed result but was, in my view, both a courageous choice and the right one. Hitler's response, though savage, undermined the Nazi cause almost as much as did Heydrich's death. The Germans had set out to destroy all evidence that Lidice had ever existed; but within weeks of the massacre, the names of towns and neighborhoods in the United States and a dozen other countries were changed in its honor. Allied soldiers painted the name on the side of their tanks, and the secretary of the U.S. Navy, Frank Knox, declared that "if future generations ask us what we were fighting for in this war, we shall tell them the story of Lidice."

Hollywood responded with two movies, both released in 1943. *Hitler's Madman* starred the lanky John Carradine as Heydrich. The more interesting—*Hangmen Also Die!*—was the work of two German refugees, Bertolt Brecht and the incomparable director Fritz Lang. Although based but loosely on the facts, the script focused on the assassin's true-to-life moral quandary: to turn himself in or to remain at large while Czech hostages were executed. The film concludes with a song, "Never Surrender," and a promise: it is *not* the end.

Today, Lidice is still widely memorialized in movies and books, while Hitler's plan to build a special memorial for Heydrich was never taken up. In 1945, the wooden marker on his grave disappeared; it has not been replaced.

# 17

# *Auguries of Genocide*

E arly in 1942, Jan Masaryk told an audience in New York City, "This is the most crucial year in the history of the human race." His listeners could but agree. The previous December, Japan had attacked Pearl Harbor, the United States had declared war on Japan, and Hitler had done the same toward the United States. The enemy had left no choice: America was in the war. Churchill rushed to Washington, where he secured a promise from Roosevelt to accord priority to the clash in Europe. Given that the United States had just been assaulted in the Pacific, this was no small commitment.

Jan Masaryk had been in America since the end of 1941, speaking and granting interviews up and down the East Coast. Known for his irreverence in youth, he now played the part of preacher. He said that the United States must assume the role of Moses because no one else had the power and credibility to lead. He pled the cause of the small countries, especially "the lovely old land of Bohemia," observing that Jesus, too, had come from a nation of modest size. He shared his feelings, as well, about Germany.

Like Beneš and other Allied leaders, Masaryk paid homage to the icons of German humanism. Past glories, however, could not excuse the "periodic ethical and moral blackouts" that scarred the country's cultural heritage. "It is not Hitler who made Germany," Masaryk asserted, "it is Germany who produced Hitler." "Until the war is over," he added later, "I do not know any nice Germans. . . . We must eliminate the people who believe that . . . aggressive warfare has any place in the eyes of God."

In London, Beneš had been in frequent contact with the antifascist Sudeten leaders, who were also in exile. Their question: would the Beneš government be expansive enough to include them? The answer, it became clear, was no. As early as 1940, Beneš had begun pondering the need to remove Germans from Czechoslovak soil. He had originally thought that some territorial concessions might be appropriate. However, the combination of Heydrich's terror and the destruction of Lidice had convinced most Czechs—including Beneš—that Germans had forfeited the right to bargain; they were culpable both individually and collectively for crimes of war. From that time forward, Beneš insisted that a massive deportation be part of the postwar reckoning. "Every Sudeten German who was not actively opposed to Nazism," he insisted, "must go and go immediately." Even though the president made an exception for people with proven anti-Hitler credentials, the ethical dilemma remained. Bruce Lockhart, the supportive British emissary, noted drily, "President Beneš has found his own solution to the problem. He has borrowed it from Hitler. It is an exchange of populations."

Although denying their requests, Beneš spoke respectfully to the Sudeten German leaders who had fled to London. He had reason to. Most of them despised Hitler for his crimes and for despoiling the reputation of their people. Forced to decide between collaboration and exile, they had chosen the honorable course, only to find that they could not win. A Nazi victory would spell disaster; an Allied triumph would leave their people without a home.

THE ANNOUNCERS ON my father's broadcast team did not use their real names when on the air for fear that reprisals would be taken against their families. That precaution extended to my father, but I doubt, in his case, that it mattered. After I became secretary of state, I was shown a copy of a document prepared during the war by the Prague command of the German secret police. The paper requested that the Czech citizenship of the "Jew Korbel," his wife, Anna, and daughter, Marie Jana, be revoked on the grounds that my father had "made himself available to the illegal Czech government in London." Even years later, I find it disturbing that the secret police had a correct listing for my family's address in England.

The stars of the London broadcasts, Beneš and Masaryk, were the opposite of anonymous; they wanted their names to be associated intimately with the national cause. Beneš spoke periodically, especially to mark anniversaries and other important events. The foreign minister, when he was not traveling, was on the air every Wednesday. His talks were unconventional, eschewing political rhetoric for lively anecdotes. He dismissed Mussolini as a "puffed-up gangster," Hitler as a "Viennese paperhanger," and the Nazis as "pagans," who, in contrast to the ancients of Troy, waged war on behalf not of a beautiful woman but of an ugly man. Aware that his audience back home had a tendency to gloominess, Masaryk spoke in soothing terms, predicting that "those who worship force must eventually be exhausted"; his favorite word was "decency," and his closing advice "Chin up and carry on!"

Beneš alone directed the government, but the younger man helped him to navigate the choppy waters of British politics. Masaryk had promised his dying father that he would do all he could to assist Beneš; his loyalty was unquestioned. Still, they made an odd couple—the smallish, reserved diplomat next to the jocular, six-foot-two-inch force of nature. Beneš represented Czechoslovak interests, but his colleague was the nation's emissary to the world.

Through the war years, Masaryk appeared before groups all around the British Isles. My father sometimes accompanied him. Decades later, he described one such moment:

Masaryk entered the hall, tall and slow, with a shy expression and markedly uncertain eyes. Those who had not met him must have felt from the first second that a personality had entered their midst. He greeted acquaintances and quickly became informal, jovial, sparkling with wit, and sharing his beautiful smile. In these moments, he sensed intimately the nature of his audience—its special interests, worries, and weaknesses. And then he spoke: tenderly about people who deserved or needed caressing, but most brutally about those who had violated the basic laws of humanity. From time to time, his hand rose as if he wanted to give with his fine aristocratic fingers the last touch to his thoughts. One asked: was he a tribune, a passionate speaker, an actor? He was all of these.

Masaryk was fond of saying that he loved England "because my charwoman keeps her hat on when she's scrubbing the floor and because the plumber who comes to mend the plug in the bathroom offers me a Player's cigarette. He'd offer the king one. That's democracy." Such phrases charmed the British, but Masaryk was brash enough, when warranted, to speak bluntly. He told the British Empire League, whose members had endorsed Munich, that the Nazis may have started in Prague but would not stop until they had endangered Ottawa, Sydney, Delhi, Johannesburg, and every outpost of the empire.

He was also frank in expressing outrage at the crimes being perpetrated against Jews. My father was with him one afternoon when Masaryk met with a group of Jewish émigré children. A girl, clothed in a traditional costume, presented him with a copy of the Torah. In reply he spoke, as he often did, about Hitler's effort to deprive Jews of their dignity and compared Jewish aspirations for a homeland to those of his own Czech people. "My dear children," he concluded, "by this holy book of yours, I solemnly swear not to return home myself until all of you are at home again."

IN OCTOBER 1942, the Korbel household was transformed by the arrival of Kathy, my baby sister. I was no longer the center of attention, but that was fine with me. The month before, I had reached a whole new level of accomplishment: enrolling as a kindergartener in the Kensington High School for Girls, about a ten-minute walk from our apartment. As required by the dress code, I wore a gray tunic and pleated skirt topped by a cherry red blazer and beret with an accessorized gas mask.

Having nothing with which to compare, I didn't appreciate how fortunate I was to be with my family. Thousands of other refugee children could communicate with their parents only by bouncing thoughts off the moon. Many were shuttled from foster home to foster home; a few were well provided for on large estates, while others functioned as unpaid servants in households that were economically hard pressed. In contrast, my father walked me to school each day and was home for dinner whenever he could. A few times, I visited his office, where I proceeded to disrupt serious work and shake hands with members of the government in exile. I also attended the occasional reception and was introduced to Beneš, who was kind but, even to my inexperienced eyes, stiff and formal.

My father, too, was a formal man but gregarious nonetheless. He loved to tell stories and didn't mind when I climbed on him even if it meant that he had to put aside his newspaper and ever-present pipe. My mother was friendly with those she knew and less inclined to be strict when I misbehaved. Now in her early thirties, she had a glorious smile and dark brown hair that she wore in a roll around her head. I was fascinated by how she managed the roll, placing a cut-off top of an old stocking on her head and tucking her hair in around it.

One morning when my mother had to stay home with Kathy, my cousin Dáša took me by the hand and escorted me to the Ealing Studios. The well-known Czech director Jiří Weiss was making a short movie about Czechoslovak refugees. I can't recall the script, but I must have been given a good part because I received in payment a pink stuffed rabbit. Not long after, I was in a crowd watching Czechoslovak troops parade through London before going overseas to fight. A passing soldier paused to scoop me up, and the next day a picture appeared in the paper together with the caption "A father says goodbye to his daughter." My mother thought the mix-up hilarious; my father not so much.

Often when my father was away on a trip, my mother seized the opportunity to indulge her spiritual side by attending a séance. Because I was born soon after my maternal grandfather had died, she thought that perhaps his spirit had been reborn in me. In 1941, she returned from a séance one summer's day with the feeling that her beloved sister

*Marie "Máňa" Spiegelová*

Marie, or Máňa, had succumbed to the kidney disease that had plagued her for years; the accuracy of this sad premonition was soon confirmed. The following year, when sister Kathy entered the world, my mother felt that Máňa had been reborn in her.

In accordance with Czech custom, our family observed both birthdays and name days—the day set aside to honor our particular saint. For me, that meant a celebration on May 15 and another in August, also on the fifteenth, the Feast of the Assumption of the Blessed Virgin Mary. As for Christmas, we had a traditional dinner, a carefully decorated tree, and gifts. Dáša, separated from her immediate family, wrote to her parents that she had received "a stamp album, a manicure set, a book, a gold pencil, bath salts, perfumed soap and a new blouse." She closed on a wistful note, "Perhaps we will see each other soon."

The letters that Dáša received from home must have disturbed her, even though the tone of her mother's words was reassuring. Her father, Rudolf, had been prohibited from practicing medicine, and his office equipment had been confiscated by the Germans. "Daddy is always at home," wrote Greta, "and in boredom helps me a bit. We went swimming three times during the whole summer. We always have to go to the lake, with the pool it does not work. We don't even go to the park, only for a walk by ourselves." Dáša's family was able to remain in Strakonice, but, deprived of income, they had to move to a smaller house.

GERMAN PROPAGANDA WITHIN the protectorate portrayed the exiles in London as captive to Jewish financial interests. Beneš was often referred to as a white (or honorary) Jew. The argument was damaging because more than a few Czechs acquiesced in the notion that Jews, especially those of German extraction, were at least partly to blame for the war. One message from the Czech Resistance informed London:

> To our own Jews, people are extending help wherever they can, prompted by sheer humanitarian motives. Otherwise we do not wish their return. We feel alienated from them and are pleased not to encounter them anymore. It is not forgotten that with few exceptions Jews have not assimilated and that they sided with the Germans whenever this was advantageous to them, causing damage to the Czech people.

Beneš alluded rarely and then only in general terms to atrocities committed against Jews and other minorities. When the Association of Czech Jews asked him to speak more forcefully, he declined, citing "the reasons of higher interests." That official reticence, however, did not extend to the government's radio broadcasts. Early in the war, Ripka authored a message titled "We Think of You" to Czech and Slovak Jews. In it, he condemned a long list of Nazi outrages, including discrimination and the confinement of Jews in ghettos and forced-labor camps.

In June and July 1942, news bulletins included accounts of the systematic execution of Jews in eastern Poland. The reports were so sensational that many dismissed them as Allied propaganda. The sources, after all, were hardly objective: the Polish government in exile and the World Jewish Congress. Who could believe that 700,000 Jews had been murdered outright and that a comparable number had been driven to their deaths by hunger and disease? Surely not even the Nazis could be shooting or gassing prisoners at the rate of a thousand a day? At first the Allied leaders were skeptical, but in December, twelve governments (including Czechoslovakia) and the French National Committee joined in a formal condemnation of the Nazis' "bestial policy of cold-blooded extermination." In London, Foreign Secretary Eden confirmed to a hushed session of Parliament that the terrible reports were indeed accurate. He said that Jews were being transported from occupied countries to "the principal Nazi slaughterhouse in Poland," where they were worked or starved to death or "deliberately massacred." Edward R. Murrow referred to the reports as "eyewitness stuff" concerning "a horror beyond what imagination can grasp."

The fearsome tidings were conveyed to the protectorate in a special broadcast. My father's colleague and friend Jaroslav Stránský urged Czechs and Slovaks to do whatever they could to assist the Jews who remained in their midst. "All the help and relief that you grant them will be for your honor and glory." Ripka warned doctors not to cooperate in Nazi medical crimes such as the forced sterilization of Jews. More generally, Beneš himself vowed that "every crime, every act of violence, every murder committed by the Nazi henchmen in Czechoslovakia . . . must and will be revenged and atoned for a thousand times over."

The words of these men were passionate because time was short, the tragedy was unfolding in the country each called home, and the crimes—though indeed beyond imagination—were real.

# 18

# *Terezín*

In February 1997, my sister, Kathy, and brother, John, visited the Czech Republic to research our family history in light of the revelations concerning our Jewish ancestry that had appeared in the *Washington Post*. With the help of friends, they were able to verify much of the information, but one important piece remained elusive. The *Post* story had mistakenly identified the first name of our maternal grandmother as Anna, instead of "Rose," or Růžena. The records kept by the Federation of Jewish Communities in Prague showed that several Anna Spiegelovás had been sent to Terezín from the area around Kostelec nad Orlicí, but no one named Růžena Spiegelová.

The puzzle remained until John remembered seeing, years earlier, a picture of me as a toddler with a middle-aged woman whom he had not recognized. Scrawled on the back of the photo was the name of an old spa town east of Prague, famed for glassmaking. By coincidence, John had previously visited the town without being aware of any connection to our family. Now he suggested to Kathy, "Why don't we look to see if Grandmother came from Poděbrady?" And so they did.

ON JUNE 9, 1942, the day that Heydrich was buried and Lidice destroyed, Růžena Spiegelová boarded a train in Kolin, near Poděbrady, en route to Terezín. Years earlier she had been a shopkeeper assuring customers that her family's coffee was the finest in all Bohemia. At the time of my birth, she had helped to care for me and was the first to call me "Madlen"—hence the picture that John had remembered. In the frightening days after Hitler's invasion, she had taken me in again while my parents moved about Prague, devising a plan for escape. She had been with her husband when, in 1936, he had died and also with her daughter when, five years later, Máňa had lost a battle to kidney disease. Aside from photographs, I have no memory of Růžena; I was too young. When growing up, I rarely thought of my grandparents; on the occasions I did, I imagined them being very old. As an adult, I had the opportunity to see my father and mother play with my children; that made me visualize someday becoming a grandparent myself. I understand now that, when she arrived in Terezín, she was but fifty-four, not old by any means—in fact, five years younger than I was upon becoming secretary of state. I have also remembered a detail: as a child, I loved to swim in cold water. When I did, my mother used to exclaim, "You are just like your grandmother."

I only wish that her fortunes had more closely resembled mine.

The train that carried Růžena Spiegelová to Terezín and—a few

*Růžena Spiegelová*

days later—farther east was one of three directly linked to Nazi vengeance for the assassination of Heydrich. Precisely what happened to the passengers at the end of their journey is not known, except that there was no evidence of survivors. Almost certainly, the train stopped in eastern Poland, where its occupants were taken off and executed. Terezín records suggest the location as Trawniki, the site of a forced-labor camp established in 1941. The facility was used by the Nazis to teach Soviet and Ukrainian prisoners of war how to become concentration camp guards; as part of their education, the students were required to shoot other captives.

FOR MY PATERNAL grandfather, Arnošt Körbel, the fruits of a lifetime's labor were now forbidden. Thrust prematurely into retirement, he had no income and his bank account ceased to draw interest. Since September 1941, those identified as Jewish had been required to wear the infamous six-pointed yellow star, with the word "Jude" inscribed in black. Their ration books were stamped with a "J," which meant no meat, fish, fruit, or dairy products. They were not permitted to have telephones or radios or to travel around the country. For Arnošt this meant no more excursions to the Dalmatian coast, where he had loved to vacation not so much with his wife, Olga, but with his beloved dog, Drolik, so named out of fondness for France and the French word *drôle*, which also means "little man" or "buffoon." While Arnošt was at the seaside, Olga would take her granddaughters Dáša and Milena to the mountains. Such trips, too, were now forbidden.

My father adored his mother, but Olga and her daughter-in-law Mandula did not always get along. Perhaps it was because they had never had a chance to know each other well; my parents had spent much of their married lives in Belgrade or London. It was family legend that one day my mother made a mistake in knitting a sweater that Olga agreed to correct. My mother, watching, couldn't summon the nerve to tell Olga to stop until the entire sweater had been unraveled.

Arnošt was kindly, but the dog did arouse his passion. Dáša never forgot the afternoon she decided it would be more fun to read than to take the aging, wobbly-legged fox terrier for his daily walk. She tied

Drolik to a doorknob and opened her book. When Arnošt returned and came across the scene, he was furious. Without a word, he secured the leash to a table and put the other end around my cousin's wrist, then took the dog for a very long run. "Let that be a lesson," he told her later, "of what it's like to lose your freedom." But soon all was forgiven. For her eleventh birthday, Dáša received from Arnošt an eight-volume set of the *Encyclopedia Masaryk*, a compendium of information about everything; seventy-two years later, the books still occupied an honored place in her apartment.

IN 1940 OR 1941, Arnošt and Olga were forced to move into a tenement shared by other Jewish families. Month by month, the community was being squeezed. Venturing outside, they were blocked at every

*Arnošt Körbel with Drolik and Alena Körbelová*

turn by the sign *Juden nicht zugänglich* (No Jews allowed). As their liberties and property were taken away, Jewish families had little to do but remain in contact with one another and wait for their names to come up. Terezín? Everyone had heard rumors, but no one knew with confidence what life there would be like.

For my grandparents, the waiting ended on July 22, 1942, at 9:45 a.m. The summons was delivered by the Jewish community leadership. Arnošt and Olga had a week to put their affairs in order. My grandmother wrote immediately to her daughter, Greta:

> I have to get used to the thought that we are actually leaving. I am going to wash my hair . . . do some shopping and . . . clean the house. In the evening, I will prepare dough for the bread to bake in the morning. . . . I hope that once I get [to Terezín], I will calm down. I am not calm right now. In fact, I haven't been calm for a long time. . . . I would like to ask you, my dear Gretichka, not to waste your strength worrying about us. You will need it for yourself. I promise that I have a very strong will to survive. Somewhere, in some foreign land, we will meet again.

She added that she hoped to be able to work with children but only as a supervisor, "because otherwise it would wear me out." Arnošt had been ordered to take Drolik to the pound, where the pets of Jewish families were being collected. "Father will have a heavy heart," she wrote. "He will be miserable, but it also makes me very sad."

My grandparents did all they could to prepare for the new chapter in their lives. They received a steady stream of farewell visits from friends, including some who expected to be fed dinner. Allowed to take about a hundred pounds of luggage, they chose carefully, trying to pack as many items of warm clothing as possible. The day prior to departure, Olga wrote again to Greta:

> We have had visitors all day long. It is now 10:30 in the evening. There is chaos in the apartment. I have taken care of everything.

. . . Gretichka, my only daughter, keep healthy. I bless you, my dear Rudolf and Milena. Remember, my first and last thoughts will be with you, my children. I am strong and believe that somewhere we will meet. I kiss you warmly, your Mother.

There was one last-minute sliver of good news: the dog was safe. A neighbor who had been taking Drolik for walks after the Jewish curfew had sworn to authorities that the animal was his.

THOSE SUMMONED TO Terezín from Prague were instructed to assemble in an old military barracks within a fairground, six blocks from the train station. There they were put through a bureaucratic ordeal that lasted, typically, two full days or more. The deportees, young and old alike, slept on straw mattresses when they weren't standing in line or filling out forms. Clerks ordered them to hand over their identity cards, house keys, ration coupons, and valuables.

On the morning of July 30, the train designated "AAv" pulled away from the station in Prague. There were 938 people on board; Olga's number was 451, Arnošt's 452. The journey from one universe to another took two and a half hours past fields of hops, lines of fruit trees, and the round-topped Mount Rip, where long ago the mythical Father Čech had promised his people "a land wet with sweet milk and honey." The passengers arrived at Terezín in a pouring rain, gathered their belongings, and trudged two miles to the ghetto entrance.

The prison experience began with more bureaucracy: further paperwork to fill in and also more hands probing through luggage in search of contraband and other objects of value. Eventually housing assignments were made. Grandfather Arnošt was sent to the old military barracks along with other men. Olga was to find space in a house, L-304, that was set aside for women. Moving in must have been traumatic, for their lodgings were impossibly crowded.

Throughout the summer, trains had been arriving—a few from the protectorate but many more from Germany and Austria. The German-speaking passengers included those whom even the Nazis could not kill without having to answer inconvenient questions—Jews who were

*Journey to Terezín*

acquainted with prominent members of the Reich, who had names that were known in business, the arts, and the professions, or who had earned medals fighting for the Fatherland in the Great War. Among their number were former government officials, barons, countesses, singers, actors, the granddaughter of Franz Liszt, the younger sister of Franz Kafka, the son of Oskar Strauss, and the former sister-in-law of Thomas Mann. The rush of new arrivals swelled the prison population from 21,000 in June to 51,000 in August, ten times the camp's reasonable capacity. The influx also increased the average age of the population by fifteen years.

Many of the German newcomers had been lured into signing contracts guaranteeing them admission to "the spa," where they were promised a life of comfort, ample meals, and rooms with a view. Instead they were greeted by shouting guards, robbed of their luggage, fed swill, and packed into barracks crawling with vermin. In a matter of weeks, rooms for four people became warehouses for twenty, then forty, then sixty. Triple-decker bunks stretched from wall to wall and floor to ceiling, with two inmates sharing every mattress. When the supply of habitable rooms was exhausted, prisoners were jammed into window-less attics, cellars with dirt floors, and dust-ridden supply closets and storerooms. The shortage of eating utensils was more than matched by the lack of food. In July, the sewers backed up. There was not nearly enough clean water. Privacy was nonexistent. The living conditions created an intense physical strain and psychological burden, especially on those already weakened by age, illness, or despair. Organization gave way to chaos.

Gerty Spies, the daughter of a prosperous Berlin merchant, arrived at Terezín on July 20, ten days before Arnošt and Olga Körbel. She wrote:

> After they looted our hand luggage, we were led through the village. Incredible! Where was the senior citizens home, the residences of which they had spoken to us? Where were the clean houses, where everybody would have their own well-furnished room? . . . They took us to our quarters. But we could not live

*Sleeping quarters at Terezín*

here! It was a shed in the back of a courtyard.... There was nothing in the shed. No furniture. No oven, no stove.... Each person was allotted a living space about two feet wide . . . enough to sleep with bent knees. For this large community there were two toilets.

The ghetto's Jewish Council of Elders had decided early on that food rations and housing assignments should favor the young, thinking it best to tilt the odds of survival in the direction of those whose potential future contributions were greatest. The decision was defensible, but the death toll among the older population was high. The overcrowding caused contagious diseases (pneumonia, typhoid, tuberculosis) to spread rapidly. There were no gas chambers at Terezín, for it was not an extermination camp. It was a killing ground still, for the deaths from supposedly natural causes were due to the unnatural conditions. Burial space was limited, and so was wood for coffins. A crematory was built and became operational in September; from the beginning its four

*The crematory at Terezín*

large black ovens were kept busy. Ashes were retrieved, labeled, and stored at first in wooden urns, then in cardboard boxes.

Autumn arrived. The foliage in the surrounding Bohemian countryside turned crimson and gold. The air cooled, beginning to hint at the bitter chill to come. Inside the prison walls, Terezín's population had—on September 18, 1942—reached 58,491, more than on any other day. This was also the date on which the most prisoners died, among them my grandfather Arnošt Körbel. At the age of sixty-four, he succumbed to bronchial pneumonia. A funeral service was almost certainly held, but it would have been a memorial for the many, not for Arnošt alone. If Olga wrote to share the sad tidings with her daughter Greta, she would have been allowed but a single postcard and a maximum of thirty words, written in block letters and in German.

ONLY THE NAZIS would think to improve their public image by establishing a concentration camp. Terezín was a prison disguised as a town. In addition to the post office, there was a coffeehouse with a band called the Ghetto Swingers, but the "coffee" was made from a blend of herbs and turnips. The one food consistently available was mustard. There were shops, but most of the merchandise had been confiscated from prisoners. The joke went round that Terezín's boutiques were the world's finest, because only there could you buy a shirt that already had your own personal monogram. In an era when nicotine addiction was commonplace, cigarettes—although supposedly prohibited—were used for bartering everything from slices of bread to kisses on the cheek. The Germans even supplied the ghetto with its own currency, featuring a caricature of Moses holding the Ten Commandments.

Administering the ghetto was a nightmare that the Nazis were pleased to leave to the Jewish Council. The elders had to cope with a population that was divided between Zionists and assimilated Jews, Communists and democrats, young and old. The Germans and Czechs, in particular, did not always get along. The Czechs resented the German Jews for being German; the Germans were upset with the Czechs for their prejudice. Both accused the other of haughtiness.

Adding to the mix was a significant minority of practicing Christians, who petitioned successfully for the right to hold services.

Because the Nazis delegated so much, they were able to direct the fate of Terezín with a contingent of just two dozen Germans. These were assisted by 150 Czech gendarmes under the command of Theodor Janeček, a sadist who bullied inmates and reported every infraction to his bosses. The typical Czech guard, however, refrained from gratuitous cruelty; fourteen were imprisoned for smuggling contraband to inmates or for illegally taking letters out.

To supplement the German and Czech security forces, the Jews formed a police unit of their own, the *Ghettowache*. These officers had the authority to arrest and punish prisoners for minor offenses, including stealing and slander. More serious violations were passed to the Czech police or ultimately to the Nazi overseers. The *Ghettowache* was also responsible for ensuring that every inmate was accounted for each night. Especially in the early months, escape from Terezín was relatively easy—one could rip off the yellow star and catch a bus. But to where would one escape? In the north was Germany, to the south occupied Bohemia. About twenty men did leave to join the antifascist resistance, but most saw no better option than trying to wait out the war in Terezín.

With so many prisoners and so few guards, Terezín had an abundance of rules that were frequently broken. Despite the risk of getting caught, there were still tunnels that could be found for private assignations;* young Zionists carved out space in a bakery attic to install a suitcase-size radio tuned to the BBC; and gardeners and kitchen workers devised ways to conceal food in their clothes. One twelve-year-old farmworker was able to hijack a single cherry, which she presented to her parents. Her father, the former chief of medical services at Prague's Jewish hospital, painstakingly cut the fruit into three equal portions.

---

* For fear of losing their belongings, inmates at Terezín liked to keep them close. This tendency moved one poetically minded suitor to complain: "My darling, I'd love to kiss you so / But you're all wrapped from head to toe. / Five panties, two dresses, a cap and a hat / How can a chap get his arms around that?"

Everyone between the ages of sixteen and sixty-five was obliged to work if physically able. Inmates were commanded to labor in mines and construction, to grow food and tend livestock, to mend German military uniforms and split mica for insulation in electronic devices.

By the end of 1942, life at Terezín had begun to develop a unique identity. The Germans had done all they could to rob the Jews of their dignity, and certainly the miserable conditions had a Darwinian effect on behavior. Those who adapted quickly and who learned how to curry favor and scrounge food survived longest, but amid the horror and death there also emerged an astonishing display of life.

From a standing start, the ghetto's Jewish administrators were able to improvise a rudimentary system of public services, including electric power, sanitation, security, law, and, with respect to shelter, making the best of a bad lot. As for education, Germans in the protectorate had scoffed that—in their brave new world—Czechs would have no practical use for schooling beyond the eighth grade, while Jews would have no need for it at all. Following this logic, academic classes were banned in Terezín, but the prohibition was overwhelmed in practice by the prisoners' desire to learn and to teach. Whatever their preferred language, inmates respected knowledge; many were scholars, and some possessed world-class expertise. The pool of qualified lecturers and instructors ran deep.

Although classes might be interrupted at any time, they were generally conducted for several hours a day in dormitories, cellars, attics—whatever was available. A lookout was posted to warn of approaching SS. In the event of an inspection, students were adept at hiding their lesson papers and pretending to be engaged in a permitted activity, such as singing, drawing pictures, or cleaning their cluttered rooms.

Academics at Terezín were not just a means of therapy. The imprisoned children were among the most broadly educated in the Reich. Their counterparts in Prague, Vienna, or Berlin were taught only what the German authorities thought essential. The educators at Terezín had nothing more to lose. If all teaching was illegal, why not provide instruction in the history of Judaism, Greek ethics, moral philosophy, and the poetry of Heinrich Heine? Why not organize study circles that delved into Russian and Latin? Why not devote evenings to cultural events that featured lectures, poetry, plays, and songs sung not only in German but in Hebrew and Czech? Why not enjoy plays based on Sholem Aleichem's tales of Tevya the Milkman?

Especially in the first two years of the ghetto, expectations were high that the children, at least, would survive. Yet even when such optimism became harder to sustain, the classes continued with undiminished vigor. To teach a doomed child about moral choices was itself a moral choice—and a brave one.

Health care too presented a paradox. There, deep in the Valley of the Shadow, heroic efforts were made to prevent infection and to treat injuries and disease. It helped that the camp possessed some five hundred doctors, albeit many who were elderly. The health care system was short of medications and chronically overwhelmed by demand, yet the survival rate for such diseases as scarlet fever and diphtheria was well above 90 percent. The camp also benefited from surgical equipment scavenged from the protectorate's Jewish hospitals, now closed. Thousands of dental, eye, and other operations were performed.

THIS WAS THE environment that prevailed when, on November 26, 1942, Rudolf Deiml, his wife, Greta, and young Milena arrived at the ghetto in company with most of the Jewish population of Strakonice. The trip had been a cold one. According to a neighbor, the snow had been so heavy that "most people couldn't carry their packages and put them, one on top of the other, on the side of the road. . . . In the railway carriage the seats were covered with a thin layer of ice." Still, at fifty-two, Rudolf was better prepared than his father-in-law had been to survive the rigors of bitter weather and of Terezín. He was also an outgoing man with a physician's skills that were always in demand. Greta, for her part, hoped to devote her time to caring for children.

I can only imagine the emotions that churned through my aunt and uncle as they exchanged their familiar surroundings for the uncertainties of Terezín; the same is true of the mixed feelings with which I suspect they were received by Grandmother Olga. Anywhere else, she would have been overjoyed, but to see them and especially Milena consigned to prison existence would have generated much anxiety and sadness.

Still, there was no choice. Before long Rudolf was supervising health care in a barracks for women and young children. Greta, although separated from her husband, was able to see Milena regularly, as she was assigned to look after girls in the room next to her daughter's. Like most, the rooms were crowded; forty or more lived in each. Greta and other women played with the girls and made sure they washed and tidied their beds each morning. Meals, prepared in large vats, consisted of watery soup, bits of potato, stale bread, and the occasional spoonful of marmalade.

Surrounded by squalor, Olga, Greta, and Milena must have drawn strength from one another. Terezín, however, was the enemy even of small comforts. An epidemic of typhoid fever broke out among the children, roughly 125 of whom were infected in January and 400 more in February. Parents became frightened. The Germans, who fretted about the dangers of contagion, also worried. A prominent Nazi doctor was called in from Prague to analyze the crisis. Himmler, scheduled to visit, suddenly found he had pressing engagements elsewhere.

Frantic efforts were made to trace the epidemic to its source. The children's kitchen was suspected, but none of the inmates who worked there was sick. The toll mounted. Twelve-year-old Helga Weissová wrote in her diary, "Lilka's sister has died. Lilka has typhoid, too. Vera, Olina, and Marta are in the infirmary. Milča was taken to Hohenelbe barracks yesterday. They say she's dying."

Two of the adult women who cared for the children also contracted the disease. One of them was my aunt Greta Deimlová. She died, after a ten-day illness, on February 15, 1943.

MILENA NO LONGER had a mother to care for her, and her father was still with the men in the barracks. Under the rules, she was assigned to a three-story house, designated L-410, which provided lodging for about 360 girls ages eight to eighteen, the majority of whom were Czech. There she was exposed to a new routine. The day began at 7 a.m. Those who awakened first sprinted for the bathroom to avoid standing in line. Hands were given a squirt of Lysol beneath the watchful eye of an older woman, whose constant refrain was "Wash your hands before you eat / And when you get off the toilet seat."

Beds were then aired and sheets shaken out in a largely vain effort to prevent infestations of bedbugs and lice. Next came the roll call and the assignment of tasks—to clean, mend, fetch, pick up food, run errands. Some of the girls were part of an organization called Yad Ozeret ("Helping Hand" in Hebrew). They volunteered to assist older prisoners by carrying luggage, reciting poems, and enlivening birthday celebrations.

Before and after chores, there was plenty of time for school. Milena was among those receiving instruction from a forty-four-year-old protégée of Paul Klee, Friedl Dicker-Brandeis, who in the 1930s had moved from Vienna to Prague, where she had maintained a children's art studio. At Terezín, her students included the girls of L-410.

"You didn't have to draw well," recalled Helga Weissová. "That was not what really mattered; the crucial thing was that you developed your talents, that you learned to see. To recognize colors. To play with colors." Dicker-Brandeis taught the girls to draw in response to stories, wishes, ideas, even rhythms rapped out on a table. "One day, she would offer a theme," wrote Weissová, "an animal in a landscape, or would simply say, 'Storm, wind, evening—paint it!' Another day, she would sketch a fantasy story in a few sentences or would say nothing more than 'Paint where you would like to be now.'"

Nothing came easy in Terezín. Dicker-Brandeis was primarily a German speaker; art supplies and paper were scarce. Nevertheless, the children produced more than four thousand drawings in pencil, crayon, and watercolor; the subjects included virtually everything except what was not permitted—life as it truly was inside Terezín. Many of the illustrations survived; when the ghetto was liberated a pair of suitcases was found in one of the children's rooms, each crammed with pictures, among them many of Milena's. From the evidence, my young cousin loved to try her hand at portraits, trains, houses, baby carriages, and a variety of animals, including dogs, pigs, horses, and camels. The sun is almost always present, sometimes smiling, sometimes not. Today, a

*Drawing by Milena Deimlová*

selection of art from the children of Terezín, including one of Milena's, is on exhibit in the Jewish Museum in Prague.

Around the corner from L-410 was L-417, formerly a municipal school, converted into a dormitory for Czech boys. It was there that one of the more remarkable literary creations of Terezín was produced. Each week, the residents of Terezín put together several magazines, including *Vedem* (In the Lead). Since reproduction was not possible, only a single copy was created. On Friday nights, the boys gathered to read their contributions aloud. Selections included poems, satire, essays on prison governance, and interviews with such ghetto celebrities as the cook, the chief engineer, the nurse, or the head of police. The editor was Petr Ginz, an improbably precocious fifteen-year-old son of a Catholic mother and Jewish father. Possessed of a boundless appetite for self-improvement, Ginz was to be seen almost every evening sitting cross-legged on his bunk, surrounded by writing and painting supplies.

For a short time, Petr kept a journal in which he vowed to devote greater effort to drawing, bookbinding, increasing his weight, the study of Buddhism, linocuts, stenography, English, Russian, Plato, and Balzac. As for maintaining a diary, he had second thoughts. "I hereby declare," began his entry for February 16, 1944, "that keeping a diary is stupid because you write things in it that one should keep forever to oneself."

In one of the essays Ginz wrote for *Vedem*, he compared the attitude of many at Terezín to a Manchu expression, *"Mey fah zu,"* or "It can't be helped":

> Manchuria is not the only place there are Manchus. There are plenty of them here, too. Are we in Terezín? *Mey fah zu*. Are we sweating like pigs? *Mey fah zu*. They take everything as given, unpleasant to be sure, but unchangeable. Is there favoritism here? Can't be helped. Favoritism is as immutable, as natural as the rotation of the earth or gravitation. It was so in the past, it will be so again. *Mey fah zu*.

Inside the ghetto, hunger was constant and so was filth—laundry privileges came along but every six weeks. It was a common sight to see men on their way to the crematory pulling carts filled with human bodies. The same wagons were used for transporting loaves of bread. Most wretched were the aged, who, deprived of equal rations and often without family to send packages from the outside, were just alive enough to shuffle about the camp scrounging for food. This was an image of shrunken humanity, skeletons barely covered with skin and sores, unable to clean themselves or to converse intelligibly. To ghetto residents, such an existence must have seemed worse than death.

The central uncertainty looming over the camp was embodied in the mysterious transports to the east, which started, then stopped, then started again. Not even the Jewish Council had much notice of when the trains would depart, nor did its members know where they went, though the prevailing guess was to vaguely defined work camps in Poland. The more realistic prisoners understood that whatever the destination, it was probably worse than Terezín. Some who could not imagine such a place volunteered for the transports—especially if loved ones had already been commanded to go.

The Germans were intent on implementing the Final Solution but not on publicizing it. To the end, they would insist that they were sending prisoners to locations where the inmates could expect to survive and even to live together as families. They were generally indifferent as to which Jews went on the transports—although, for a time, persons married to Aryans and those holding German war decorations were exempt. With characteristic sadism, the Nazis left to the elders the responsibility for deciding who would go, dictating only the number of passengers and whether they should be young, old, possessed of certain skills, or of a particular nationality.

The task of selecting victims created a hideous moral dilemma for Jewish leaders. Names were added, then scratched out, according to such subjective criteria as ideological affinity, family connections, language, nationality, and degree of personal hardship. Each time an exception was made, another prisoner had to take his or her place. The most vulnerable were orphaned teenagers, who had no one to speak for them. Inevitably, the power wielded by the council caused resentment. Its members had more comfortable housing, fuller dinner plates, and cleaner clothes; they were also in a position to help their friends. Inmates referred derisively to the importance of vitamins B (*Beziehung*, or connection) and P (*Protektion*). Gonda Redlich, the council member responsible for youths, complained, "The elders will never agree to cutting back on a morsel of their rights."* He went on to ask, "Is a man who is given two portions of food fit to judge a thief who is given only one portion, when he tries to take a second from the kitchen?"

---

* Redlich maintained a diary from January 1942 until his death early in October 1944. His acerbic and often moving words, written on sheets of office calendars, were not discovered until 1967, when workers found them—stashed in a woman's purse—in an attic in Terezín.

How to discern, in such a place, the difference between right and wrong? Vera Schiff, only seventeen, worked in the Terezín hospital. One night a renowned surgeon hurried in with a bundle in his arms. Vera pulled back the blanket to discover a newborn child, whom the doctor begged her to kill. The infant's mother had arrived at Terezín only a few days previously and had managed to conceal her condition. To give birth at the camp was a capital crime. The doctor knew the mother and wished to save her life, but to harm the newborn would violate his Hippocratic oath. He prepared a syringe and implored Vera to use it. Her reaction:

Although I was not bound by any oath, I found it impossible to calmly take the syringe, inject the baby, and walk away. We were both unnerved by the deliberate act of extinguishing a life, even the life of a baby who was doomed to die, even if it was to try to save the life of the mother.

We exchanged a pained and embarrassed glance. Then the boy began to whine, making Dr. Freund's flesh creep. Coldly and tensely, he snapped that we would do it together. Before I could say anything, he grabbed my hand, pushed the syringe into it, and with his hand wrapped around mine, he forced the needle into the baby's thigh.

According to official records, the child had never existed. The doctor had betrayed his oath and implicated an innocent girl in his crime, all to do what was—in his judgment—the right thing. Surely the blame rests not with those forced to make such choices but with those responsible for creating the circumstances in which such choices must be made.

# 19

# The Bridge Too Far

Hitler's fateful decision to invade Russia had left his troops exposed to the same three indomitable warriors—October, November, and December—that had foiled Napoleon more than a century before. In January 1943, the German Sixth Army surrendered to Soviet forces after the failure of a prolonged and bitter siege of Stalingrad. Allied troops, having finally prevailed in the deserts of North Africa, prepared to pressure Hitler from the south through Sicily, then the Italian mainland. Churchill and Roosevelt, meeting in Casablanca, vowed to demand Germany's unconditional surrender. In England the mood was beginning to brighten despite the ongoing blackout. "Hitler and his lot are moaning at present," wrote a woman from a village near Coventry, "and we—well, we are feeling much bettah!"

Beneš, too, was upbeat. "Our cause is internationally assured," he told his advisers. "Our government in exile has been recognized by all the democratic countries. We have a treaty of alliance with Britain; we have renewed the French-Czechoslovak treaty with de Gaulle. The allied powers have declared the Munich agreement null and void. The time has come to sign a treaty with the Soviet Union."

Establishing a firm bond with Moscow was an essential element in Beneš's postwar strategy. Munich would not have happened, he argued, had the British been less distrustful of the Soviets. If his people were to be safe, the partnership between Russia and the West must continue. Yes, the Soviet leaders were totalitarian, but that was to be expected, given the country's czarist tradition. Prolonged exposure to the West would surely have a liberalizing effect, a process that Czechoslovakia, with its democratic values, could help to speed. Whether or not that hope was realized, Beneš believed that his country needed a powerful friend. Even in defeat, Germany, Hungary, and Austria would still loom, encircling and menacing. He no longer had faith in Western promises; Moscow was to be courted.

The prospect of such a romance stirred little excitement in Great Britain. The Foreign Office wasn't overly concerned about the future of Czechoslovakia, but Poland, much larger and possessing a 200,000-man exile army, did command attention. If the Soviets and Czechs arranged a separate peace, where would that leave the Poles? Like Beneš, the Polish leaders wanted to restore the prewar boundaries of their country. The difficulty was that although Czechoslovakia had been occupied by Germany, Poland had been gnawed on by both sides. The Germans, when beaten, could be forced to give back what they had taken, but the Soviets were allies and would have to agree voluntarily.

To complicate matters further, in April the Nazis discovered the bodies of four thousand Polish military officers in Katyn Forest near the Russian town of Smolensk. The Wehrmacht blamed the killings on the Soviets, who indignantly denied the allegation and blamed the Nazis. This dispute between the kettle and the black pot inflamed the curiosity of a junior British diplomat who investigated and then informed his superiors that the Germans, for once, were right; Moscow had been responsible for the Katyn executions and for many others. Privately the English agreed that Stalin was an appalling butcher but one who had murdered so many of his own people that doing the same to a few thousand Poles should not be considered a shock; publicly they said nothing for fear of offending him. Across the Atlantic the Roosevelt administration refused even to evaluate the evidence.

Squabbling allies are a threat to any war effort. Beneš, at British urging, hoped to quiet matters by endorsing the idea of a postwar federation that would tie his country and Poland together and that would also have the diplomatic blessing of the USSR. To that end, early in 1942, he signed a statement of principles and began a series of discussions with his Polish counterparts. The project stalled when the Soviets refused to consider returning any of the territory they had stolen, a position the Poles found impossible to accept. As the weeks slid by, Beneš became anxious. He did not want his country's security held hostage to a negotiation that would never succeed; instead, he would make his own arrangements based on Czechoslovak interests.

About that time, Beneš considered and rejected an invitation from Stalin to transfer his base of operations to Moscow. The Soviets hinted that Beneš should move if he wanted to accompany the eastern contingent of the Czechoslovak army when it liberated his homeland. According to Soviet propagandists, this force was expanding rapidly and would soon reach 20,000 members; in fact, it was still modest and of little military value.

The Soviets sought to control Beneš, hence the welcome mat in Moscow, but the president could not have made such a shift without betraying his London-based democratic supporters, including my father. Because of the war, the rivalry between members of the Communist and other parties was subdued; everyone was fighting on the same side. However, within the exile community, there were two broadcasting centers, two sets of soldiers, and two groups of politicians trying to position themselves for the future. Back home, there were vastly different ideological tendencies within the Czechoslovak underground. Competition between the factions was inevitable. Beneš, who was now acknowledged by all as the nation's rightful leader, was determined to preserve his status by positioning himself above the fray. He decided to go to Moscow, not to remain but to sign a treaty of friendship that would secure one pillar of the diplomatic structure he wished to create.

THE FOREIGN POLICY of every small country begins with one question: how can we survive? The issue is particularly acute if the country is in possession of resources that others prize or is located in a place of interest to larger powers. This vulnerability explains why smaller states are often the most vocal in supporting institutions—such as the United Nations—that are designed to protect the rights and sovereignty of all.

Beneš, in 1943, could not rely on the hope that the UN of the future would succeed; he had just seen the League of Nations fail. Instead, he had to cope with the reality that preserving a small state often requires at least a limited dependence on a major power. In Czechoslovakia's case, it demanded even more: a friendship with the USSR that the West would not find threatening and a warm relationship with the West to which Russia would not object.

Having decided to go to Moscow, the president needed to forge a parallel bond with the West—and by the West he had in mind more than his ambivalent relations with the British. One did not need to be as keen an observer as Beneš to know that the United States would have greater postwar influence, even in Europe, than would the overstretched authorities in London. He did not want any misunderstandings with Washington to arise and so thought it prudent to reintroduce himself and explain his intentions to officials in that capital. Never before having flown across the Atlantic, he summoned his nerve, wrote a new will, and boarded a plane headed west.

Four years earlier, when Beneš had arrived in the United States, he had been the deposed leader of a disintegrating country. Now there was a war on, and he was a significant, if not a leading, member of the Allied team. On May 12, 1943, he was received at the White House with elaborate military honors. At a reception on the South Lawn, he was treated to the Marine Band's rendition of "Where Is My Home?" and after an official state dinner, conferred privately with President Roosevelt until 2 a.m.

In the words of a U.S. intelligence report, FDR found his visitor's plan for the future of Europe "right interesting." Beneš foresaw "a Russian sphere of influence in East Europe and . . . [another] in Western Europe under the leadership of England." He offered himself to Roosevelt as someone who could serve as a courier between Moscow and Western capitals. He also made plain his desire for friendship with Stalin, pointing out that his country and the Soviet Union would be neighbors. It was therefore inevitable that the Soviets would have extensive influence in the postwar period. Beneš assured the Americans that the Czechs, although Slav, were basically Western in culture. They

*Beneš and Roosevelt, Washington, D.C., 1943*

would not become subservient to the Communists but would seek to have cordial relations with both sides.

In a victory of pragmatism over principle, the two men agreed that Poland's attempt to blame the Soviet Union for the Katyn Forest killings was ill advised and that the Kremlin's territorial demands in Poland would have to be honored. FDR encouraged his visitor to develop strong ties with the Russians and asked whether he, too, should meet with Stalin. After a lengthy discussion that touched on France, postwar institutions, and the future of Germany, Beneš pressed for and was grateful to receive implicit U.S. support for his goal of expelling Sudeten Germans from Czechoslovak territory at the war's conclusion.

In succeeding days, Beneš had a series of meetings with immigrant groups, held long discussions with legislative leaders, and addressed a joint session of Congress during which he referred to his country as a "godchild of the United States." He also spoke to enthusiastic audiences in New York, Detroit, and Chicago and to an emotion-filled gathering in the recently renamed town of Lidice, Illinois. Before leaving, he was told by Harry Hopkins that "Roosevelt esteems your sound advice and judgment on European matters. Although he himself carefully follows European affairs, he cannot know all the details and he would appreciate keeping in constant contact." Beneš returned to England satisfied that Roosevelt both endorsed his policies and valued his role. He felt that he could go to the Soviet Union without fear of alienating his democratic friends.

ON THE NIGHT of November 23, Beneš began an eventful pilgrimage to the East. Like T. G. Masaryk in World War I, he would visit Russia in aid of his country's independence and freedom. Due to poor weather, his plane could fly no further than the picturesque Caspian Sea port of Baku, from which the president embarked on a four-day train ride across the Caucasus to Moscow. As the countryside flashed by, he had an opportunity to view what little was left of cities and villages that had caught the brunt of the German invasion. "I passed . . . demolished hamlets, railway lines and stations, bridges and roads," he wrote later, and "endless dumps of destroyed tanks, motor cars, planes, railway wagons, and of weapons of all kinds. One beautiful bright night I went through Stalingrad and saw the incredible destruction wrought by the Germans; demolished houses of which only the four main walls were left pointing to heaven like dreadful and warning fingers."

Beneš believed he was on a mission of historic importance and was therefore eager to return with news that would validate his trip. That impulse, along with a lack of negotiating leverage and his decision to travel without senior advisers, made him an easy guest for the Soviets to handle. It didn't help that Zdeněk Fierlinger, his ambassador in Moscow, cared less about defending Czechoslovak interests than about currying favor with the Soviets.

After being received by the Kremlin with highest honors, Beneš engaged in wide-ranging talks with Stalin followed by a VIP tour of factories, scientific institutes, military installations, and theaters. Like many visitors to Moscow, he was lured into adopting a favorable view of the Soviet system in part because he felt among friends. Wherever he went, he came across hardworking citizens who seemingly adored communism and—in contrast to Londoners—were well informed about the Czechs' plight. He was genuinely impressed by the USSR's resilience in rebuilding rapidly despite the massive blows it had endured; the Red Army had lost half a million men in the fight for Stalingrad alone. He also credited the revolution with transforming the country from a nation of illiterate peasants into a modern industrial society. "Masaryk refused to accept that the Soviet regime would last," he remarked to one of his advisers. "I wonder what he would have said now. A regime that can improve the living standards of 90 percent of the people is bound to maintain itself. That is what so many in the West fail to realize."

Beneš was convinced that countries and their leaders could be transformed by events. He saw that happening with Stalin and Russia. In his view, the Soviet strongman was devoted entirely to the defeat and dismemberment of Germany, goals that were also paramount to Czechoslovakia. Given the modest nature of his own postwar ambitions, Beneš felt sure that the spirit of cooperation would continue. He assured the Soviets that, once restored to office, his government would conduct foreign relations in a manner fully acceptable to them and that he envisioned close, even intimate, economic and military collaboration. He asked only that Moscow support his desire to expel Germans and that Stalin refrain from interfering in his country's internal affairs. To those requests, the dictator agreed without a moment's hesitation. On December 12, 1943, the two men signed a treaty pledging mutual nonaggression and friendship for a minimum of twenty years. Beneš delivered his speech in Russian with a skill in pronunciation that Stalin joked was at least "better than yesterday."

The Czechoslovak leader's pride in a job well done was evident in the cable he sent back to London: "I consider all our negotiations as wholly successful. . . . It can be regarded as certain that all [Soviet] treaties and agreements not only with us but also with the British and with America will be kept." He was convinced that "a new Soviet Union will come out of the war," one that would be more tolerant of others and cooperative in dealing with the West.

Before leaving the Russian capital, Beneš met with the Czechoslovak Communist exiles who had gathered there, among them their leader,

*Beneš and Stalin, Moscow, 1943*

Klement Gottwald, whom he had not seen since the contentious days of Munich. Gottwald, forty-seven, had been trained as a toolmaker and immersed in party dogma and discipline for more than two decades. His life mission was to achieve a workers' revolution in his home country. Short and stocky, with dark hair and a broad face, Gottwald was known for wearing caps instead of hats and for his affinity, hardly uncommon among his peers, for strong drink. A self-educated and often wily tactician, he was deeply attracted to power and would never intentionally deviate from the Soviet line.

The circumstances of exile and war had left the president and Gottwald sharing the same political boat, a reality that pleased neither of them. Beneš was loyal to a nation, Gottwald to a doctrine that—at least theoretically—despised nationalism. The former was a committed democrat; the latter considered democracy a trick employed by the bourgeoisie to deny laborers their rights. Beneš was disciplined and meticulous to the point of being fussy; Gottwald was bombastic and undiplomatic, almost a lout. Yet they had no choice but to conduct business because, at the time, each needed the other.

During their meeting, the two were able to agree on the primacy of the war effort, punishing collaborators, and sharply reducing the number of Germans in their country. Gottwald insisted that the first postwar prime minister come from one of the leftist parties, and the president, having just met with Stalin, felt in no position to object. Both took it for granted that the new government would be led by the exiled leaders—that is, by themselves—in preference to the resistance fighters who were struggling to survive at home. The harmonious atmosphere dissipated, however, as soon as discussion turned to decisions made before the war. Gottwald blamed the government for capitulating and ridiculed the contention that subsequent events had shown the prudence of that judgment. When Beneš asserted that the nation would "survive the war better than anybody could have imagined," the Communist leader pounded the table and denounced "the evil moral consequences that Munich has had for our people." Beneš replied by asking him to consider what would have transpired had the Czechoslovaks gone to war alone. "I claimed the merit," he wrote with a touch of smugness, "of having foreseen in 1938 that certain things would happen and that other things would not."

Few temptations are more damaging to a leader than to act on hopes instead of facts. Chamberlain had put his trust in the reasonableness of Hitler, Daladier in the supposed invulnerability of the Maginot Line. Stalin had thought the Germans would not dare attack him; Hitler fancied himself an agent of destiny. Beneš, the leader of a small country in a dangerous neighborhood, yearned to believe in Stalin's capacity for intellectual and moral growth. Thus he regarded the treaty he had negotiated as a landmark in his nation's diplomatic history. He claimed it would safeguard Czechoslovak security, dissuade the Soviets from intervening in his country's affairs, and create a model for relations between the USSR and the rest of Central Europe. These were lofty expectations.

In his defense, Beneš knew that, after the German defeat at Stalingrad, Hitler's best chance for survival was to divide the Allies. Thus

Nazi propaganda was increasingly built around the idea of saving civilization from the Bolsheviks. Beneš worried that anti-Communist passions would undermine Western unity in the final stages of the war. He thought it important to counter that sentiment by defending Stalin's trustworthiness and by envisioning a future in which the West need not fear the East. It was sound strategic thinking, provided Beneš did not fall too deeply under the spell of his own words.

# 20

# Cried-out Eyes

Summer 1943. The cellar of the girls' home at Terezín (L-410) had become a rehearsal hall for concerts and plays. The residents, who included my ten-year-old cousin, Milena Deimlová, often found time to go downstairs to listen and watch. This was where the girls' choir practiced and the ghetto's productions of *The Bartered Bride*, *The Magic Flute*, and *Figaro* came together. A number of girls from L-410 also appeared in *Brundibár*, a children's opera. That work, written in Prague, had been performed there the previous winter by a cast of Jewish orphans. When the composer and many of the singers found themselves in Terezín, they revived the show, holding rehearsals in the attic of L-417, the boys' dorm. The opera's libretto depicts a battle of wits between an evil organ grinder (Brundibár) and a pair of impoverished siblings who sing on street corners to raise money for their bedridden mother. With help from some musically gifted animals, the children ultimately win out. The final song, "Brundibár Is Defeated," was especially popular among the many prisoners who sang of Brundibár while thinking of Hitler. Beginning in September, the opera was staged fifty-five times, always before a full house.

Like a desert oasis, culture and the arts enlivened the ghetto's landscape. There was a constant menu of lectures, readings, and plays, while musical performances were hostage only to the scarcity of functioning instruments. Residents were eager for diversions despite the physical toll of their daily routine. Even relatively humble troupes issued

invitations so that crowds would not exceed the capacity of their "theater." One read, "The Cleaning Service . . . takes pleasure in inviting you to a cabaret evening on January 12, 1943 at 8:00 p.m. in the potato peeling room of HB [Hamburg barracks]."

Gerty Spies, a prisoner who had in happier times sampled the cultural life of Berlin in all its diversity, wrote:

> Performances multiplied [, becoming] . . . more varied, more comfortable both for performers and for the audience. . . . One could choose concerts, theater (without scenery, of course), travelogues, scientific and literary lectures, evenings of ballads, and who knows what else.

The repressive atmosphere made perilous any attempt to cross the line separating art from politics. Because of their familiarity with their own language and culture, Czechs had an advantage that German artists did not. A few months before the war's end, a second children's opera was presented, this one based on "Fireflies," a well-known fairy tale. The prison audience was delighted to hear the Czech national anthem discreetly worked into the score. Karel Schwenk's satire *The Last Cyclist* was also written at Terezín. It tells the story of a dictator who blames people who ride bicycles for all his country's problems. The tyrant banishes everyone unable to prove that their ancestors had been pedestrians for at least six generations. One intrepid cyclist rebels and is placed in a cage, where he is ridiculed by the local population. As in *Brundibár*, virtue triumphs in the end.

The Nazis deprived inmates of their physical freedom but not of their capacity to think—and to do so about far more than the terrors of their situation. The prison included people eager to exchange ideas about linguistics, botany, anthropology, theology, literature—almost everything. Most popular among the lecturers was Leo Baeck, a seventy-year-old reform rabbi from Berlin who offered talks on "Philosophical Thinkers from Plato to Kant." Dignified and eloquent, Baeck inspired those around him to maintain their self-respect. Even as his body withered due to a lack of nutrition, he continued wearing his suit and tie and carefully trimmed his beard. "Never become a mere number," he said. "We bow before God, but stand erect before man."

Baeck was admired for his learning, moral integrity, and courage (four of his sisters died at Terezín). He did, however, have a secret. An escapee from a labor camp in Poland had gotten word to him about the gas chambers at Auschwitz. This meant that, for most prisoners, the summons to a transport was the equivalent of a death sentence. After reflection, Baeck decided not to share what he had learned because he didn't want to demoralize his fellow inmates further and because it was still possible to survive if one were chosen for a work detail. The right answer to his dilemma—to tell or not to tell—has been debated ever since.

WHEN, IN DECEMBER 1942, the Allied nations had denounced Nazi atrocities against Jews, they had cited reports of mass executions in the prison camps in Poland. Himmler denied that any such slaughter was taking place. Feigning indignation, he invited the media and Red Cross to inspect a labor camp and also the "model facility" at Terezín. Of course, before the ghetto could receive visitors, a few preparations would be required.

To begin, the population level was stabilized. Emphasis was placed on cleanliness to reduce health risks. The food became more palatable. New wells were dug and a sewer line built. Prisoners were given time to improve the appearance of their living quarters. Children and teenagers were allowed to form soccer teams. Streets that had been designated merely by letters and numbers were given more appealing names: L-1 became Lake Street, despite the absence of any lake.

These welcome, if largely cosmetic, changes were interrupted when, in July 1943, the record-keeping division of the Gestapo demanded office space secure from the threat of Allied bombing. Several thousand prisoners were evicted from their housing, among them my uncle Rudolf Deiml and a friend, Jiří Barbier, a professional carpenter. Together, they and a few others were able to build new quarters for themselves, complete with a table, four chairs, a wardrobe, and small stove.

But not everyone was a carpenter. The displacement created by the record keepers led to the return of overcrowded conditions. Per Himmler's order, transports had been suspended for seven months, but notice was given that, in September, a new convoy—massive in size—would go. The list of potential passengers included Olga Körbelová, Rudolf Deiml, and Milena Deimlová. With much trepidation, the three prepared to leave. Exactly what happened next is unknown, but the answer can probably be found in a note written by Gonda Redlich. In it the youth leader explained to the council that Milena's mother had died while caring for children during the typhoid epidemic and that the girl had since come down with tuberculosis. Rudolf's value as a doctor might also have contributed to the reprieve. In any case their names were stricken from the list.

On September 6, 1943, more than five thousand mostly Czech-speaking prisoners left Terezín, to be followed in December by an equal number. The passengers on these transports, though bound for Auschwitz, did not go through the usual selection process, that is, the division between inmates thought able to work and those immediately sent to the gas chambers. Instead, they were diverted to nearby Birkenau, where a "family camp" of Terezín prisoners was established. This was the purportedly humane facility that Himmler planned on inviting the world to see. Children were given their own play area and supplied with half-decent food. Adults worked, in addition to manual labor, at weaving and making clothes. In time, this camp too became overcrowded. To clear space, on March 8, 1944, the passengers on the September transport who had survived the first six months were summoned to a phony work detail. That night more than 3,700 Czech Jews were executed, by far the largest mass killing of Czechs during the war.

Once again the Polish underground sought to publicize the murders. However, three months passed before reliable reports reached the government in exile in London. The news was accompanied by a warning that the Nazis planned to liquidate survivors from the December transports on June 20, only a few days away. My father's broadcast team highlighted the report immediately, coupled with a vow to punish any

*Selection at Auschwitz, 1944*

and all of those responsible for future murders. The Gestapo responded by putting their plans on hold and by ordering prisoners at the family camp to send postcards, dated June 21, back to Terezín.

IN OCTOBER 1943, the first of several groups of Danish Jews arrived at the ghetto. Their reception differed from that of any other and would prove a test of the Nazis' capacity to deceive. Denmark had provided an instructive example of what happens when evil is confronted. Under the Nazi occupation, King Christian X and the Danes had refused to become complicit. Tipped off that Eichmann planned to deport the country's eight thousand Jews, the Danish underground had succeeded in smuggling out or otherwise hiding 90 percent of them. That September, Eichmann's thugs had rounded up those who remained and sent them to Terezín. Instead of accepting defeat, King Christian and the Danish Red Cross inquired continually about the welfare of the prisoners, showered them with postcards and food packages, and demanded that an international delegation be allowed to inspect their living conditions.

The timing of the visit by the International Committee of the Red Cross (ICRC) took many months to nail down and was repeatedly postponed by the Nazis. This afforded Himmler time to produce a counterfeit of the model ghetto about which he had been boasting. Given the ample supply of slave labor, all that was needed was some paint, building materials, playground equipment, and trust in the desire of most people to believe what they wished. Workers were instructed to create a new performance hall, refurbish the post office and bank, adorn the remodeled cafeteria with white tablecloths and flowers, and erect a children's pavilion complete with sandboxes and swings. Artists were called on to use their imaginations and draw pictures of the ghetto's supposedly carefree social life. The new Terezín featured a pharmacy, a bakery, a band pavilion, store windows crammed with tempting merchandise, a fancy conference room, improved housing, and a renovated school bearing the sign "Closed for the holidays."

On June 19, the Red Cross received permission to conduct an

inspection four days later. The delegation consisted of two Danes and a Swiss, Maurice Rossel, who represented the ICRC's Berlin office. Rarely has so much trouble been taken to impress so few. Jewish leaders were drilled on what they would or would not be permitted to say. Child performers sat in front of lamps to darken their sun-starved skin. To minimize the potential for disruptive incidents, most of the Danish inmates, who were thought to be less intimidated and therefore more likely to speak the truth, were kept out of sight. To reduce crowding, five thousand more prisoners were shipped to Auschwitz, among them many who were disabled or ill.

On June 23, at 10 a.m., the delegates arrived by limousine from Prague. Every move of the six-hour visit had been carefully orchestrated. At the bank, the visitors saw lines of customers waiting to transact business. At the laundry, smiling women were washing clothes of the highest quality. In the dining hall, inmates were digging into generous helpings of grilled meat, vegetables, and cake. Outside, young women laughed as they marched off—with rakes on their shoulders—to work in a field. When the delegation passed by a soccer match, cheers erupted to mark the scoring of a goal. The visitors reached the performance hall just in time to catch the finale of *Brundibár*. Everywhere they looked, they saw chess players intent on their game, old people listening to a concert, youths striding eagerly about. If they had paid closer attention, they might even have noticed the same caravan of well-dressed children being herded past them several times during the course of the day.

One of the inspectors, Rossel, had brought a camera to Terezín with which he snapped three dozen pictures. In the course of researching this book, I had an opportunity to review some of these images and did a double-take at one. In the photo, children, gathered in a small group, are standing with their eyes on the camera. Among them is a girl who has placed a friendly arm around the shoulders of a companion. Not only was the girl's face familiar to me but so was her dress. My family's last photo of Milena had been with her parents, taken in 1941. Although a positive identification is not possible, it appears likely that my cousin was among the children compelled to march around Terezín that June day.

*Children at Terezín, 1944, during ICRC visit*

*Milena Deimlová, 1941, with her parents*

The ICRC had supplied the delegation with two lists of questions to explore during the visit. These dealt primarily with the handling of relief packages. In keeping with its humanitarian purpose, the organization wished to acquire a reliable roster of those living in the ghetto, to facilitate mail deliveries, and to be sure that food, medicine, and clothing went to the intended recipients. During the war, the ICRC conducted more than 11,000 visits to camps where prisoners were being held. To reduce the risk of deception, the standard practice was to insist on the right to speak privately with detainees. This did not happen at Terezín.

Throughout the visit, German officials, including a close aide to Eichmann, were on hand to monitor conversations. The inmates had no chance to talk freely. The Danish representatives nonetheless detected signs of tension. They asked one inmate how long he had been living in his finely furnished room—the answer: "Since yesterday." They asked the head of the Jewish Council what he thought would happen to the prisoners. "I don't know any way out" was the reply. A number of other inquiries were met with confused statements, as were requests to see such imaginary places as the "fully equipped" maternity ward. Most crucial, however, was the answer to the question of whether prisoners were being deported to the East. No, they were told, Terezín was a permanent, self-governing community, an *Endlager*, not a transit point.

From the Nazi perspective, the charade could hardly have been more satisfying. The report presented by the Danish representatives congratulated the Jews of Terezín for what they had accomplished but was otherwise neutral in tone. Denmark's media, under Nazi control, used the findings to quash rumors that Jews were being sent to labor camps. The account of the Swiss observer, Maurice Rossel, was even more harmful:

> This Jewish city is truly astonishing. . . . One found in the ghetto foods that were almost impossible to find in Prague. The smarter women were all wearing silk stockings, hats, scarves, and carried modern handbags . . . certainly there had seldom been a people who had better medical care than those at Terezín.

On July 19, Nazis held a press conference for foreign journalists using Rossel's words and the accompanying pictures to deny that Jews were being mistreated, let alone gassed.

Certain aspects of the ICRC visit remain mysterious. Rossel was not an experienced inspector. He had been hired that February, trained in March, and had never before conducted an inspection without the accompaniment of a more senior employee. His superior in Berlin, Roland Marti, had been in discussions for almost two years about a visit to Terezín. When it was finally arranged, Marti went on vacation. To this day, the Red Cross has no explanation for why this occurred. In a message to me, an ICRC research officer speculated that Marti had known that a serious inspection would be impossible and therefore had withdrawn out of concern for his future credibility.

What are we to make of the delegation's reports? My first instinct is to ask how the inspectors could have been so blind, to question their integrity and, in the case of Rossel, his attitude toward Jews. My second thought is to wonder how well I would have done in their shoes. In 2011, when visiting Terezín, I was shown a washroom in the Little Fortress. There were two long rows of gleaming white sinks, a shower, and a water closet. How civilized, I thought, who could complain about this? Then the guides explained. The washroom was so clean because it had never been used. It had been built solely for the Red Cross visit, to be displayed if the inspectors insisted on touring the Little Fortress—which they did not.

After much thought, I cannot fault the inspectors for being impressed by what they saw and were told; I do blame them for failing to probe beneath the surface. Tens of thousands of Jews had been sent to Terezín in the thirty months prior to the inspection. Where were they? The ICRC knew the names of many who should have been in the ghetto; why weren't interviews demanded? If the ghetto was such a showplace, why had the Nazis postponed the visit so often? The inspectors had no means to verify the answer they had received to the key question of whether Jews from Terezín were being sent to camps in the East. Yet their credulous accounts helped to sustain Himmler's lie.

There is a lesson in this for those who conduct inspections in our day, whether of prisons, sweatshops, refugee camps, polling places, or nuclear facilities: do not trust—push; control your own schedule; do your homework. Remember the adage that a little knowledge can be dangerous. The truth is more likely to be served by a canceled or aborted inspection than by a whitewash.

In the case of Terezín, the tragic consequences of the flawed inspections were felt far beyond the prison boundaries. Himmler had promised the Red Cross an opportunity to visit a labor camp in Poland. After the delegation's visit to the "spa," the issue was not pursued. That meant that the family camp at Auschwitz-Birkenau no longer had a purpose to serve. In December and May, a total of 11,000 Jews had been transported there from the ghetto. In July, some were selected for work detail; twins were sent to the infamous doctor Josef Mengele; most were murdered.*

The following month, the Germans decided to capitalize further on the cosmetic changes at Terezín by making a film titled *Hitler Gives the Jews a Town*. The scenes of casual and prosperous living that had been created for the Red Cross were replicated before the cameras. Women

---

* One of those sent from Terezín to the family camp at Auschwitz-Birkenau was nineteen-year-old Jiřina Smolková. In July she took her place among the prisoners lined up for "showers" in the gas chamber. Unaware of their impending fate, the prisoners were not in an especially fearful mood. When a German guard smiled at Jiřina, she gave him a wary smile back. A few seconds later, she found herself pulled from the line. Shortly thereafter, she was sent to a newly opened subunit of the Neugraben concentration camp. Women there were assigned various duties, including home building, pipe laying, and clearing the rubble caused by Allied bombing. In February 1945, the SS transferred the women to another subcamp (Hamburg-Tiefstack), then on to the grotesquely overcrowded women's prison at Bergen-Belsen. On April 15, the camp was liberated by the British army, which found 60,000 prisoners, many desperately ill. Among those who had perished during the previous month were Anne Frank and her sister, Margot. Jiřina was not in Bergen-Belsen long enough to succumb to the lack of sanitary conditions and rampant disease. Once again, she survived. After liberation, she met Vilém Holzer, also a Czech and also a survivor. Vilém had been arrested at the outset of the war and sent to a labor camp in Plzeň. In the autumn of 1939, he had been forced to take part in a German experiment that involved the injection of typhoid bacteria. He was one of the few who did not die. He spent most of the war in Buchenwald. Jiřina and Vilém Holzer made a new home for themselves in Argentina. Their granddaughter, Mica Carmio, now works in my Washington office.

were once again outfitted in fancy clothes and made to stroll through the summer heat. Girls walked down a block nibbling pieces of fruit, which, as soon as they turned a corner, were ripped half eaten from their hands.

A grimly satiric piece in *Vedem*, the boys' magazine, captured the mood:

"Now then, gentlemen, you with the long nose, you Fatso, you four-eyes, line up for filming. Look pleasant, self-satisfied, as if you'd just dined on goose. What, you stinking Jew, what sort of a look is that? Here's a slap in the face for you"—and the blows begin to fall, elbow jabs, kicks administered by a gentleman in green to the head of a helpless old man. A whole company of old ladies are commandeered to go bathe. . . . An old lady, who doesn't even know how to swim, has to get into the water. . . . Orthodox Jews and rabbis were sent to the municipal orchestra and had to jump up and down to the rhythm of a jazz band.

The spectacle was obscene, but then so was it all. As Redlich observed in his diary, "Even the kings of Egypt did not film the children they wanted to kill."

New inmates continued to arrive; in late summer, more than two thousand came from the Netherlands. With the ghetto population rising once more, the Nazis began to worry about the possibility of rebellion. Their solution was to resume transports, with an emphasis on able-bodied men. Ghetto residents were told that the deportees would be sent to nearby Dresden to work on building projects. The news gained credibility when only men between the ages of sixteen and fifty-five were scheduled to go. The range was just wide enough to snare both Petr Ginz, the promising young writer, and Rudolf Deiml, the father of my cousins Dáša and Milena.

By that time Deiml had been appointed health commissioner for the whole camp, with responsibility for inspecting kitchens and testing water and food. Earlier in the war, a person of such prestige could easily have avoided being sent away. No longer.

But was the summons really bad news? Council elders were convinced that the Nazis meant what they were saying about the new transports. As the passengers boarded, an official statement was read advising the inmates not to be concerned, that the food would be better and the work a source of satisfaction. Unlike earlier leave-takings, the mood was expectant. Perhaps even Milena and Grandmother Olga were reassured. When, a few days later, the Nazis announced that relatives would be allowed to follow, several hundred volunteered. A feeling had begun to take root that a critical corner had been turned, the war was nearing an end and the Nazis—now desperately short of manpower—really did need their help.

The transport designated "Ek" left Terezín on September 28 with 2,500 men on board. Rudolf Deiml was among them, as was his friend Jiří Barbier, the carpenter. Shortly after departure, they were given postcards to send to their loved ones saying that all was well. When the train reached Dresden, it stopped so the guards could collect and mail the cards, then continued on its way. "As far as Dresden," Barbier recalled, "no one had any doubts, but after leaving in the direction of the East (toward Auschwitz), we realized what was happening to us."

The journey consumed two days and nights. Barbier and Deiml sat together, sharing bread and tinned meat, reflecting quietly. They promised that if one survived and the other did not, to get word to each other's families. On September 30, at five in the morning, the train reached its destination. According to Barbier:

> We had to get out without baggage and wait for further orders. Meanwhile prisoners came and began to unload our things. They told us to give the valuables we had to them, but we didn't feel we could trust them. They told us during inspections to deny being ill and to say that we were workers.

There on the platform Barbier urged Deiml not to admit his profession but to say that he was a carpenter, too, and that the pair of them worked together. Deiml was noncommittal. The inspection was conducted by Mengele and by a second doctor, Schwarz. Each of the prisoners was asked the same questions. Deiml went ahead of Barbier.

"How is your health?" asked Schwarz.

"Good," replied Deiml.

"What is your job?"

"I am a doctor."

Deiml was sent to the left, toward the gas chambers; Barbier, the carpenter, to the right. "With that we parted," wrote Barbier of his friend. "His last look will remain forever in my memory."

IN EARLY OCTOBER 1944, the camp received welcome news: there would be no more transports. A memo was posted to this effect and must have caused celebration. However, the Nazis were beginning to panic. Some officials wanted to refrain from more crimes in order to avoid future punishment; others sought to achieve the same goal by killing witnesses. Thus the decision to suspend transports was soon reversed. Within days a new round began, and before month's end eight more trains would leave, carrying a majority of the camp's remaining population and all of its senior Jewish leadership. The Nazis still insisted that the transports were going to a new work facility, a lie buttressed by receipt of the postcards from Dresden. Many who boarded the trains in October—including Jiří Barbier's wife and Friedl Dicker-Brandeis, the art teacher—expected to be greeted by family members who had left previously. Still, the October transports did not go as smoothly as those in September. The Nazis had taken over the job of selection, and it soon became evident that many of those summoned were too old, young, or sick to perform physical labor.

The surge in deportations and the accompanying chaos disrupted all other aspects of camp life. It seemed that everyone was going, waiting to go, or helping others to pack. There was no way to obtain an exemption from the Jewish elders. The sole means of appeal was to the mercurial camp commandant, Karl Rahm, who had a well-deserved reputation for brutality but who sometimes granted reprieves even while the boarding process was under way.

Finally, again, the summons came. Sixty-six-year-old Olga and twelve-year-old Milena were required to report in two days to the assembly point, called the sluice gate, in the Hamburg barracks. The date of departure was Sunday, October 23. Many girls from L-410 were among those slated to go; the group was a mixture of young and old.

For each transport, a team of healthy inmates was assigned to help the ill and elderly board the train. One such aide, Alice Ehrmann, recorded the scene:

October 23, 1944: Night time in the sluice-gate. At nine-thirty getting people into the cars. The sick, the sick, the sick, stretchers without end. And all this, including loading luggage, is done by forty people with white caps. Luggage everywhere. Luggage in front of the sluice-gate, luggage in the sluice-gate, on the platforms, in the cars. And everyone has so ridiculously little, and even that will probably be taken from them. . . .

Small children, three to ten. Screaming. Each has a little backpack. . . . There is not a person here whose history is not a tragedy; all have been abandoned. . . . One stares peculiarly at those with cried-out eyes. One is brave. Those who walk have turned to stone; those who remain swallow their tears. In the end, the luggage remained; there was no space.

The train made poor time, stopping frequently to allow other, higher-priority trains to chug past. A survivor reported that it arrived at its destination in the middle of the following night, the passengers greeted by barking dogs, shouted orders, and powerful flashlights shining in their faces. Ordered to abandon whatever belongings they had, the prisoners tumbled out of the carriages and lined up in the yard. Of the 1,714 on board, 200 women and 51 men were put onto trucks and driven to a labor camp. The rest, including Olga and Milena, were condemned to the gas chamber.

AT TEREZÍN, THE Nazis were determined to leave as little evidence as they could. The final transport to Auschwitz left on October 28, 1944, five days after the train that had carried Olga and Milena. Two weeks later, the commandant ordered the disposal of urns from the crematory. The job consumed four days. The work was done primarily by women and children, who were paid in sardines. Beginning at the mausoleum, long lines formed and inmates passed the wood and cardboard containers along like water buckets in the manner of an old-fashioned fire brigade. Each bore a label with a name (such as Arnošt Körbel or Greta Deimlová) and the dates of birth and death. The makeshift urns were loaded into trucks, transported, then unloaded—again hand to hand—before being turned upside down, their contents dumped into the Ohře River. An acre of ash floated along the surface.

The prisoners suspected that the disposal was but the first phase of a strategy to bury the truth of what had transpired at Terezín. They were right. In subsequent weeks, the Germans ordered Jewish engineers to build a vegetable storehouse and an enlarged poultry farm. Construction teams began work, but the engineers soon grew suspicious. Why design a storehouse without ventilation and with doors that could not be opened from the inside? Why surround a poultry farm with a wall eighteen feet high or build an enclosure big enough to hold the entire camp population? Why hoard the supply of toxic chemicals ordinarily used to kill bedbugs? As the prisoners speculated, they also asked: why, at this point in the war, should we do what the Nazis ask? The engineers decided to confront Rahm. We quit, they announced. Angrily, the commandant hit their spokesman several times with a pistol but—to their surprise—did not order the men shot. Instead, he retreated to Prague the next day for consultations. By that time, the Red Army had begun to come across German death camps and gas chambers. The horrible truth had reached every front page. Eichmann reportedly told subordinates, "I've had enough." Plans to kill the 15,000 prisoners who remained at Terezín were scrapped.

BETWEEN 1942 AND 1944, at least twenty-five members of my family were sent to Terezín; none survived. On the side of my paternal grandparents—in addition to their daughter, Greta, son-in-law, Rudolf, and granddaughter, Milena—the toll includes three of Arnošt's six siblings, a sister-in-law, a brother-in-law, two of his nieces, and a nephew. On the maternal side, my grandmother, Růžena, her sister, brother-in-law, and nephew perished, as did Grandfather Alfred's brother, sister-in-law, two nephews, and a niece, her husband, and two children. Some, such as Růžena's brother-in-law, Gustav, died in the ghetto, but most were sent to the east; Gustav's wife, Augusta, their children, and a grandchild lived for a time in the family camp at Auschwitz-Birkenau. My relatives had been among the first to arrive at Terezín and the last to leave; Arnošt's brother Karel and his wife were on the final transport. At the time, their son Gert, twenty-six, was in a labor camp near Auschwitz. Early in 1945, when the camp was evacuated, he was sent on a forced march back to Czechoslovakia. Enfeebled by malnutrition and typhoid fever, he died in a barn just a few days before liberation. Like that of so many others, our family tree had been stripped bare.

I THINK SOMETIMES that there are really only two kinds of stories, one ending in hope, the other despair, although it is not always obvious

*The author's paternal grandfather, Arnošt Körbel (back row, left), with his parents and siblings. Marta (back, middle), Irma (front, left), and Karel (front, second from left), also died in the Holocaust.*

which is which. There is no deeper cause for despair than malicious hope (Hitler proved that), and few traits more valuable than sadness and anger at suffering. The distinction that matters is not whether a story concludes happily but whether there is at its core an affirmation that life has meaning. That is why this book of remembrance and war will end in hope—as does this particular story:

One morning in the middle of June 1942, thirty men climbed into the back of a battered green truck and stood crammed together. Tools were piled on, then two barrels of lime. The truck and its cargo lumbered past a succession of towns and farmyards with ducks and geese running this way and that, little ponds with children splashing, and old people sitting peacefully in front of their cottages. The passengers crowded into the back of the truck could derive little pleasure from such bucolic scenes, for they were inmates at Terezín. Suddenly, up ahead, they were startled to see shoots of flame accompanied by plumes of thick black smoke.

A few minutes later, the vehicle drove by a half-fallen yellow signpost bearing the name "Lidice." The truck groaned to a halt; the men jumped out and glanced around. What remained of the village was still burning. Bullet-riddled bodies were piled haphazardly in front of the execution wall of stacked mattresses and pallets. Using the silver handle of his whip, the German commander gouged into the ground a rectangle. "Twelve meters long, nine meters wide, and four meters deep! You understand, you sons of bitches from Jericho, you pig eaters, you criminals?" The workers began to dig. The hours passed, the heavens darkened, the guards lit torches. Sweating and half naked, the men wielded their shovels throughout the night. Around dawn, Lidice's smoldering church broke apart, the walls collapsing as the tower bell plunged, ringing and echoing for the last time. One of the workers was, by trade, a composer; when the church crumbled, he began quietly to sing. The melody was that of Antonín Dvořák's *Requiem*; the words *Dies irae, dies illa, solvet saeclum in favilla* (The day of wrath, that day, will dissolve the world in ashes) and *Agnus Dei, qui tollis peccata mundi* (O Lamb of God who takest away the sins of the world). Hours later, their digging complete, the men were ordered

*Lidice, June 10, 1942*

to strip the corpses of money and identification papers, then drag the bodies into the grave and cover it with dirt, all 432 cubic meters.

There followed the long drive back to Terezín. Before surrendering to sleep, the exhausted men took time to join other inmates in singing Kaddish. Among those who had endured the ordeal was a barrel-chested thirty-seven-year-old journalist named František R. Kraus. Years later, he recalled his emotions at the conclusion of those two miserable days:

> I sink back. . . . Outside the night is of deepest black. And beneath me, on the lower bunks, the composer sings quietly: *"Requiem aeternam dona eis, Domine, et lux perpetua luceat eis."* (Rest eternal grant them, O Lord, and may light perpetual shine upon them.) A spray of tiny stars glitters outside beyond the bars of the barrack windows.

The Nazis had tried to destroy the Czech spirit by obliterating a village and by forcing inmates from Terezín—also an instrument of destruction—to participate in that unspeakable crime. They had sought, in so doing, to deprive their enemies of a future. But the story does not end so simply.

Two and a half years earlier, František Kraus had been on the first train from Prague to Terezín; he was one of those whose hard labor had helped prepare the ghetto for its new and terrible role. In the fall of 1944, he was sent to Auschwitz but was again selected for a work detail and survived to raise a family—fully Jewish and fully Czech. Since 1991, František's son Tomáš has been head of the Federation of Jewish Communities in the Czech Republic. In 1997, he helped me learn about the fate of my own family, including that of my grandmother Růžena, who was transported east during the very hours that František Kraus and his fellow prisoners were burying the victims of Lidice and singing softly of perpetual light.

# 21

# Doodlebugs and Gooney Birds

In 1943, my parents, Kathy, and I moved from our apartment on Kensington Park Road to Walton-on-Thames, a picturesque town in northwest Surrey about thirty miles south of London. There, at 22 Stompond Lane, we shared a redbrick four-bedroom house with a Czech couple, the Goldstückers. Out back was a garden, in front a strange prickly plant called a monkey puzzle tree. Each weekday morning, my father and Mr. Goldstücker walked half a mile to the newly electrified train line on which they commuted to London. The Ingomar School, where I was enrolled in the first grade, was en route, so I pedaled along beside the men on my bicycle. Like my father, Eduard Goldstücker had attended Charles University. He was a scholar with a specialty in German literature who worked in the Education Department of the government in exile. He was, I later learned, a Communist, but quite a nice one—I never heard him argue with my parents about politics.

At school, I felt a proper little English girl in my brown-and-white uniform, which included a tie and a straw hat with striped headband. For lunch every day I ate cold meat and bubble and squeak (a mixture of leftover potatoes and cabbage all fried up, so named because of the sound it made in one's stomach after being consumed). I loved going to Ingomar because it made me feel grown up and because I have always been eager to learn. According to my report cards, I possessed "the ability to do well, but must try to be a little more steady." In arithmetic, I was admonished to avoid "careless slips" and in drawing "not to rush" my work. Even for students of my tender age, the arts were not neglected. Little Madlen apparently had "an excellent sense of rhythm" and was "full of enthusiasm" when letting loose on such classics as "Camptown Races" and "The Lass of Richmond Hill." The school offered instruction in piano, to which I readily took, falling in love with an old Austrian (not German) named Mozart. In geography, my first-term grade was D minus, which did not bode well for a career in world affairs, but the next semester I improved to B, so there was a chance after all.

I was just six years old, but impressing my parents was already a preoccupation. As was typical of an English public school, the student body was divided into teams and one scored points by succeeding in various activities. When I first earned credits for my team, my father indicated approval. Wanting more, I began to make up exploits for which additional marks were awarded, including, as I recall, pulling my teacher out of a rosebush. Before long, I had tallied so many imaginary points that I decided to invent a special award, telling my parents that I had won the "Egyptian Cup." They asked to see the trophy, which obviously wasn't going to happen. Instead I devised a whole new set of fibs about how awful everyone was being to me. "They even make me sit on needles!" I exclaimed. My mother insisted on going to school to find out what was happening to her poor child. As Hus had predicted long before, truth prevailed, and I was duly punished. In later years, whenever a story I was telling seemed at odds with the facts, my parents had only to murmur, "Egyptian Cup," and I stopped.

Dáša was usually away at school in Wales or else staying in Berkhamstead with our aunt and uncle. That made me the big sister. There was a small grocery store four blocks down and across a bustling street to which I pushed Kathy in her green pram. I brought along the shopping list and ration book, but there was little in the way of fresh meat or fruit and a strict limit on purchases of milk. This didn't seem unusual because it was the only reality I had ever known. I was also entrusted with the task of giving water to the noisy chickens we kept out back. The first time I just grabbed an empty milk bottle and filled it about halfway. My mother suggested that a bowl might be a better strategy and asked, "How did you expect the poor chickens to drink?" I gave the matter some thought: "They have long necks."

It was at Walton-on-Thames that I first indulged my fantasy of becoming a priest, which for a young Catholic girl was certainly a sign of ambition. Each night, I prayed to the Virgin Mary, fashioning a bedroom altar complete with candles and a silver cup that I used to douse the flames. Whether it was the smell of burning wax or the prayerful rhythm of the words that most appealed, I am unsure, but the religious experience was more meaningful to me than to my parents, particularly my father, who rarely attended services and—when he did—complained about the requests for money.

Looking back, I do not know what went through the minds of my mother and father as they worried about what might be happening back home. I do know that they did all they could to make life for Kathy and me seem as normal as possible. We went on family outings, including to the beach, where we swam, ignoring the huge barriers that were there to keep the Germans from invading. On weekday afternoons, I sometimes slipped between the hedge branches to share high tea with our English neighbors. On Sundays my parents often hosted a gathering of Czechoslovak friends. After dinner the women cleared the tables while the men—most of whom were in the Beneš government—paced up and down alongside our garden in earnest discussion. They walked with their hands clasped behind their backs, as European men do, my father usually with his pipe and a puff of smoke around his head. In the evening the men drank beer, filled ashtrays with cigarettes, and played Mariáš, a distinctive Czech and Slovak card game with thirty-two cards divided into four suits (hearts, bells, leaves, and acorns) and rules more complicated than those of bridge. The wives drank coffee, traded news, and laughed.

IN THE SPRING of 1944, I had my first look at American troops. There were many of them roaming about the city streets, where they were besieged by children seeking handouts of candy and asking, "Got any gum, chum?" Others could be seen on the country roads, casually driving their camouflaged jeeps, trucks, and peculiar-looking amphibious vehicles known as ducks. Even the enlisted Americans had well-cut uniforms the color of a green olive, and, in contrast to those of the British, their guns were shiny and new. For a time, "Yanks" were everywhere; then, in an eyeblink, they were gone.

On the morning of June 6, Operation Overlord, the largest amphibious landing in the history of war, took place on five beachheads along a fifty-mile stretch of the Normandy coast. Despite a cold northwest wind, 160,000 troops crossed the Channel. The skies were thick with 11,000 airplanes, and there were so many ships and boats that it seemed almost possible to walk from England to France. The early-afternoon news confirmed that D-Day had arrived. Announcers, delighted to have such an event to discuss, told the exciting story of how the men and equipment had been assembled, the guns and ammunition packed, and the critical element of surprise maintained. The local French population had been warned only an hour ahead of time to evacuate the area; a devastating bombardment of roads, rails, and bridges was under way. An evening broadcast featured the king's speech, followed by a prayer service led by the archbishop of Canterbury. By nightfall, 9,000 Allied fighters had given their lives. In succeeding weeks, soldiers continued to swarm into France, penetrating more deeply every day; the decisive struggle to reclaim Europe had begun.

WE HAD LIVED in Walton-on-Thames for more than a year. The blackout was still on, but we didn't worry much about bombs anymore. After D-Day the Germans were on the run—or so we thought.

On June 13, at 4:13 a.m., an explosion was heard in Gravesend; something big had landed in a lettuce patch. Ten minutes later, a second explosion damaged a railway bridge in London. For months there had been rumors that the Germans were developing a secret weapon to be unleashed if and when the Allies invaded France. Some said it was an invisible beam that would lay waste to all in its path, others an ingenious method for spraying poison gas, still others a superbomb more devastating than any previously conceived. After three days of investigation, the government confirmed that the Nazis were deploying a pilotless flying bomb called the V-1, or *Vergeltungswaffe* (vengeance weapon).

Shaped like small airplanes, 27 feet long and 17 feet across the wings, they zoomed along at 350 miles per hour. Most were launched from ramps near the French and Dutch coasts, some by low-flying bombers.

Churchill at first adopted what Cadogan dubbed "the light-hearted bulldog view," deeming the V-1s but a nuisance, the last gasp of a desperate regime. He paid more heed as the casualties and damage mounted. Many Britons were outraged at the authorities for not being better prepared. The attacks were harder for some civilians to cope with than the Blitz. When the future had seemed dark, danger was less daunting; now, on the threshold of victory, the prospect of death from the sky seemed a cruel trick.

One hundred or more of the bombs were sent toward England each day, and it was several weeks before defenses were able to intercept any at all. When we heard the loud buzzing, we held our breath. The sound meant that the rocket still had fuel and would rapidly pass by; when the noise stopped, so did the bomb, which would fall almost straight down. The V-1s were called "doodlebugs" because of their resemblance to the large buzzing insects of the same name. "One can see its little black body hurtling along," wrote a Briton who was there, "with something flapping behind it which looks like gauze being blown out from an electric fan." At night "this exhaust flames like a meteor, and one can follow the thing as it hurtles through the sky like a falling star . . . unswerving, vindictive, and horribly purposeful."

The scream of sirens became common once more, and again I had the experience of sharing long hours with neighbors in an air-raid shelter. To pass the time, we sang "A Hundred Green Bottles" from start to finish, then again from the beginning. On some days the entire school-age population of Walton-on-Thames was confined to the shelter. The town, being south of London, was located in the middle of "bomb alley," the flight path from the V-1 launch sites to the capital. In addition, British double agents—that is, English spies pretending to be Nazi spies—were instructed to tell Berlin that the bombs were overshooting London. The Germans shortened their aim, sending fewer bombs toward the densely populated city and more crashing down around us. On June 19, a V-1 shook the ground just a few blocks from our house;

then another landed and another—a total of eighteen on our town. "Why are the doodlebugs in such a hurry?" went the joke. "You would be, too, if your behind was on fire!"

My father was a volunteer air-raid warden. That meant that when he returned from London each evening, he had to rush through the neighborhood making sure that everyone had their curtains drawn, even though the flying bombs had no pilots and couldn't see. One night in the pitch dark, he ran into a brick pillar in front of our house, bloodying his nose and breaking his glasses. I was sympathetic but by then had my own problem: sleepwalking. Whether it was caused by the V-1s or not, I have never known, but for weeks I went to bed without being sure where I would end up. My parents and the Goldstückers, worried that our residence might be hit, bought a Morrison table, named after Home Secretary Herbert Morrison. Overnight, the heavy rectangular steel object became a center of our lives. We ate on it; the adults had their coffee and drinks while sitting around it; Kathy and I played on top of it; and when the siren sounded, all six of us dived under it.

One doodlebug fell outside the BBC headquarters, shattering windows and damaging the building. The midday Czechoslovak broadcast, which was about to begin, went on as scheduled despite injuries to a secretary and two announcers. Another time my father and Mr. Goldstücker arrived at Waterloo Station a few moments after a bomb hit. Ambulances had not yet arrived, and passengers were throwing down their briefcases to administer first aid to dazed and bleeding victims. On Sunday, June 18, one of the bombs smashed the roof of Guards Chapel in central London shortly after the weekly worship service had begun. Rubble buried the congregation, a mix of military and civilians, killing 121 and sending 140 more to the hospital. Two days were required to dig out the dead.

The doodlebug raids lasted three months, taking the lives of more than 5,500 people and injuring thrice that number. As during the Blitz, tens of thousands of children were evacuated. The whole country celebrated when, in September, Allied troops overran the V-1 launch sites along the northern coast of Europe and Mr. Morrison announced that the second Battle of Britain had been won. By then, Rome and

Paris had been retaken, as had Brussels. Returning evacuees clogged the train and tube stations in London while in Walton-on-Thames the centuries-old bells in Saint Mary's Church resumed their cheerful ringing.

AUTUMN 1944: The Germans were being driven back on all fronts. Allied forces were advancing from the south, west, and east while firing relentlessly from the skies and at sea. In September, Luxembourg was free and Holland, too. The Red Army was moving through Poland, Romania, and the Balkans, surging via the dense Carpathian woods to the border of Slovakia. In November, partisans liberated Greece, and Churchill indulged in a well-earned day of speech giving and wreath laying as the guest of honor in a celebratory and inebriated Paris. German troops, hampered by the bombed-out rail lines and shortages of fuel, were taken prisoner by the gross. A captured codebook enabled the Americans to decipher enemy U-boat communications, leading to the destruction of nearly three hundred Nazi submarines in just a few months.

For Polish democrats, the prospect of being liberated from the Nazis by the Soviets had the appearance of a prayer only half answered. In August, the main resistance force, the home army, learned that Soviet troops had reached the eastern suburbs of Warsaw. Rather than wait to be freed by the Communists, the Poles resolved to do the job on their own. A force of 40,000 fighters (but with only 2,500 firearms) attacked German military outposts and civilian installations. A fierce struggle commenced as the entire city and surrounding area became a battleground. The beleaguered German commanders called in reinforcements and issued orders to exterminate every Pole. The rebels had hoped that panic caused by the uprising, coupled with fear of the approaching Soviets, would cause the Nazis to flee; it didn't happen.

British, Polish, and South African pilots sought to save the fighters by flying in supplies from bases in the United Kingdom and Italy, but the quantities were not sufficient. A more effective airlift could have been mounted had the Soviets granted permission to use airfields under their control, but Stalin flatly refused. After several weeks of fighting, the momentum shifted in favor of the better-equipped Germans. In the end, the uprising became a debacle. An estimated 200,000 Poles were killed and another 800,000 captured or expelled from their homes. The Nazis looted every portion of the city before burning what was left. When, in January 1945, the Red Army and their Polish partisans finally entered Warsaw, they found much of it destroyed.

A pro-Soviet governing council was established in Lublin, the center of Poland's southeast region. The London-based Polish democrats counted on the Allies for help in securing a role for themselves in any postwar settlement. President Beneš was not sympathetic. Past border disputes had left a bad taste in his mouth, and the most capable Polish leader, General Władysław Sikorski, had died two years earlier in a plane crash. From a diplomatic perspective, Beneš was convinced that the Polish exiles had played a weak hand badly. "Poland in this war has made one fundamental mistake," he told a friend. "She can achieve her aims only with the help of Soviet Russia and the collaboration of the three great powers. . . . The Poles [think] . . . that they are strong enough to discuss the future as equals. . . . I learned long ago that the big nations always settle their questions among themselves at the expense of the small."

The question Beneš did not address is whether the Poles ever had a chance. Unlike Czechoslovakia, their country had deeply felt grievances against the Soviets, who had inflicted terrible atrocities against its people and seized more than a third of its territory. Sikorski had, in fact, sought to mend relations, only to be rebuffed when he demanded, quite reasonably, that there be an international investigation of the Katyn Forest massacre. Stalin was unlikely to accept any arrangement that gave leverage to Polish democrats because he had already murdered so many. He also wanted his country to have the widest possible territorial buffer against Germany.

The Czechoslovaks, by contrast, could approach the Soviet Union as a friend. There was no history of conflict between the two and no apparent reason why they could not go on as partners. Beneš certainly favored that scenario. By his calculation, once the Nazis were defeated, the Russian military would dominate Central Europe. For every Western soldier in the region, the Russians had three. The British were too weak and too thinly spread to counter them, while the Americans and Canadians would want nothing more than to return home. What Stalin desired in the region, Stalin would get; the only course open was to influence what the Soviets felt they needed.

Accordingly, Beneš continued to make himself agreeable. Otherwise the Soviets might do in Czechoslovakia what they had already done in Poland and were preparing to do elsewhere: choose a replacement set of leaders and back them with money and arms. The Red Army would soon be in Czechoslovakia. If Beneš were to be there, too, he would need its help. Although London was the official center of the government in exile, the Czech and Slovak Communists in Moscow would have to be included in the postwar government. They lacked a leader of genuine stature, but their ideology gained in popularity every time Hitler denounced it.

Even as he was reassuring Stalin, Beneš had no wish to be seen as a Soviet minion. He insisted that his country could avoid having to choose between East and West. Hadn't he signed a treaty of friendship with England as well as the USSR and journeyed to Washington before going to Moscow? Hadn't Stalin promised to refrain from interfering in Czechoslovak democracy? Communist ideology called for the liquidation of bourgeois capitalism, but the Soviet government had just spent years fighting in a grand alliance with the West. Surely the shared interests that had brought the Communists and other parties together would still be present after the war?

Beneš was far from the only one who held this view. Aside from Stalin's victims—of whom there were millions—people tended to have affection for old Uncle Joe. He was jovial, smoked a pipe, didn't rant like Hitler, and seemed far more practical than ideological. More than that, when the outlook had been bleakest, his country had delivered. For years, every victory by the Soviet armed forces had been cause for cheers and sighs of relief in the West; that kind of experience makes a difference.

Harold Nicolson, a British parliamentarian who was close to Churchill, explained:

People say to me, "but why, when you cursed us for wishing to appease Hitler, do you advocate the appeasement of Stalin?" I reply, "[F]or several reasons. First, because the Nazi system was more evil than the Soviet system. Second, where Hitler used every surrender on our part as a stepping off point for further aggression, there does exist a line beyond which Stalin will not go. On a purely material basis, Stalin needs economic help from the Americans and so will not go too far to alienate. [My] feeling [is] that if we are patient, the Soviet tide will recede."

In the Czechoslovak context, this analysis was buttressed by the reassuring rhetoric of the Communist leadership. Speaking from Moscow, Gottwald delivered scores of broadcasts about the importance of a united war effort. For inspiration, he cited not Marx and class solidarity, but Hus and patriotism. As a goal, he extolled the return of freedom, saying nothing about world revolution. In London, Vlado Clementis and the other Communist leaders spoke of their objectives in terms that could scarcely be distinguished from those of the democrats.

Beneš was what we might call today a democrat with socialist leanings. He favored state control of basic industries, a vigorous trade union movement, and generous public services. He had some sympathy for communism's egalitarian ideals but doubted that the ideology could be applied in Czechoslovakia the way it had been in the USSR. He felt sure that most of his people would remain loyal to the democratic model established prior to the war by Tomáš Masaryk. However, the senior Masaryk, far more than Beneš, had been an articulate and convincing critic of communism's flaws. There was a reason why Lenin had called Masaryk his most serious intellectual opponent in Europe. Beneš, who sought to please almost everyone, found himself in the position of trying to champion democracy, implement leftist economic policies, mollify the West, and butter up Stalin all at the same time.

The frenetic maneuvering made some of the president's democratic followers uncomfortable, but there was little they could do. Beneš responded to every expression of concern with an assurance that he knew exactly what he was doing. Jan Masaryk, for one, was not about to challenge his boss on matters of political ideology. Even his love for democracy was tempered by concerns about the fickleness of public opinion; after all, Hitler had adoring crowds and the British public had applauded Chamberlain after Munich. For Masaryk, the Communists' greatest sin was that they took themselves too seriously. He told a friend, "Lenin said that people must stop listening to Beethoven, because he has the power of making people happy. He was afraid lest they would become too soft to make the revolution. There you have the whole of Lenin."

LESS THAN A week after the last of the V-1s had taken to the air, V-2 rockets were unleashed. "What I want," Hitler had demanded of his munitions designers, "is annihilation—[an] annihilating effect!" German engineers, long skilled in rocketry, set out to craft a technology powerful enough to destroy whole cities and thereby alter the outcome of the war. What they produced was a 9,000-pound, 46-foot-long missile, launched from mobile sites in Holland and France, that soared fifty miles into the sky before crashing without much precision into the urban neighborhoods, towns, and—more often than not—the empty fields and vacant lots of England. Unlike the doodlebugs, whose buzzing advertised their approach, these rockets were supersonic; their impact was felt before any sound could be heard. The doodlebugs had been tracked by radar, crippled by balloons, and shot down by fighters and antiaircraft guns. The V-2s flew too high and fast for such countermeasures. The payload was no match for Hitler's fantasies but heavy enough to leave bus-size craters; the missiles were far more destructive than a conventional bomb. The British, who had a nickname for everything, called them "gooney birds," another term for the albatross—big, awkward, and, as in Samuel Taylor Coleridge's poem, very bad luck.

Fortunately for the Allies, the V-2s were expensive to make and tricky to launch. The Germans were able to fire four a day, then six, but never more than thirteen. That was still enough to worry British leaders, who feared a panic if word spread about the heavy rocket. For almost two months, landing sites were quickly roped off and the huge explosions attributed to malfunctioning gas mains or sabotage; the government denied that the gooney birds even existed. The charade ended in November, when Churchill felt he could no longer pretend that what his people were seeing and hearing was imaginary.

Between September 1944 and the following spring, more than a thousand V-2s hit British territory, about half in and around London. The deadliest crashed into the Woolworth's in Deptford right at the peak of Christmas shopping. As frightening as the explosives were, they did not have an "annihilating effect." They didn't carry enough weight, nor could they be manufactured rapidly enough, to make a strategic difference. The weapon certainly didn't cause British morale to plummet to the extent that the Nazi leadership had hoped. In fact, the Germans would have been wiser to have invested in a more capable fleet of long-range bombers. Still the development of rockets that could travel a distance, deliver a large payload, and remain roughly on target was sobering to those thinking ahead. "Every time one goes off," wrote George Orwell, "I hear gloomy references to 'next time' and the reflection, [that when war comes again] they'll be able to shoot them across the Atlantic."

Two additional aspects of the V-2 story are worth noting, the first for what it says about Nazi depravity, the second for what it reveals concerning American pragmatism.

The rockets were produced in a construction complex about 130 miles southwest of Berlin. The workers—or more accurately, the slaves—included French, Soviet, Belgian, Dutch, and German prisoners who labored underground for twelve hours each day. Their bodies were nourished only by coffee, thin soup, and bread. The combination of malnutrition, impure water, poor ventilation, harsh climate, and physical abuse killed so many that replacements had constantly to be shipped in from nearby concentration camps. In April 1945, immediately before surrendering the construction site, the Germans locked 1,046 of the workers into a barn and burned them alive.

The director for technical development of the V-2 was Wernher von Braun, a thirty-one-year-old SS major who personally provided Hitler with updates on the project. Through a combination of his own fast talking and America's willingness to overlook the faults of those with useful skills, the young Nazi was spared postwar punishment. He later became a prominent figure on the "Tomorrowland" segment of the Walt Disney television show, and—more valuably—an architect of the U.S. space program. In recognition of his accomplishments, von Braun was publicly congratulated by President John F. Kennedy, whose older brother, Joseph Jr., had died while on a 1944 bombing mission against a doodlebug launch site in France.

# 22

# Hitler's End

The air was so cold that my feet turned numb and I could not feel the stairs, so cold that I thought my bones would freeze, until King Albert II of Belgium gallantly offered to share his blanket. From 1993 until 1997, I was the U.S. ambassador to the United Nations. During those years, I felt as if the world were celebrating the fiftieth anniversary of everything; not that I minded. On December 16, 1994, I had the honor of representing my government on a day of commemoration and thanksgiving.

The setting was Bastogne, a small Belgian city just west of the German border. The frigid temperatures were apt because the battle waged there half a century earlier had been fought in subzero conditions, in the ice and snow of the dark woodlands of the Ardennes. From the podium, I could see beyond the steamlike condensation of my breath to the colors that had guided the U.S. Army through its sternest test: among them the banners of the V Corps, the 7th Armored Division, the "damned engineers" of the 291st Engineer Combat Battalion, and the 30th Infantry Division. There, too, were veterans from the 101st Airborne Division and elements of the 9th and 10th Armored Divisions that had made their stand not far from where I now sat.

From the landing at Normandy in early June 1944 until the middle part of the autumn, Allied troops advanced more rapidly than anticipated. This was good news, but it overextended supply lines, requiring a pause before the final push. The Germans chose that interval to mass their forces and launch—on December 16—a desperate counterattack aimed at compelling the enemy to seek a negotiated and more favorable peace. Thus began the epic clash known in Europe as the Battle of the Ardennes and in the United States as the Battle of the Bulge. The German generals had given the assault little better than a 10 percent chance of success, but the element of surprise led to a promising start. They had timed the action to coincide with poor weather so the Allies would be unable to generate air support or track hostile troop movements from above. The Nazis pounced where the opposing lines were most thinly spread and caught their foes at a moment when many units were short of equipment and experienced manpower. Suddenly the U.S. Army leaders were on the defensive, for most a novel experience.

A turning point was reached when the 101st Airborne Division found itself surrounded in Bastogne. The Germans had expected the position to fall on the assault's second day; it was now day ten. Asked by a Nazi messenger if he would care to surrender, General Anthony McAuliffe said just one word: "Nuts." The courier asked, "Should this be interpreted as a positive or a negative response?" The answer: "Negative, and it means 'Go to Hell!'" In the days before Christmas, General George Patton's 4th Armored Division prepared to break the encirclement. Because air support would be critical, his chaplain led a prayer for better weather; when skies cleared early the next morning, the churchman was rewarded by Patton with a Bronze Star.* On December 26, the Allied tanks drove and dodged their way through a maze of enemy artillery and minefields to relieve the trapped division.

Skirmishes continued for several more weeks, but with the 101st back in action, fewer clouds, and a frenzied commitment of fresh men and equipment, the Allies quickly regained the upper hand. Eisenhower had told his men, "By rushing out from his fixed defenses, the enemy may have given us the chance to turn his great gamble into his worst defeat." So it proved. The German attack stalled, and with their reserve forces now exhausted, the defenders of the Reich began their

---

* Patton's exact words were "Goddamn! That [Chaplain] O'Neill sure did some potent praying."

final fallback. In a month of bloody encounters up and down the front, roughly 90,000 U.S. and 1,400 British troops were killed, wounded, or captured. More than 3,000 Belgian citizens died. The U.S. Army lost nearly 800 tanks—more than it had even possessed prior to the war. Hitler's final charge left the Wehrmacht with 60,000 casualties and four times that number taken prisoner. All in all, a fearsome price.

As I shivered that morning in Bastogne, I could see a silver-haired army of old soldiers who had crossed the Atlantic to relive memories and recall lost friends. After greeting our hosts in French, I directed my words to those brave Americans:

> You, the veterans of this conflict, may have felt that you were fighting only for yourselves, your buddies, your unit, and your family. When the scourge of war is visited upon us, it is not countries that fight, it is people, and the emotions of conflict are intensely personal. But your skills, courage and sacrifice were enriched and ennobled by the cause for which you fought. Let us never forget why this war began, how this war was fought, or what this war was about. . . . History did not end here in these fabled woods; it did not end with the Nazi surrender or the fall of the Berlin Wall. Each generation is tested; each must choose.

EARLY IN FEBRUARY, Churchill, Stalin, and the ailing Roosevelt met in the czar's former vacation home in the resort town of Yalta on the coast of the Black Sea. There, in palatial surroundings, they planned for the finish to the war and thought about what would come next. Three topics headed their agenda: the future of Germany, the creation of a new world organization (the United Nations), and the question of what to do about Poland.

Given their deservedly honored status in the history of the twentieth century, one yearns to imagine the U.S. president and the British prime minister at their best in this moment of triumph. We might envision them going briskly about the world's business informed by a shrewd strategy and clear principles, their thoughts seasoned by profound insights concerning the legacy of war and the substance of future hopes. We might wish to see them pressing Stalin hard not only on the terms of agreement but on the precise meaning of each word and phrase, understanding that a quarrel over interpretation could drain the value from any pact. Alexander Cadogan, who admired both leaders, was ordinarily a sympathetic observer. Still, he wrote that at Yalta, Churchill behaved like "a silly old man" who rambled on at variance with the positions of his own government, while Roosevelt "did not look well and was rather shaky. I know he is never a master of detail, but I got the impression that most of the time he hardly knew what it was all about." The spectacle was, he added, "rather disturbing."

In this company, particularly with his army but sixty-five miles from Berlin, Stalin did not have to work very hard. He began in disarming fashion by pledging to atone for the Soviet Union's mistreatment of Poland, a state that he agreed should be fully independent, with leaders chosen through free elections. As to the border, he proposed that the Poles be given a large chunk of Germany in return for the 30 percent of their country that he referred to as "western Ukraine and Byelorussia."

*Churchill, Roosevelt, and Stalin at Yalta, 1945*

He also accepted the idea of broadening the Soviet-installed provisional government in Lublin to include democrats; this would ensure that balloting was fair. Churchill, whose mind possessed a vast library of martial phrases, vowed that Poland must be "free and sovereign, captain of her soul." Stalin purred that he had precisely that intent; elections, if all went well, could be held within a month.

Returning to Washington, FDR declared that the Yalta Conference had put an end to the kind of balance-of-power divisions that had long marred global politics. His assessment echoed Woodrow Wilson's idealistic and equally inaccurate claims at the end of World War I. In London, Churchill told his cabinet that "poor Chamberlain believed he could trust Hitler. He was wrong. But I don't think I'm wrong about Stalin." Soviet-British friendship, Churchill maintained, "would continue as long as Stalin was in charge."

There had been other business done at Yalta: the Soviet Union had agreed to participate in the conference to be held in San Francisco that would establish the United Nations; it had also consented to join the war against Japan, a move thought likely to hasten an end to the still bloody conflict in the Pacific. In the British capital, however, it was Poland that generated the most controversy. The London-based Polish exiles denounced the proposed border settlement agreed to at Yalta and criticized what they termed the legitimization of their rival, the pro-Russian Lublin regime. When the House of Commons debated the issue, a group of twenty-five parliamentarians sided with the exiles and offered an amendment condemning the allegedly unfair treatment. The proposal failed but not before it caused significant trouble—for my father.

The Czechoslovak government in exile needed the support of the governments that had participated in Yalta; it was natural, therefore, that Beneš—whose route home would take him through Moscow— endorsed their decisions. It followed that my father would use his broadcasting platform to explain his government's point of view. In a program on March 3, 1945, he did just that but with perhaps too much enthusiasm. Instead of offering a straightforward defense of the Yalta Agreement, he stuck verbal pins into the British parliamentarians who opposed it, labeling them as past supporters of the Munich pact and as "members of the British-German Association." The targets of this criticism, led by a Mr. Maurice Petherick, MP, thought it outrageous that they should be insulted by a foreigner—let alone a guest of the BBC. They demanded Korbel's immediate dismissal. My father rarely cast aspersions without meaning every syllable, but in that case he prudently agreed to an apology, conceding that "the offending passage should have been more accurately and carefully worded."

The next act in this drama took place in California, where delegates had gathered to write the rules for the United Nations. The question arose as to who should represent Poland. The USSR pushed for recognition of the Communist provisional government. The British and Americans preferred to wait until a more democratic body could be assembled. For the first time, Czechoslovakia was forced into a zero-sum choice between acquiescence to Moscow and alignment with the West. Sadly, the motion to approve the Soviet position was made by none other than Jan Masaryk. "We want a strong and democratic Poland," Masaryk explained, "but only a Poland that will collaborate with the Soviet Union." He later confided to U.S. Envoy Charles Bohlen:

> What can one do with these Russians? Out of the clear blue sky I got a note from Molotov saying Czechoslovakia must vote for the Soviet proposition in regard to Poland, or else forfeit the friendship of the Soviet government. . . . You can be on your knees and this is not enough for the Russians.

The Soviets could not have done more in this period to expose the naïveté of their apologists. In March, Stalin invited sixteen Polish democratic leaders to Moscow for talks—promised at Yalta—about forming a unified provisional government. No sooner had the democrats arrived than they disappeared. For six weeks, the Russians claimed not to know where their visitors might be. The British and Americans demanded to know what was going on; the Polish democrats in San Francisco were incredulous. Finally the Soviets admitted that the Poles were in Lubyanka prison, having been arrested for "organizing diversionary tactics in the rear of the Red Army." The prisoners were tried, and most were sentenced to jail terms, thus embarrassing all those who had vouched for the trustworthiness of Stalin. The betrayal was just beginning. Despite the promises made on the shores of the Black Sea, the USSR imposed a totalitarian government on Poland that would not relax its grip for more than forty years.

IT WAS APRIL 1945, the European war's last month. Prisoners were being released by the thousands. The days lengthened, and the weather warmed. Streetlamps were turned on. Motorcar headlights were unmasked. Now almost eight years old, I learned with glee what it was like to live in a house without ugly dark curtains draped over the windows. Fresh meat and produce reappeared in the grocery store; the return of coconuts made the news. Our radio was almost always on, and for the first time in six years, it was considered safe to have weather reports. I listened whenever I was trapped inside by spring rain. When the sun came out, I worked in the garden with my mother, planting vegetables that we did not expect to be around to eat. I spent much time playing with Kathy because my parents were often on the phone. The Sunday-afternoon gatherings in our backyard were filled with laughter and a sense of anticipation. My father was soon to have a new job. Best of all, we would be going home—a grand adventure, to be sure, but one I had no way to envision.

Patton's army, with all its equipment, had pushed its way across the Rhine and into Germany. The British and Canadians were linking up from the north. The Soviets had reached and entered Austria. The Allied bombing of Germany was relentless, the destruction horrible. On Friday the thirteenth, we learned that President Roosevelt had died of a cerebral hemorrhage while having his portrait painted in Georgia; the West bowed its head at the loss of a man who, while sitting in a wheelchair, had raised democracy from its knees. Two days later, U.S. soldiers stormed into the Nazi death camp at Buchenwald. Edward R. Murrow was with them, and his account of the human devastation was broadcast by the BBC. Never has the need to confront evil been more graphically demonstrated.

Still the bulletins came. Berchtesgaden, the mountain retreat where Hitler had met with Chamberlain, had been bombed. The Allies had taken Munich. On April 28, there was a snow- and ice storm in England; so much for the vegetable garden. Now Berlin was encircled by Soviet, U.S., and British troops. In London, people prepared to celebrate. Large quantities of liquor were imported but held in reserve for the special day. The air-raid warning system was turned off. Flags were everywhere on sale. Bunting was purchased in the dimensions of Buckingham Palace. The Nazis had nowhere left to run.

It took a day for the tidings to reach her, but the news penetrated even the thick walls of Terezín, where, on May 2, Eva Ginzová, the fifteen-year-old sister of Petr Ginz, wrote in her diary, "Apparently, Hitler has croaked."

# PART IV

❧

# *May 1945–November 1948*

*No man shall dictate to me what books I shall read,
what music I shall hear, or what friends I shall choose.*

—JAN MASARYK

# 23

# No Angels

In April 1945, Edvard Beneš returned to the homeland he had left six and a half years previously. Escorted by the Red Army, he crossed by train from Ukraine into the town of Košice in the eastern part of Slovakia. He was welcomed by joyous villagers garbed in national costumes, waving flags, cheering, and tossing flowers. A young girl offered him the traditional greeting of bread and salt. From London, the president had communicated with his people only by radio. On the BBC broadcasts he had made no secret of his identity, but over the clandestine network used by the Resistance, he had been referred to by his code name: Mr. Comeback.

BENEŠ HAD LEFT London four weeks earlier in the company of Jan Masaryk and representatives from the various factions of the London government in exile. The entourage, large enough to warrant three Royal Air Force bombers, stopped for refueling in Tehran. Waiting to greet them was the government's double-dealing ambassador to Moscow, Zdeněk Fierlinger, who informed Beneš that the Communists had decided to put him forward as prime minister of the postwar government. Beneš was surprised because Fierlinger had no domestic policy experience but relieved that the Communists hadn't chosen one of their own. As sympathetic as the ambassador was to Moscow, he belonged to the Social Democratic Party. The West would be more comfortable with him than it would have been with the Communist leader, Klement

Gottwald. Beneš also thought that Fierlinger's lack of popular support would make him relatively easy to control.

The presidential party arrived in Moscow on March 17; meetings to organize the government began five days later, chaired by Gottwald. Beneš didn't attend on the grounds that, under the Constitution, he should stand above the parties and await their recommendations. This was a miscalculation. By far the most popular man in Czechoslovakia, he could have used his political capital—acknowledged even by the Soviets—to mold the state institutions for which he would bear ultimate responsibility. Instead he left it to the democratic party leaders to look after their own interests, which they were poorly prepared to do. For years they had done little without his consent; now they were on their own. They saw their mission as one of restoration, to re-create a pluralist system in which an appointed government would soon be replaced by an elected one. To the party leaders' way of thinking, the country's future direction would be determined at the ballot box, not by temporary arrangements decided on in Moscow.

The Czech and Slovak Communists, by contrast, were intent not on restoration but revolution. They saw the war as a catastrophe brought on by capitalist decadence but also a rare opportunity to create a totalitarian state by democratic means. They were patient enough not to seek victory all at once but determined to set the country on what they intended to be an irreversible course. Stalin advised them to accept Beneš as president but otherwise to increase their leverage at every point. Their thirty-two-page draft program served as the basis for discussion because the opposing parties had neglected to fashion an alternative. The deck was stacked further by an informal alliance between the Communists and Social Democrats, the party of the moderate Left led by Fierlinger—a man whose heart and marrow, the Communists knew, belonged to them.

Gottwald also sought to take advantage of the perpetual tension between Slovaks and Czechs. He promised the Slovak representatives that their region would have full autonomy. Beneš and the Czech democrats saw no alternative but to accept this, albeit with much regret. Beneš in particular believed that Slovak nationalism had no racial or linguistic basis and that the country could not prosper without a robust central authority. He had tried to enlist Stalin's support for these arguments, but the dictator had no interest. This meant that the Slovak national dream would survive—but to what end?

THE WARTIME EXPERIENCE of independent Slovakia had failed the vision of its champions. In place of its junior partnership with Prague, the nation had chosen a disgraceful subservience to the Reich, although in fairness many Slovaks resented German domination and struggled to lessen its hold. The country's Communists opposed Nazism ideologically; many Protestants were still drawn to the Czechs; and a fair number of Roman Catholics took offense at Hitler's perverted view of scripture. One U.S. diplomat compared the relationship between Bratislava and Berlin to that of a puppy on its master's leash, pulling always in one direction or another, unable to shake itself loose. The Slovak leadership, unforgivably, made little attempt to break free.

Father Tiso, the jowly, crew-cut president, was an enthusiastic collaborator. He had sent his country's boys to fight and die alongside German soldiers on the eastern front and was generous in funneling food and mineral resources to the Nazi war machine. More damning still, his government had hastened to rid its homeland of Jews. This was not because Tiso and his advisers agreed with Nazi racial theories—as Slavs, they couldn't have done so without betraying their own ethnic heritage. Instead, their policy was poisoned by a witch's brew of greed, revenge, and bigotry: greed because Jewish properties were a tempting target for plunder; revenge because many Slovak Jews were of Hungarian extraction; bigotry because, in the words of one Catholic publication, "The source of the tragedy of the Jewish people is that they did not recognize the Messiah and prepared for him a horrible and infamous death on the cross."

The parliament approved anti-Semitic laws comparable to those in the Reich but with more power for the president to grant exceptions—which Tiso routinely did for Christian converts and the wealthy. Despite those reprieves, 60,000 Slovaks (about three-fourths of the Jewish population) were deported, supposedly to labor camps. In July 1942, the Vatican alerted the government that the deportees were in fact being systematically murdered. When the Slovaks asked Germany for permission to visit the sites where the exiles were allegedly working, they were turned down. The cabinet and parliament put pressure on Tiso to suspend the transports; so too did Monsignor Angelo Roncalli, the papal representative in Constantinople, known later as Pope John XXIII. In time, the president gave in.

Slovakia's partnership with Nazi Germany was purely a marriage of convenience. The Germans exploited the Slovaks; the Slovaks gained the right, within certain constraints, to govern themselves and were initially convinced that they had backed a winner. Hitler, the bully, seemed sure to remain the dominant force in Europe. After German fortunes soured, Slovak attitudes began to change, and those who had never been comfortable with the Nazis grew more assertive.

Toward the end of August 1944, four weeks after the start of the antifascist uprising in Warsaw, the Slovak resistance mounted its own attack. There, as in Poland, the organizers hoped that the approach of the Red Army, coupled with setbacks elsewhere, would cause the Nazis and their collaborators to give ground. Instead, German forces poured into Slovakia and, in two months, suppressed the rebellion. A major reason for the coalition's collapse was the lack of coordination among its eclectic components, which included pro-Beneš democrats, nationalists, liberated Jewish prisoners, Communists, and renegade units of the Slovak military. Again, as in Poland, the Red Army was of little help, either because Stalin didn't want the uprising to succeed (as anti-Communists later insisted) or because he had other legitimate priorities (as argued by apologists for Moscow). In victory, the Germans were typically ruthless, executing thousands of rebels and sending a last trainload of Jews to Auschwitz.

With the war winding down, the Slovaks found themselves in a unique position. The Allies had demanded Germany's unconditional surrender, but they had also been persuaded by Beneš to reject Slovakia's claim to independence. This meant that in the eyes of the West, the country was still part of a unified Czechoslovakia—and therefore on the winning, not the losing, side of the war. That considerable piece of luck did not dampen the Slovaks' desire for separation. When they gathered to celebrate the war's end, they filled the air with Slovak flags and a scattering of Communist banners; the emblem of a united Czechoslovakia was almost nowhere to be seen.

THE OUTCOME OF the talks in Moscow was made public in Košice on April 4, 1945. The interim government would consist of three representatives from each of the four major Czech parties and the two Slovak ones. There were six appointees without party affiliation, plus Beneš. Although superficially equitable, the distribution of power gave the Communists virtually everything they had sought. Directly or indirectly, they controlled the prime minister and most of the key ministries. A new position of state secretary for foreign affairs was also created, to be filled by Vlado Clementis, a friend of my father's but a Communist nonetheless. His assignment would be to keep a close eye on his nominal boss, the foreign minister, Jan Masaryk.

In Moscow, Masaryk met Gottwald for the first time. The two men had in common a deep love for Czech folk songs and an instant distaste for each other. They talked for an entire afternoon without agreeing on much. Gottwald complained that the foreign policy advocated by the London-based exiles had been insufficiently pro-Soviet. This, he insisted, must change; total cooperation would be required. Gottwald said that he doubted Masaryk understood, to which Jan replied that indeed he did but would not promise to acquiesce. Summarizing the exchange in a memo to Beneš, Masaryk pointed out that when his father had been president, no one dared attack him, preferring instead to lambaste Beneš, the foreign minister. Now that Beneš was president, he, Masaryk, was in the post—and had a target on his back.

AFTER ANNOUNCING THE government program, Beneš remained in Košice, waiting for the final stages of the fighting to play out. Although the nation's acknowledged leader and back on his own soil, he was hardly in a position of command. The Soviets took charge of his security, sticking him in a house surrounded by guards and preventing him from communicating directly with London or Prague. Citing safety

concerns, Moscow refused to allow British or U.S. diplomats to accompany him on his journey from London or to join him in Košice. If
Beneš wanted news, he had to appeal to the Russian ambassador, who
passed along only what he deemed prudent. To his aides, the Czechoslovak leader complained about this degrading treatment; with the Soviets, he held his tongue.

Earlier in Moscow, Stalin had treated Beneš to a victory dinner enlivened by traditional music, storytelling, folk dances, and toasts. In
his remarks, he had emphasized the two countries' shared interest in
thwarting German ambition and disclaimed any desire to promote
Soviet-style communism across Europe. Stalin could make a visitor feel
like a king simply by telling him what he wanted to hear. That evening,
he also tried to prepare Beneš for what was to come. "Our soldiers will
be going into your country," he said. "Do not judge them too harshly;
they are tired by a long war and have become a little uncontrolled.
Anyway, Red Army men are no angels."

Indeed. Like a Carpathian storm, the army marched through Slovakia and westward into Moravia, greeting happy throngs with a hearty
"Hitler kaput!" and giving the retreating Germans a firm shove. The
Soviets, welcome as they were, often failed to heed the distinction between liberation and conquest. Relatively few were professional soldiers; most were half-trained farm boys who, having gone through hell
with poor equipment and worse food, were now eager to indulge their
appetites. As liberators, they were given as much to eat and drink as the
hard-pressed citizenry could spare. The Russians found this arrangement to their liking and urged their hosts to dig a little deeper. The
men were particularly fond of watches, fabrics, carpets, and clothes—
especially boots. They drank vodka, of course, but also wine, beer, medicinal alcohol, and, in one notorious incident, methylated spirits that
had been used by a museum to preserve animal specimens.

Their officers were little better. They requisitioned houses for their
own use and upon departure carried off all the valuables they could.
They also tried their hand at pilfering cars, making two attempts to
hijack that of the British ambassador, once from the driver alone and
once with the envoy, mightily distressed, sitting in the backseat. In

September 1945, the Russian military broke into some sugar refineries and began stealing the contents. This was too much for Beneš, who, without consulting the cabinet, ordered his own army to intervene, causing the Soviets to back down.

Far worse, in Czechoslovakia as elsewhere in Eastern and Central Europe, the men of the Red Army raped thousands of women and girls without the least sign of disapproval from senior officers. The Czechs and Slovaks who encountered such behavior were disgusted but also afraid. Not everyone reacted as my father did: "They have liberated us from lice and set onto us leeches." Instead, some sought protection by turning for help to members of the local Communist Party or by signing on themselves. In that way, Russian brutality became something of a boon to party organizers. More significantly, the Communists benefited from the fact that it was the Soviet Union—not the United States—that had liberated Prague.

AS EARLY AS the 1943 Tehran conference, it was understood by the Allied leadership that the Soviets would be responsible for securing Central Europe, including Czechoslovakia. Military planning was done on that basis. The Americans had no objections to this, and neither—at the time—did the British.

Circumstances change, however, and Churchill concluded that it might indeed make a difference which Allied army marched where. His faith in Stalin's intentions had vanished abruptly after the Big Three's brief flirtation on the shores of the Black Sea. In mid-April 1945, the British urged the United States to dispatch its forces to Prague. Having received no response after two weeks, Eden sent a second note:

> In our view the liberation of Prague and as much as possible of the territory of western Czechoslovakia by US troops might make the whole difference to the postwar situation. . . . On the other hand, if the western Allies play no significant part in Czechoslovakia's liberation that country may well go the way of Yugoslavia.

The State Department was persuaded by the argument and recommended that U.S. forces proceed to the Vltava Valley. However, Truman, just starting out in his presidency, was loath to meddle in arrangements previously agreed to by Allied military leaders. The situation changed only slightly when General Patton's Third Army, moving into Austria, required protection on its northern flank. The supreme allied commander, Dwight D. Eisenhower, asked the Soviets for clearance to send troops to southern Bohemia. This was granted, and a new understanding was reached: U.S. forces could penetrate as far east as Plzeň, some fifty miles from Prague. They did so without enemy opposition during the first week of May, setting off a wild celebration and causing impatience to build throughout the Czech lands.

From one direction, Soviet troops were heading toward the capital; from the other, U.S. forces were crossing the border. Victory was within sight, yet the ignominy of foreign rule went on. German soldiers were still standing on Prague street corners. Insulting the führer remained a crime. The Gestapo continued to round up and shoot partisans while political prisoners sat in jail, at risk of execution. It was little wonder that, in cellars and attics, people monitored the radio nonstop, hoping for word that the Germans had quit. According to foreign news broadcasts, Hitler had killed himself; his top advisers were dead or in flight; the Third Reich was collapsing; so why wasn't the enemy going home?

In the early days of May, the people of Prague and other urban centers decided to wait no longer. Acting spontaneously, they began to take back their country, ripping down German signs and replacing swastikas with Czech banners. Shopkeepers and tram conductors refused to accept reichsmarks, while German soldiers were harassed and, when possible, disarmed. On the morning of May 5, the main radio station broadcast a plea: "Come help us everyone! We are fighting the Germans!" When Nazi troops rushed the station, the previously docile city police challenged them. All afternoon, the two sides fought. Reinforced by a detachment of guards who had crossed to the radio station on rooftops, the Czechs cornered the Nazi unit and forced its surrender. Rebels also seized the loudspeaker system and telephone exchange.

Late that afternoon, a U.S. intelligence team arrived in jeeps. The commander, a Lieutenant Fodor, agreed to return to Plzeň and convey a request for assistance.

That night, the local SS commander wired his superiors that half of Prague was in the hands of insurgents, who "are fighting unexpectedly well." Tragically, the Germans were not about to put down their weapons; they needed to control the capital to protect their overall retreat. Possessing both firepower and troops, they struck back, using incendiary bombs to destroy apartment buildings and armor to break through barriers and kill as many people as possible. The rebels, expecting U.S. help to arrive at any minute, would not yield. Whole families joined in erecting barricades made of trash barrels, sandbags, torn-up cobblestones, pieces of timber, and mattresses. To retain control of the streets, they retrieved munitions that had been concealed in floors, gardens, even coffins. Women disguised as Red Cross nurses went to the railway station, where a cache of arms had sat undetected since the occupation had begun; there the women picked up baskets that were labeled "bandages" but that required, in the carrying, every ounce of their strength.

The Czechs broadcast repeated pleas for help. Churchill cabled Washington, urging that the Third Army move. Briefed by Lieutenant Fodor, Patton was eager to march into Wenceslas Square. Eisenhower informed the Soviet high command of his readiness to send his fighters east. The Russians replied: do not proceed beyond Plzeň, lest a possible confusion of forces be created. At that decisive moment, the American general acquiesced, adding only that he presumed that "the Soviet forces [would] advance rapidly for the purpose of clearing up the situation in the center of the country."

This exchange meant that the Third Army would not be going to Prague; the Russians, meanwhile, were not there yet.* The Czechs

---

* Four U.S. tanks entered Prague on May 7, but this was to convey news of the German surrender in Berlin to local Nazi officials. The Americans said that U.S. troops would not be liberating the capital. That disappointed the Germans (who were terrified of the Soviets) as much as it did the Czechs.

*At the barricades*

stayed on the barricades, fighting desperately. On May 7, the rebel leadership exhorted their followers to "stand firm and strike still harder. Let every shot find a target, let every blow avenge the death of your brother, sister, father, or mother. Tonight let all men, women, boys and girls build still more and bigger barricades which no tank can penetrate, no shell can pierce."

For a full twenty-four hours after the Nazi capitulation in Berlin, the battle raged; streets were torn apart and buildings damaged, including the Old Town Hall, where a decade earlier my parents had been wed. Before the uprising was over, some 1,700 Czechs lost their lives. Finally a cease-fire was negotiated, allowing the Germans a secure retreat. On May 9, the first units of the Red Army appeared; one witness recorded the scene:

*Russian soldiers being welcomed*

People streamed into the streets to cheer, to welcome, to embrace their liberators, asking them into their homes, offering them every good thing they had. Pretty girls covered the tanks with flowers and climbed onto the armored trucks. The Russians laughed good-naturedly and took out their accordions. The world was full of fragrance, and music and joy.

In later years, many writers, including my father, cited Eisenhower's failure to send U.S. troops into Prague as a sign of Western indifference. This is not entirely fair. Beneš had never advocated liberation by the Americans and, on the contrary, had made clear his warm relationship with Stalin. Further, the Allies had no role in planning or encouraging the last-minute outbreak of violence. Eisenhower was in the middle of orchestrating the surrender of Germany—this to relieve everyone's suffering, including Prague's. Victory was imminent only because the Soviet army, which had two million men committed to the battle, had prevented Hitler from sending more of his troops to fight in the West. The Allied effort had proceeded smoothly, moreover, because all the participants, including the USSR, had abided by their agreements. With the war in the Pacific still undecided, a decision to break faith with the Kremlin at that critical juncture would have carried extraordinary risks.

In any case, the responsibility for making policy was not Eisenhower's. The general had been ordered to destroy the German military and bring the war to an early and victorious end, not to concern himself with the postwar political balance. Still, the record is clear that Ike was prepared to unleash Patton and would have done so had the Soviets not objected. The blame for what happened in Prague rests properly with Moscow.

There is, however, little fair about the creation of national myths. Symbols matter, and some quests—however quixotic—cannot be ignored without paying a price. The Prague uprising made little sense tactically but possessed its own rationale as an expression of bottled rage, coming as it did from a people denied earlier opportunities to fight. The rebellion was not about logic but about courage and honor, or what my father referred to in the context of Munich as "national ethos." So the legend was born that the United States had turned away from the Czechs at their moment of greatest need. For years to come, Communists would exploit the perception that Americans had "sat in Plzeň drinking Pilsener" while the people's quest for freedom was drowned in blood.

That perception lingers. When anniversaries of the uprising are marked, politicians still refer to Eisenhower's failure. This is true even in Plzeň, where, as I can bear witness, the local population has preserved many of the U.S. jeeps and trucks that Patton's men left behind. In 2010, Václav Havel told me that an American liberation of Prague would have made "all the difference." Havel, whose family spent the war in the countryside, remembered the end of the conflict as a time of uncertainty. The Germans were being driven out; Soviet troops were running around with half a dozen stolen watches on each forearm; and people were popping out of the forest claiming to be resistance fighters when, in some cases, they were brigands. A Czech pilot returning from England landed his plane in a meadow not far from Havel's house. The whole town threw him a celebration; they had deviled eggs with ketchup and salad.

One of the principles outlined at Košice was that the new Czech and Slovak armed forces would be trained and equipped along the model of the Red Army; that meant, in practice, that the Russian-based exiles would form the core of the new military while the soldiers and airmen who had fought with the British would be shunned. The Communists wanted a monopoly on wartime heroes and so redefined the London-based military as a tool of capitalist oppression. Within a few years, the majority of the men who had fought so bravely with the RAF were either forced again into exile or—as with the pilot who had been feted by Havel and his community—in jail.

# 24

# Unpatched

In July 1945, I returned to the land of my birth, flying across Europe in the belly of an RAF bomber. I was eight, my sister, Kathy, only three; we huddled between my mother and my cousin Dáša, now seventeen. The seats—actually hard benches—were in bays where the air force usually kept bombs. The noise was deafening, the plane rattled and shook; many of the forty or so passengers became ill, and I was petrified of flying for years afterward. Our stomachs churned when the pilot zoomed in low over Dresden, the German city destroyed by Allied firebombing earlier that year. Inexplicably, the U.S. Air Force had dropped 150 tons of munitions on Prague at the same time; no military targets had been struck, but five hundred people had been killed. Evidently the pilots had mistaken the Czech capital for Dresden.

My father, who had returned in May, met us at the airport. He was distressed by how far from the reception facility the plane had been directed to discharge its passengers. The British pilots were allowed to use only the most remote areas, a worrisome sign of how pervasive the Soviet presence had become. But at least we arrived safely; two months later, a similar flight bearing returnees crashed, killing all aboard.

When the war ended, Dáša had been at sixes and sevens about what to do. She had spent the previous semester at a school in Wales specially established for Czech students. After being pressed to speak English for several years, the youngsters were now encouraged to brush up on their Czech in preparation for the return home. Understandably confused, they invented a language that was half one, half the other, calling it "Czechlish."

When the time came to go, Dáša hesitated. Uncle Honza and Aunt Ola invited her to remain with them in England rather than face the uncertainties of postwar Prague, yet one of those uncertainties was never far from Dáša's mind. Like other exiles, she had made nerve-racking and often frustrating visits to the Red Cross to discover what she could about the fate of her family. There had been no letters for several years. News reports about Czech and other Jews were horrifying. One sad day, she learned of the deaths of her mother and sister, but she was also shown a list of survivors that included the name Rudolf Deiml. That settled the issue; she would go back to Prague to be there when her father showed up. "I had to go and wait for him," she told me much later, "because we knew that everyone else was dead."

IN PRAGUE, OUR family was provided by the government with a flat on the second floor of a seventeenth-century house overlooking Hradčany Square. The apartment was spacious, with large bright rooms, a fireplace, nice furnishings, and ivy-covered balconies. Dáša had her own bedroom; I shared one with Kathy. I adored the apartment but was unsure, at first, what to make of Prague. Walton-on-Thames, where I had had friends and playmates, had been pretty and green. Here, I knew no one. The streets were often too crowded for my comfort, and then there were those Russian soldiers.

Still, it was not long before the city had me under its spell. Across the street was a small park dedicated to the fourteenth-century Saint John of Nepomuk, a man as revered by Roman Catholics as Jan Hus is by Protestants. While Hus had been martyred for challenging the Church, John had been thrown off Charles Bridge for defending the Seal of the Confession against the sacrilege of a secular ruler. In statues and portraits, the holy man's head is typically encircled by a halo of five stars, representing the heavenly witnesses to his drowning.

When I wasn't in the park, I was happily wandering across the square (actually a long rectangle) to the famous castle where Beneš and his wife now lived. The guards there were in fine uniforms, and nothing was more amusing to an eight-year-old than to make odd faces at the men in hopes that one would smile, which they never did.

During the war, German engineers had confiscated more than 14,000 church bells, intending to melt them down into artillery and tanks. The faithful now rushed to return them to their rightful towers and spires. Prague had seen fighting, especially in those final days, but most of the glorious baroque architecture, the ornate palaces, and the slate-roofed apartment buildings remained intact. The pavement that had been torn up was soon repaired, and trams began again to rumble through the rabbit warren of streets from Old Town to New. The various shops retained their distinctive hand-wrought signs, separating the boot maker from the apothecary and the butcher from the baker. Whether from one of the bridges or the castle heights, I enjoyed gazing down on the flowing water and swooping gulls, at the anglers in their small boats and the cargo vessels carrying who knew what to who knew where. Every evening at sunset, lamps were aglow amid the trees and blossoming shrubs on the river banks. This, I thought to myself, is what my parents had been talking about when they looked forward to returning home.

Of course, I didn't understand that Czechoslovakia had been through an ordeal that had changed it forever. Three hundred and fifty thousand of its people had died, including roughly 80 percent of the Jewish population. Tens of thousands of homes had been destroyed. Many of the larger factories had been bombed, and the country's rail and highway network had been ripped asunder. Food was scarce. In the capital, the intense fighting of the final days had left its mark. On the streets, women who had survived the concentration camps wore long sleeves to cover scars and tattoos. The new government moved to reclaim offices and ministries that still "stank of the Nazis." The Gestapo torture chamber, where Ata Moravec and thousands of others had suffered, was preserved, its guillotine caked with blood, draped now by a Czechoslovak flag and small wreath. The ethnic German and Hungarian minorities, once welcome participants in Czechoslovak democracy, faced the prospect of expulsion by presidential decree. From the top down, the victors rushed to punish wartime collaborators, thus advancing the cause of justice but also generating abuses and creating opportunities for political mischief.

In his writings, my father described a country divided among returning exiles from London and Moscow, resistance fighters, "sit-it-outers," "comrades" (who talked the most), and former concentration camp inmates (who said the least). So much had happened that the sense of national solidarity had all but drained away. Too many people had grown used to taking orders. The Czechs who had survived the occupation resented their countrymen who had been "safely out of it" in England. Many of the exiles who had served under arms questioned the bravery of those who had remained at home. The gulfs separating these groups, lamented my father, "were deep, always emotional, sometimes rational, and rarely bridgeable."

As in Walton-on-Thames, my father walked me to school. The difference was that, in English schools, I had thrived. In Prague classrooms, I spent most of my time in the corner. When my parents asked why, they were told I had been arrogant. How so? My teacher said I had complimented her on a dress that she was wearing, a simple courtesy in England but too familiar an utterance for a child to make in the more formal atmosphere of Czechoslovakia. At least that is how I remember the story; in any case, the school was too strict for my taste.

The Foreign Ministry, where my father worked, was only a few blocks from our apartment, housed in the imposing Černín Palace. Years later, I would have the opportunity to compare the palace to the U.S. Department of State. From the outside, Černín is a striking example of seventeenth-century architecture, while the State Department resembles an oversize discarded box. A visitor to the palace is greeted by a massive hall with vaulted ceilings, exquisite tapestries, antique furniture, and a dramatic sculpture of Hercules slaying the Hydra. The State Department welcomes its guests with metal detectors and a security desk. To be sure, the U.S. diplomatic reception rooms are gorgeous, but they are hidden away on the eighth floor and used only for special occasions or tours. Both buildings offer a spectacular view. The State Department overlooks the Lincoln Memorial, while from its Czech counterpart one can see a historic church, wherein reside the remains of Saint Starosta, a Portuguese princess who rebuffed the advances of her unwanted husband by—with God's help—growing a beard.

My father assisted both the foreign minister, Jan Masaryk, and his deputy, Vlado Clementis. Each of these men had a person assigned to public relations, and there were two secretaries, one who handled business in Czech and the other in Slovak. The office manager was a veteran civil servant of "frail stature [with] red cheeks, thin lips, a pointed nose, mousy hair and small grey eyes." That flattering portrayal comes courtesy of Hana Stránská, a young woman who had been on my father's staff in London and whom he had recruited to help out in Prague. Stránská, twenty-seven years old, worked mostly as a translator of English and also handled the overflow of paperwork in Czech.

My father's duties included the organization of what would become a rapidly growing Foreign Ministry and the oversight of day-to-day political activities, a heavy burden given that Masaryk spent much of his time abroad. Dealing with important visitors also consumed considerable energy; among the guests that busy summer were the two military icons of the West: General Eisenhower and Field Marshal Bernard Montgomery. My father was entrusted with these responsibilities in part because he was one of the few who enjoyed a good relationship with both Masaryk and Clementis. The two diplomats, though thrust together professionally, were barely of the same species. Unlike the informal Masaryk, Clementis was habitually serious and businesslike, with an intellectual and ideological commitment to communism. Masaryk disliked ideology of all kinds, thinking that it made people forget their humanity in the vain pursuit of foolish goals. As a child, I knew both of them: Masaryk with his round face, big belly, and joking manner, Clementis with his stern eyes and deep voice.

When asked to help manage the Foreign Ministry, my father was told by his superiors that the job would be temporary; in fact, Clementis urged him to stay longer than planned. Despite his relatively young age (thirty-six), my father was in line to become an ambassador. The logical assignment, given his experience, was a return to Belgrade as our country's minister to Yugoslavia. Dáša chose to remain behind

*Masaryk and Clementis, 1946*

with a great-aunt to finish school and await news of her own father. The rest of us, once again, would be on the move.

For months, Dáša clung to the hope that Dr. Deiml was indeed still alive. There were many rumors, including the possibility that former prisoners had been sent to the Soviet Union. Finally, in February 1946, she received a letter from Jiří Barbier, the carpenter who had known her family at Terezín and who had accompanied Rudolf on his final journey. Barbier, who had obtained Dáša's address from the Red Cross, apologized for being the bearer of shattering news but thought that perhaps she had already discovered the truth. She had not.

Dáša's distress was unknown to me then because of my age, self-absorption, and the fact that we had spent just two months in Prague before leaving for Belgrade. Looking back, I can barely conceive her pain; but I have come to realize that she was not alone among the members of my family in experiencing and wrestling with grief.

MY MOTHER WAS not ordinarily one to hide her feelings. If upset, she would say so; if sad, her tears would flow. But when we returned to

Prague after the war, I am convinced in hindsight that she made a courageous effort to conceal her pain. She had loved her mother and sister dearly, but I saw no sign of the agony she must have felt. My father, too, showed no outward signs of mourning. I did not wonder about this, for I was told only that my grandparents had died. Alfred had succumbed before I was born; Růžena, Olga, and Arnošt were names I barely recognized; I could not remember ever calling anyone Grandmother or Grandfather.

Fifty years later, when I learned the circumstances of their deaths and those of so many other relations, I again wondered what my father had felt. I could imagine the depths of his sorrow but had no evidence. Now I have. Rummaging through the boxes in my garage, I found a manila folder containing a document, 123 pages long, triple-spaced with narrow margins. The text is typed neatly with a few revisions in pencil. This was my father's attempt at a novel. He had mentioned it to me once, but I had not taken him seriously and in any event had heard no more about it. A professor and historian, my father had a genius for making the past come alive, but he also dealt in facts. In his books and articles, he wrote to develop a thesis and prove a point. Why would he try his hand at fiction? What did he care about profoundly enough that he felt compelled to write but could not do so in his customary way? I picked up the folder, removed the paper clip, and turned to the opening page.

"The plane was about to land," the story begins. Peter Ptachek,* a young diplomat, is returning to Prague after six years in London, where he had directed wartime broadcasting for the Czechoslovak government in exile. Unmarried, he fantasizes about a long-awaited reunion:

> He will slip quietly through the backyard and open cautiously the door. Careful, one tile in the corridor is loose and makes [a] noise. . . . There she is. Bent over the oven. . . . He covers her temples with his hands, and. . . . Maybe she will not be at home. She just went to buy something for supper. Maybe she is in the

---

* My paternal grandmother's maiden name was Olga Ptačková.

country and [has] left a letter behind. It will be under the second glass from the right side of the kitchen cabinet. So it always was in the old times.

Once on the ground, Peter takes a taxi from the airport to the Alcron Hotel, where returning officials have been assigned rooms. His route takes him past the castle and cathedral, down the steep street named for Jan Neruda, across the bridge toward Wenceslas Square. "Centuries have passed and centuries will pass," thinks Peter to himself. At the hotel, he is greeted cordially by the concierge, who had probably spent the previous six years repeating, just as courteously, "Heil Hitler." In the lobby, he overhears snatches of conversation among other return-ees: "Yes, I have found them all" or "I haven't found anybody."

Peter takes streetcar number 1. Disembarking, he walks—then runs—to the familiar door. Behind it, instead of his mother, he finds a stranger, who says that she has been living in the house for three months; before that a German family had stayed there. She has no in-formation. Stunned, Peter decides to hike to the home of his mother's younger sister, Martha, who lives with her husband, Jan, and two chil-dren. As he walks, he thinks of the many Friday evenings before the war when he had joined Martha's family in playing chamber music. He arrives and knocks. The door opens—again, not a familiar face but a stranger:

Peter introduced himself and asked about his uncle.

Yes we knew him very well, answered the man. We were good friends of both Jan and Martha. We also knew your mother, poor soul.

What happened? Where is she? muttered Peter in fear.

My dear friend, if I have to be the first person to tell you. She is no more. It happened in May 1942. They took her away, and two weeks later Jan got a note about it. Then came Jan and Mar-tha's turn. Before they came for them, Jan asked us to move into their apartment.

What happened to the boys?

God in heaven. They took them too, two days later. I have got some pictures. Would you like to see them? Your mother's picture, too.

No, I don't think I would. Not now. I'll come later.

Darkness of the night swallowed Peter's sinking body. Heavy, slow steps carried him through streets and squares. Prague, his cradle, suddenly became to him the strangest city in the world. Charles Bridge was leading him to the bank of the unknown. As he was crossing it a woman, standing under the statue of Christ, emerged and asked: Would you buy me a drink, darling?

Peter went on and looked down to the river. The life on and under the bridge obviously continued unchanged, he thought.

It was well after midnight when he reached the hotel. He tottered to his room. Tense and exhausted, he fell on the bed. His face dropped into the grave of the pillow. "God give me one, I beseech you, give me at least one," he sobbed. Stones of tears were falling through a hole in the pillow cover. The war was over. It left behind many holes. Some could be patched up. Some couldn't.

So there it was. My father was no stoic; the opposite, in fact. The emotions were there and had been tugging at him for years. Probably he had begun writing with an idea to publish, but he must have concluded that it was not something he could do. Why had his mother and cousins been taken away?

Later in the story, Peter goes to the countryside to visit his boyhood home. The door is answered by a remarkably short man. Peter explains himself and asks if he might come in and look around. The man shrugs in embarrassment, then opens his mouth and makes unintelligible sounds. After a moment, Peter realizes that his host is deaf and dumb. Following some awkward attempts to communicate, Peter politely takes his leave. "I am grateful," he thinks to himself as he walks away; the encounter must have been a sign: "The past was to be deaf and dumb to him. It was to be neither heard, nor spoken."

# 25

# A World Big Enough to Keep Us Apart

During the war, Beneš had sought diplomatic backing for his plan to banish ethnic Germans from Czechoslovakia save only those who could prove their resistance to fascist occupation. In 1944, he submitted a memorandum to the great powers (United States, Great Britain, and the Soviet Union) proposing to expel about two-thirds of his country's Germans. Those transferred would be entitled to take with them all movable property and receive compensation for the rest. He pledged that the process would be completed within two years. "Our people," he declared, "can no longer live in the same country with Germans."

This policy of confiscation and expulsion—embodied in what became known as the Beneš decrees—reflected a passion that had seeped into the bones and blood of virtually the entire Czech population. Throughout their lives and through the inherited memory of their nation, Czechs had shared space with their German neighbors. Each people had frustrated the ambitions of the other, and the two had maintained separate identities despite mixed marriages, personal friendships, commercial ties, and countless common experiences. It was never inevitable that this intimate relationship would terminate in war, but the war had happened and dug a deep well of bitterness. In May 1945, most Czechs had no interest in defining a new relationship with Germans; they wanted to *end* that relationship.

When, on May 17, one of the major democratic parties held its victory celebration in Prague, our friend and former neighbor Prokop Drtina was among the principal speakers; he would soon become minister of justice and a central figure in the new government. For the event, in front of a large and enthusiastic assemblage of future voters, his political antennae were fully extended. Writing later, he confessed that denouncing Germans to a Czech audience in 1945 was "a demagogic opportunity" too favorable to pass up. Getting rid of Germans, he said in his speech, "is the historic task of our generation. . . . Our new country cannot be built except as a pure state. . . . One of us must leave—either the Germans or us—and since this is a Czech country and we are the winners, they are the ones to go!"

The Communist leader, Klement Gottwald, stoked the same fire by suggesting that expulsions were merited for reasons deeply rooted in the past. "Now we will definitely compensate for White Mountain," he promised. "And not only that; we [will] reach even further into our nation's history: by confiscating the property of the Germans, we will rectify the mistakes committed by the kings of the Přemyslid dynasty who [invited] . . . German settlers: now they will be expelled from our land once and for all." These and similar declarations evoked ovations of such foot-stomping rowdiness that any talk of orderly procedures and equitable compensation was soon forgotten. The assurances that had lent a veneer of reasonableness to Beneš's diplomatic memos were cast aside in the first flush of victory.

During the late spring and early summer of 1945, an unknown number of Germans were shot, lynched, or beaten to death. The citizens of Brno gathered as many as could be found, roughly 20,000, and forced them to march toward Austria. Due to a lack of food, an outbreak of dysentery, and an almost total lack of organization, an estimated 1,700 died. There were other abuses. According to one report:

In Nový Bydžov, 77 captured German soldiers were executed; in the mountain town of Špindlerův Mlýn, 30 German civilians were murdered along with 50 soldiers; near Přerov 265 were killed, including 120 women and 74 children younger than fourteen years of age. In Postoloprty, a Czech investigative team later unearthed the corpses of 763 Germans who had been concentrated in the area and liquidated.

Due process, especially in the first weeks, was widely neglected. In some cases, alleged collaborators were simply killed; in others, they were hauled off to makeshift prisons to be interrogated and tortured. In many towns, the maiming of local Germans became a spectator sport, as crowds gathered to jeer. To the local guardians of security, the rough treatment was not lawlessness but justice. Germans were given the same rations that Jews had been allotted during the war and were prohibited from entering hotels, restaurants, and shops. They could no longer speak their language in public. In some towns, they were required to wear specially colored armbands; in others, swastikas were painted on their backs. Their schools were closed and many of their businesses seized. Czech women who had a reputation for fraternizing with Germans were humiliated. Not surprisingly in this environment, horrible mistakes were made. In early May, an elderly man was beaten to death in a Prague hospital after he cited as his hometown a village in the Sudeten region. The killers assumed that he was German when, in fact, he was Czech. In any case, he was not causing harm to anyone from his hospital bed.

A few weeks after the war's end, Hana Stránská (the twenty-seven-year-old who worked in my father's office) went on an excursion to the American-occupied resort town of Marienbad. She found the streets crowded with easygoing, wisecracking U.S. soldiers and was upset to see some of them walking "arm-in-arm with dirndled Sudeten-German Frauleins." Hana could not forget the camp survivors she had seen on the streets and trolleys, with their haggard faces, scarred bodies, and hair just starting to grow back.

As she walked along, her senses were attracted by the sound of a Czech love song and the sight of a group of men singing and dancing in the middle of the street. These were not, she soon realized, ordinary

*Germans surrounded by angry Czechs*

singers; they were, in fact, German prisoners being ordered to perform by a contingent of Czech soldiers. Every so often, when the Germans stopped or lost their place, the soldiers screamed at them to resume. Hana smiled.

A U.S. serviceman, standing nearby, was not so content. He yelled at the Czechs to stop. "The war is over, so halt your bullying!" he shouted. Some of his buddies agreed.

That was too much for Hana. "How dare you?" she demanded of the American. "Where in the States are you from, anyway?"

"Mississippi," he said.

"Miss-iss-ip-pi?" said Hana, drawing out the syllables sarcastically. "I see. So you've come all the way from Miss-iss-ip-pi to tell us in Czech-o-slo-vakia how we should treat our traitorous Nazi scum, our prisoners. You find it too much if we humiliate those dregs of humanity by making them sing Czech folk tunes? Where have you been all this time? Do you know what they have done? Do you know they tortured and killed millions? Or haven't you heard? Or maybe," said

Hana, drawing a deep breath, "you sympathize with them because you float dead Negroes down your river?"

Her words caused a commotion: furious and indignant soldiers gathered round; Hana's own phrase was thrown back at her: "How dare you?"

Another American intervened. "She's absolutely right," he said. "I've just come from those camps where we've been liberating the inmates. You should see it. Besides, these Germans are not being harmed in any way." Turning to the first soldier, he said, "Let's you and I keep out of it, okay?"

Like many Czechs, including my parents, Stránská had lived with Germans most of her life. She had known them at school, spent summers with them, learned their language, shared dinners and social occasions. But the war had changed her. She later summarized her thoughts on that day in Marienbad:

> I will not take a German's word for it that he is innocent. Who would admit guilt as colossal as that? As far as I am concerned, they were guilty until proven innocent. And they would so remain in my eyes for the rest of my life. I vowed that I would never utter a word in German unless no other language would serve . . . never willingly set foot in either Germany or Austria ever again . . . not buy German products, large or small . . . not talk to a German or Austrian, not even to ask the time of day. The idea that a German might smile at me gave me goose-pimples. The world is big enough to keep us apart.

The first time I heard this story, I thought to myself: who is right—the soldier who intervened or the one who said it wasn't America's business to try to sort out disputes that arise among others? It is a question that—in this context and many more recent ones—I still ask.

As reflected in Hana's account, the U.S. troops who occupied Plzeň, Marienbad, and other parts of southwest Bohemia generally did not allow the abuse of Germans. The Soviets, who were in control of the

rest of the country, encouraged and joined in it. This discrepancy filled
many Czechs with resentment—toward the Americans.*

My father, a student of history, came to appreciate the remarkable
and deeply felt efforts made by Germany to atone for the most unspeakable chapter in its history. My mother's reaction was comparable
to that of Hana. She never wanted to hear a good word on Germany's
behalf. Years later, when I first told her of my love for a man named Joe
Albright, she asked me to repeat the name. When I did, she sighed,
"Thank God it's Albright, not Albrecht."

OFFICIALLY, THE GOVERNMENT'S plan called for the division of ethnic
Germans into three categories: (1) collaborators and opportunists; (2)
those who had been arrested or persecuted by the Nazis; and (3) others.
People in the first category were required to leave, those in the second
could stay, and those in the third could reapply for citizenship. By presidential order, 270,000 farms covering more than six million acres were
confiscated.

The Nazi racial laws had been hard to implement because people
of even arguably one blood were the exception, not the rule; the Beneš
decrees ran into a similar problem. Many Czech and Sudeten families
were culturally mixed or had bounced back and forth between the
two nationalities based on which was more convenient at the time.
Even Hana Stránská, who had attended German schools as a child,
had to go to great lengths to prove herself a Czech. Less successful
was a man, Emmanuel Goldberger, who in 1942 had escaped from
a concentration camp and ultimately joined the Czechoslovak exile
army. Because he had been raised in a German-speaking family, the
Defense Ministry denied his application to return home. Goldberger
was accused of having chosen a Czech identity during wartime "in
order to remain hidden and avoid attention," not out of "authentic"

---

* I should note that U.S. soldiers were otherwise wildly popular. Unlike the Soviets, they
had their own supplies and could be generous in handing out such novel items as Her-
shey bars and cans of flavored soda. For a time, the most popular tune in Czechoslovakia
was "Chattanooga Choo Choo."

national loyalties. The fact that he was Jewish was not considered a sufficient extenuating circumstance.

To its credit, the Beneš government soon took steps to curb the excesses. It called for an end to extralegal violence, jailed thousands of people on charges of plunder and theft, and established a structure for adjudicating cases of alleged collaboration. In mid-June, Beneš declared that the transfer policy would not go forward except with international cooperation and in an organized way.

IN JULY, NINE weeks after V-E Day, the leaders of the United States, Great Britain, and the Soviet Union met in Potsdam, a riverside city once home to Prussian royalty, fifteen miles southwest of Berlin. Of the trio that had convened in Yalta five months previously, only Stalin would remain through the conference. Roosevelt's place was taken by Truman; Churchill had to excuse himself after a few days to return to England, where elections were under way. To his chagrin, British voters decided that, with Germany now crushed, they no longer had need for his services.* His chair in Potsdam was filled by Clement Attlee, the comparatively colorless head of the Labour Party. After discussing the future administration of Germany and Austria and the organization of war-crimes trials, the leaders found time to approve the "orderly and humane" transfer of ethnic Germans from Czechoslovakia.

The Allied governments accepted Beneš's basic argument, but they also asked him to slow the pace. Prague should not be deporting Germans until the occupation authorities were prepared to receive them, a period of waiting that would consume several months. The trains finally began to run in December. The deportations were implemented by the army, which secured the perimeter of villages one or two at a time, then notified residents that they would have to leave. Under the rules specified by the Allies, families were not to be divided and food and clothing allowances were to be adequate. For the deportees, that

* When, in the first chapter of his history of World War II, Churchill quoted Plutarch ("Ingratitude towards their great men is the mark of strong peoples"), he did so in reference to France; it is possible that he had another country in mind.

still meant leaving behind their land, dwellings, livestock, farm equipment, and the graves of their ancestors. There was no right of appeal. During the yearlong exodus, more than 1.2 million were sent to the American Zone of Germany and another 630,000 to the Soviet Zone. Several hundred thousand more had been pushed out before the program officially started. In the end, only a quarter-million Germans remained in Czechoslovakia, less than 10 percent of the prewar level.

IN JUSTIFYING HIS policy, Beneš argued that the conditions existing prior to 1939 could not be repeated after the war. The Sudeten minority had served as a pretext for Munich, which had in turn destroyed the republic and imperiled the very existence of the Czech people. Further, Sudeten support for the Nazis had been enthusiastic and widespread; such a population could never be at home in Czechoslovakia. Finally, the German presence was a provocation; if it remained, people would likely be killed out of a desire for revenge.

In defense, most Sudetens said that they had been unaware of the extent of the Nazi atrocities. They were, they insisted, just ordinary citizens—butchers, farmers, shopkeepers, tailors, factory workers. They had not known about the death camps; they had never liked the Nazis; they had become party members out of fear; they had only been protecting their families; they could not fairly be blamed. The Czechoslovak government responded that it was impractical to evaluate the behavior of every individual. Lists were drawn up of Germans with proven antifascist credentials; they could stay, but others were required to go.

The expulsion of ethnic Germans remains controversial. Was it a legitimate response to the crimes of war or an overzealous reaction grounded in vengeance? Was it flawed in conception or merely in implementation? Did it help to make Czechoslovakia a better country?

Certainly a case could be made for deporting individuals who were shown, after an objective legal process, to have joined in persecuting their neighbors. Under the Beneš policy, however, a German and a Czech who had acted in the same manner would be judged differently. Passive obedience on the part of a Czech or Slovak was acceptable; in a German, it was not. Undoubtedly, many of the Germans who were expelled deserved their punishment, but many who were not culpable also lost their homes.

My views on Czechoslovak policy in this period are colored by my experiences as an adult far removed from the passions of the day. As a diplomat, I sharply condemned ethnic cleansing in Central Africa and the Balkans and championed the creation of a war crimes tribunal to ensure that individual rather than collective responsibility would be assigned for humanitarian crimes. Collective punishments, such as forced expulsions, are usually rationalized on the grounds of security but almost always fall most heavily on the defenseless and weak. According to the Czechoslovak government's own figures, 80 percent of the Germans targeted for eviction were women, children, or elderly. It seems revealing, as well, that under the Austro-Hungarian Empire, the Czechs had been among the leading advocates of minority rights. Beneš had personally helped establish the legal protections that were in place under the League of Nations. That devotion to principle was no doubt sincere, but it had been consumed in the fire of Nazi atrocities.

Legal philosophers have long debated whether it is better to have a system in which some who are innocent are punished along with the guilty or one in which the innocent are held harmless but some who are guilty escape. I tend to favor the rights of the innocent, but my parents—whose values I inherited—supported the expulsion policy. When my father wrote about it, which he did only briefly, he admitted that its execution was "sometimes accompanied by excesses of brutality which no decent man can condone." He blamed the abuses on the Communists, but in truth the mob actions were a product of passion, not ideology; non-Communists were equally to blame.

It would not be until the 1990s and the presidency of Václav Havel that the Czech people would be challenged to revisit this chapter in their history. Speaking in 1992, Havel said, at considerable expense to his personal popularity and political standing:

> The disease of violence and evil spread by Nazism ultimately afflicted even its victims. . . . We accepted the principle of collective

guilt and instead of punishing individuals, opted for collective revenge. For decades we were not allowed to admit this, and even now we do so with great reluctance. But just as the Germans have been able to reflect upon the dark sides of their history, so must we.

The most damaging inference one might draw from the Beneš decrees is that the presence of a German minority within Czechoslovakia was a primary cause of World War II. It was not. The Sudeten German situation was exploited by Hitler, but that was his responsibility; it cannot be blamed on T. G. Masaryk's dream of a viable multinational country. Without Hitler and the economic setbacks that drew people to his cause, the Czechoslovak Republic could well have succeeded. Over time, the presence of an industrious (if still occasionally quarrelsome) German minority might have been recognized as a major asset. I underline this point because of its relevance to an understanding of history; even more because multiethnic cooperation remains critical to the success of democracies everywhere. Defending this principle matters if one believes, as I do, that the emergence of Czechoslovakia in 1918 was a cause for celebration not because the new country was Czech and Slovak but because it was democratic—and that Munich was a tragedy not because Germans triumphed over Czechs but because the Western democracies lacked resolve when faced by the evil of a racist totalitarian state.

BEFORE THE NEWLY reconstituted government could devote itself fully to the future, it needed to settle past accounts. That meant prosecuting those of whatever nationality who had been guilty of war crimes. To this end, a network of national, regional, and local tribunals was created to hold people responsible for actions taken (or not taken) during the conflict. The list of potential violations included everything from murder and torture to voicing support for the enemy. The quality of the tribunals varied widely. Some were professional; others lacked trained personnel and made little pretense at proper procedure. Many of the alleged violations, such as collaboration and fraternization, were loosely defined. There was no mechanism for ensuring that legal interpretations and penalties were consistent. Because emotions ran so high, especially at the outset, public opinion had an influence on judges. There were instances, as well, of jurists using their authority to confiscate property that then found its way into the hands of their family members and friends.

In the first weeks, tens of thousands of people were arrested. The prisons, ill equipped and unsanitary, became more wretched as new inmates were wedged in. To settle cases quickly, neither defendants nor prosecutors were given the right to appeal and death sentences were carried out within two hours of judgment or, if requested by the condemned, three hours.* Beneš had the authority to grant clemency, but with so brief a window of time, the option was rarely used. As a result, the Czechs executed almost 95 percent of 723 condemned prisoners, a higher rate than any other European country.† This created another problem: finding qualified hangmen. Professionals were few because those who admitted to having been employed in wartime (by the Nazis) were likely to be hanged themselves.

In the postwar environment, overheated as it understandably was, individuals had a power not ordinarily possessed in a democracy: to destroy others through political denunciation. Whether or not charges were truthful, the accused were placed on the defensive and could be detained for long periods, interrogated, beaten, and deprived of property. This meant that justice could be manipulated by people who were angry or opportunistic enough to cause hardship for unpleasant acquaintances, troublesome business partners, local rivals, or inconvenient spouses.

Even judges trying to be fair would find it hard to discern truth when neighbor was denouncing neighbor based on rumor, hearsay, or claims that could not be verified. How were they to define the appropriate limits of guilt by association? What about the friends and family of collaborators or people who might have shown weakness once but at

---

* In the protectorate, under Nazi rule, those sentenced to death had been granted but ninety seconds to address the court.

† The Slovak justice system, which was separate from the Czech, did include the right of appeal. Its rate of execution was 41 percent.

other times had stoutly resisted pressure? What about people who had given damaging information while under torture or because their loved ones were threatened?

In one instance a man confessed to working with the Gestapo, helping to track down anti-Nazi partisans, and stealing property from Jews. Yet the same man had sheltered a Jewish woman in his apartment, refused to betray prison escapees, and secured the release from jail of resistance leaders who later served in the Slovak government. Both villain and hero, he was sentenced to prison for thirty years.

THE POSTWAR ADMINISTRATION of law in the Czech lands was uneven and messy but no more flawed than comparable efforts in neighboring countries. As tempers slowly cooled, prosecutors began to dismiss more cases than they tried. There were many acquittals, and the pressure for long sentences and more trials diminished, especially after the completion of the most prominent cases. In these highly publicized instances, at least, it is reasonable to conclude that the interests of justice were served.

Among those receiving a death sentence was the commander of Terezín at the end of the war, Karl Rahm, the Nazi who had sent so many prisoners to the gas chambers. The trial of K. H. Frank, the Sudeten German who had worked closely with Hitler, was broadcast live over the radio. The witnesses to his execution included seven women from Lidice. Six of the Gestapo officers who had participated in the massacre of that village were also condemned to die. Fittingly, the prosecutor in these cases was Jaroslav Drábek, a friend of my father from before the war, a member of the Resistance during the conflict, and a survivor of Auschwitz.

In April 1947, the National Court in Bratislava found Father Tiso guilty of treason. Beneš favored life imprisonment but deferred to his cabinet, which—by the margin of a single vote—recommended execution. Tiso was hanged, then buried in secret so that his grave would not become a Slovak shrine.

Karel Čurda, the parachutist who had betrayed the assassins of Heydrich, was caught trying to flee during the last days of the war. Neither the reward he had received from the Nazis nor his German identification papers could shield him from trial. When the judge asked him how he could have informed on his friends, he replied, "I think you would do the same for a million marks, your honor." Precisely two hours after his conviction, Čurda, unrepentant and still telling jokes, met his doom.

As for Konrad Henlein, the Sudeten German leader who had prayed for the day when all Czechoslovakia would join the Reich, there was no need for a trial. Captured in Plzeň by the U.S. Army, he begged the Americans not to turn him over to the Czechs. When it became clear that his request would be denied, he cut his wrists with broken glass.

# 26

# A Precarious Balance

Czechoslovak democracy died with Munich and was resurrected when Beneš and his government returned to Prague. In less than three years, it would be buried again. Was this second death inevitable, or, with wiser leadership and more outside help, could democratic Czechoslovakia have survived?

I posed this question to Václav Havel, who replied that survival had indeed been possible. "The Yalta line was meant to be military, not political," he asserted. "Beneš thought that the country could serve as a bridge between East and West, but he did not frame this idea properly. In any event, he was a good diplomat and an excellent foreign minister for calm times, but he was not the best to lead at a moment of high drama."

Stalin would not have agreed that the Yalta line had been intended to be military only. In his view, where the Red Army had gone the Communist system was licensed to set up shop. Czechoslovakia would provide a testing ground for the two perspectives. Unlike the rest of Eastern and Central Europe, the republic remained free to hold meaningful elections.

Appealing to voters for support was not an unwelcome prospect for Communists in postwar Czechoslovakia. After all, their ideology offered a remedy to every ill—or so many believed. Old Europe had been held back by the artificial divisions of class and nationality; the Nazis had put people into boxes according to religion and race.

The Communists, by contrast, spoke of one another as comrades and claimed to represent workers of every background. This egalitarian mind-set meshed well with the image that Czechs and Slovaks had of their own past rebellions against German and Hungarian nobility. What could be better, after the horrors of the Second World War, than to create a society free from the scourges of poverty and privilege? Surely, the hour of the worker had finally come, when those who labored with their hands would receive their due while those who profited from the sweat and calluses of others would be brought low.

Joining the party was a way to connect with a movement pushed along by the currents of history; a means—according to a woman drawn in at the time—to achieve "victory over one's smallness." The Communists also made a claim on voters' respect. Had not their partisans been the bravest in facing down the Nazis and had not Stalin stood with the country through the critical tests of Munich? Had not the Red Army liberated Prague? Following years of Aryan savagery and Anglo-Saxon indifference, did it not make sense to look for salvation from Mother Russia, the unofficial capital and protector of Slavs?

However, this glorious new world could only be brought about through political change, and for that, discipline would be required. The workers' revolution could not arrive until its enemies—rapacious capitalists, reactionary politicians, and the decadent bourgeoisie—had been defeated. Victory would emerge as a result of meticulous planning and preparation, everywhere from the smallest precinct to the largest city. This would demand firmness and, for nonbelievers, a liberal dose of reeducation. Even while the war had been under way, Czechoslovak Communists had set out to become the country's most thoroughly organized political faction.

The program announced in Košice in April 1945 had called for the creation of administrative committees from the local to the national level. By sweeping away the old governing structures, Communists were able to secure disproportionate representation in the new ones. Gottwald instructed his aides to use these committees "to rebuild the very foundations" of the state. The party's control of vital ministries allowed its operatives to penetrate deeply into every segment of society.

This infiltration was made easier by a general political climate favorable to centralized rule; few voices were raised on behalf of capitalism. The new government moved swiftly to nationalize banks, mines, insurance companies, public utilities, and major businesses. These measures encountered little resistance because, in most cases, the previous owners were pro-Nazi and therefore in no position to protest.

In hindsight, it is reasonable to say that the nation's fate was decided in its villages. The Communists were active everywhere, helping one another to intimidate foes and mold public opinion. One of the characters in my father's unpublished novel is the owner of a Kostelec bookstore who had gladly removed copies of *Mein Kampf* from his front window, only to be strong-armed into replacing them with biographies of Lenin and Stalin. Thinking back to the golden era before the war, when great literary works had enjoyed pride of place, the shopkeeper says regretfully:

> That window used to be my greatest joy. Every morning at eight when I opened the store, I would stand for a minute in front of it, and my heart smiled. I liked to think it was a photograph of me. Now I am ashamed.

The key to the Communists' success was that when they lost a local election, they would concentrate their resources and try again. When they won, they employed every means at hand, legitimate or otherwise, to remain in office. They also used their agents within the security apparatus to harass their domestic rivals. To cite one of many examples, Vladimír Krajina had been among the most prominent leaders of the wartime resistance. The Communists wished to discredit him in order to sustain the fiction that they alone had fought back against the Nazis. They produced a deposition signed by Frank, the despised Sudeten leader, purporting to prove that Krajina had been a Nazi collaborator. During the trial, a prosecutor showed the statement to Frank, who admitted signing it but—as he was unable to read Czech—had done so without knowledge of its contents. All charges were dropped.

The Krajina case reflected the precarious balance that now came into being. The Communists dominated the security forces and thus had the authority to investigate and arrest. The Ministry of Justice was headed by Drtina, who did his best to frustrate Communist schemes. In some cases, the Interior Ministry ordered arrests based on the testimony of witnesses who had been bribed or coerced. Drtina took the cases to court but opened new investigations into the actions of the security agents who had compromised the witnesses. This produced an equilibrium of sorts, but not a sustainable one.

ON SEPTEMBER 28, 1945, my family boarded an old prop-driven Junker that had been seized from the Nazis. My father's new title was Czechoslovak Minister Plenipotentiary and Envoy Extraordinary to Yugoslavia.* After the mercifully short flight, we arrived in Belgrade, a capital that had been reduced to dust and debris by the bombs of Allied and Axis powers alike. More than one-tenth of the population had died fighting the Germans or one another. Everywhere shabbily dressed people were hard at work clearing streets and repairing or replacing the buildings that had been destroyed.

Before leaving Prague, my father had been given his instructions by Beneš, who had asked him to return home as often as he could. "Don't write down anything of a confidential character," the president cautioned. "The Soviet embassy would have it the day after your telegram arrives in the ministry of foreign affairs. You must report to me personally." Beneš emphasized his distaste for Josip Broz Tito, Yugoslavia's flamboyant leader. Like many a dictator, Tito used the trappings of power to burnish his personal legend, which, in turn, helped justify his rule. From the Slovenian woodlands to the Dalmatian coast, towns and streets were named after him and stories repeated about his wartime exploits. According to the officially approved slogan, "Tito belongs to us and we belong to Tito." Children even sang songs about him; I remember learning one myself: "Tito, Tito, Little White Violet."

In keeping with diplomatic custom, my father's first duty upon

---

* His title was later lengthened further to include "and Albania."

arriving in Belgrade was to present his credentials to the head of government. Asked to wait in a hallway, he was nearly knocked over by Tiger, Tito's affectionate Alsatian shepherd. When the prime minister finally appeared, my father found him to be shorter and plumper than expected, quick with a smile, and impressive in his military uniform and high boots. The fifty-three-year-old had regular features except for a somewhat prominent nose and was, despite his paunch, an active sportsman who maintained a stable of horses and loved to fish and hunt. My father had numerous opportunities to converse with Tito, discussing all aspects of the world situation, including the possibilities of coexistence between East and West. One night, the dictator invited my father to his home, which, like the U.S. White House, included a bowling alley. When the ambassador gripped the ball in his left hand, his host applauded and said that my father had been born a leftist. When the ball was released, however, he exclaimed, "Just look at it. It goes suspiciously to the right!"

As the child of an ambassador, I was privileged to live in a house that included both the embassy offices and our living quarters. Located

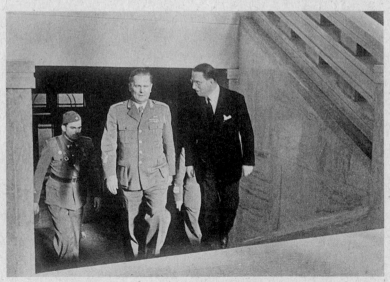

*Tito and Josef Korbel*

on a main boulevard, the building was just a block or two from the Yugoslav parliament. The front of the embassy was marked by the long balcony from which my father had observed demonstrations before the war and where we stood now for speeches and parades. A circular staircase led to the private quarters, which occupied three floors. We had been provided with a butler, a chauffeur, a cook, and several maids. In the reception area, there was a ballroom with crystal chandeliers and an abundance of marble. When we moved in, my father was appalled to discover that the outside walls were covered with pro-Tito slogans. He ordered them removed, but within days the partisan graffiti were back.

The posh surroundings belied the economic straits my family confronted. The embassy had been used by the Nazis and had, prior to our arrival, been thoroughly looted. My parents had to request furnishings from Prague, the first in an endless volley of requests for aid in coping with the expenses of the job. In those strained circumstances, we all had to do our part. Mine was to dress in our country's national costume (white blouse, pink skirt, blue apron, lots of embroidery and ribbons) and hand out flowers at parties. The costume, which was actually Slovak, not Czech, proved a survivor, residing to this day in my hall closet.

The life of a diplomat abroad agreed with my father. It's true that he had to spend much time at his desk studying and marking up documents. I know because I found thick folders of yellowing official papers from that period among his belongings in my garage. His chief interest, however, was learning more about the Yugoslav people. Whenever he could, he broke free from his office to explore the country and to meet with representatives of its many ethnic groups and political factions. No matter the audience, he loved to talk and probe, pushing people gently to be open about their disappointments, hopes, and fears. He was a skilled questioner, a sympathetic listener, and intellectually curious. He spoke to Serbs, who complained bitterly of wartime massacres committed by Croats and about the steady erosion of their national identity under Tito. He met Croats, who opposed the very existence of Yugoslavia and desired a country of their own; many Bosniaks and

*The author and her sister, Kathy, prepared for flower duty*

Slovenes felt the same. All this must have seemed both familiar and depressing to a man who had grown up amid the ethnic rivalries of Czechoslovakia. He developed a deep affection for Serbs and other Yugoslavs yet despaired at their inability to live in harmony—a shortcoming that would remain tragically in evidence during my own years in government.

Intrigued as he was by his travels, my father also had cause for disappointment. Many of the friends he had known before the war were unwilling to resume their relationship because, under Tito, contact with a suspect foreign embassy was grounds for arrest. Even the most innocent interaction could lead to a problem. For example, the French ambassadress had a dog for which she wanted to find a mate. Through inquiries, she located a Yugoslav family that owned an animal of identical breed and opposite sex. A conjugal visit was arranged, after which the police descended on the family in question, interrogating the head of the household for days.

The altered atmosphere drained the spontaneity from conversation; people either repeated the party line or restricted themselves to pleasantries. One friend eventually told my father that he had stopped visiting because he had been ordered to spy on him, which he refused to do. Other acquaintances, such as the Ribnikars, with whom our family had been close, were now partisans of Tito, whether out of sincerity or survival instincts my parents were unsure. As a result, my father could share his thoughts about the most interesting and meaty issues only with other members of the diplomatic corps or with the rare Yugoslav who didn't care what anybody saw or heard.

Early in his tenure, my father attended a session of the Yugoslav parliament at which Tito was the main speaker. Instead of just standing politely when the dictator entered, the Soviet ambassador joined in enthusiastic applause at various points throughout the remarks. His lead was followed by the representatives from Communist-controlled Poland, Hungary, and Romania. This was the beginning of what the Cold War would produce—a sad collection of compliant satellite regimes whose officials would clap in unison whenever the right rhetorical buttons were pushed, such as an attack on bourgeois capitalists or a complaint about Western imperialism. My father refused to join in this already tired game. He instructed his embassy staff that when attending events they might courteously applaud Tito's arrival, but they should listen to his speeches in silence.

This effort to place professionalism above politics nettled the embassy's lawyer, a Communist who started to cause trouble in Prague, only to find that the ambassador was no pushover. After a little investigation, my father uncovered evidence that the counselor had been smuggling currency on the black market. As that was illegal, the miscreant was promptly dismissed. There were, however, no final victories. The Foreign Ministry soon sent a replacement, who spent his time reporting on every allegedly disloyal comment he heard. This may explain an entry I was shown in 2011 from the files of the secret police complaining that my father was "not a Communist" but instead a Beneš supporter who had done little "to earn the favor of leading Yugoslav officials." Further, the embassy staff included Gottwald's daughter, Marta, likely a pipeline to her father and—because she was married to a Yugoslav diplomat—possibly passing secrets to Tito as well. Given the scrutiny my father received, it's a miracle that he lasted as long as he did, for he took every opportunity to share information with the British and U.S. embassies—disclosing nothing that would hurt his country but whatever might help the democratic cause.

To shield me from Yugoslavia's poisonous politics, my parents had asked Blanka, a twenty-year-old Czech governess, to come to live with us in Belgrade. She took charge of my schooling and helped to look after my sister, Kathy, as well. To this day, all the grammar I know in my native tongue I learned between the ages of eight and ten. Once again my parents did all they could to make life, as changeable as it was, feel normal. My father's office was connected to the residence through a passageway on the third floor. When he was not busy, he joined us for lunch, and in the afternoon we drove into the countryside in our black Tatra, a Czechoslovak car with fins on the back resembling a Batmobile. If the weather suited, we enjoyed walks in the woods or stopped at Mount Avala, where we climbed the steps to the huge World War I Monument to the Unknown Hero.

Sometimes my father invited Yugoslav government officials or journalists to join us, I think now because open-air conversations were less likely to be overheard. It may be that the Yugoslavs who were brave enough to come didn't expect to live long in any event; they were the most reckless drivers I had ever seen. I remember well my horror at watching one of their cars run over a dog. My father worried because the Czechoslovak government had given a Tatra to Tito as a gift. The old man had passed the vehicle on to his son, who drove like a maniac despite having lost an arm in the war. "Suppose there is a crash," my father said. "Who do you think Tito will blame—the driver or the car?"

STALIN WAS TRUE to his word in one respect: Soviet troops didn't stay to occupy Czechoslovakia. The U.S. Embassy helped to negotiate a mutual withdrawal so that American and Soviet soldiers departed by the end of 1945—with the Russians carrying off as much in the way of jewelry, crockery, farm implements, carpets, plumbing fixtures, toys, musical instruments, mattresses, and other loot as they could manage.

The Red Army's departure, however, did not mean an end to Soviet influence. At public events, there were as many portraits of Stalin as of Beneš. Gottwald and his comrades spoke constantly of the country's debt to Moscow and argued that the USSR was a valuable trading partner—selling grain, buying armaments, and exchanging a long list of goods.

Less openly, the Soviet interest in Czechoslovakia was piqued by what was, at the time, the globe's most sought after substance. When, in August 1945, the Hiroshima and Nagasaki nuclear explosions changed the world forever, there were three sources of uranium: Canada, the Belgian Congo, and the Jáchymov mine in Bohemia. The Russians did not have access to the first two; hence their hunger for a special relationship with Prague. Earlier in the century, tailings from the Jáchymov uranium mine had been used by Marie Curie to produce radium for health treatments and by glassmakers as a source of yellow coloring. With the dawn of the nuclear age, uranium itself became the prize.

Stalin wanted a guaranteed supply, and he would get one. His already advantageous negotiating position was enhanced by the eager cooperation of Prime Minister Fierlinger. Well before any formal talks began, Soviet security agents were allowed to inspect the mines, take samples, and send soldiers to guard the area. Bypassing the Czechoslovak Foreign Ministry, Fierlinger worked directly with Moscow. On October 7, a Sunday, he walked through the garden of his country estate to that of his neighbor, President Beneš, for a confidential discussion. He explained the Soviets' desire for uranium and suggested that a pact on the right terms would ensure Stalin's support on other matters, such as the development of oil fields in Slovakia and the clarification of minor border disputes with Poland and Austria. The president insisted that the Czechoslovaks retain a share of the uranium (the agreed amount was "up to 10 percent") but did not object either to the proposed arrangement or to its secrecy.

The treaty was approved at a closed meeting of the cabinet on November 23, 1945, the day Jan Masaryk first learned of it. Two months later, he gave a speech to the United Nations promising that his country's uranium would be used solely for peaceful purposes. Moscow had other plans. Stringent security measures were put into place, and the Red Army, although gone from the rest of the country, remained in the Jáchymov district. Labor at the mines was furnished by a combination of civilian workers and prisoners, first German and later Czech and Slovak. In the early years, union officials were sufficiently independent to protest when safety standards were ignored; after 1948, such courage went out of fashion. Although uranium deposits were soon identified in the USSR and elsewhere in Central Europe, the Czechoslovak mines made a substantial contribution to the Soviet arsenal throughout the first decades of the nuclear arms race.

IN 1945 AND 1946, communism was ascendant in Czechoslovakia. However, the democratic parties were not without assets. Nationalism remained a powerful force. This helped the democrats because Stalin, for all his popularity, could not claim to be one of us; Beneš was still the legitimate custodian of T. G. Masaryk's vision. The Sokol gymnastics organization, with its deep roots in Czech culture, rebuffed Communists' efforts to infiltrate it, as did the Boy Scouts. Gottwald railed against the West, but many of his countrymen admired its democratic values, first-rate universities, and cities that were more exciting to visit (or imagine visiting) than snowy old Moscow. Notwithstanding the trendiness of leftist economic ideas, there remained businesspeople and farmers who held more conservative views. Finally, Communist ideology was incompatible with religion. The Czechs habitually invoked God's name whether or not they believed, while in Slovakia the pull of the Vatican remained strong. That is why Gottwald and other Communist leaders assured voters that, alongside Lenin and Stalin, Christmas too had a place in their hearts.

The first—and, as it would turn out, the only—meaningful national elections in the postwar period were held in May 1946. Previously, Communists had rarely received as many as one vote in ten. They were expected to do better in this balloting because they controlled more media outlets than their rivals; the right-wing Agrarian Party had been outlawed; and the Communist minister of agriculture had been given the popular assignment of distributing confiscated farms. In addition, hundreds of thousands of alleged collaborators had been erased from the electoral register. Even the most pessimistic of democratic political leaders, however, did not anticipate that the Communists would garner 38 percent of the vote—more than any other party.

That outcome gave them a plurality of seats in parliament and the right to name a new prime minister, enabling Gottwald to take the reins from Fierlinger. The new cabinet consisted of nine Communists, three Social Democrats, and a dozen ministers from the more moderate democratic parties, a twelve-to-twelve split. The two remaining cabinet members, Jan Masaryk and the defense minister, Ludvík Svoboda, had no party affiliation. The delicacy of this political equation would have a major impact in future months.

The election results tarnished Czechoslovakia in the eyes of the United States. Americans were less likely to look with favor on a country whose people had chosen—freely, no less—to elect a Communist prime minister. American assistance at the time was limited to some agricultural credits, but even those were held up as the U.S. Embassy sought to push Czechoslovakia's economic policies in a direction more agreeable to the West.

Two months earlier, speaking at Westminster College in Missouri, Winston Churchill had declared that an Iron Curtain was descending across Europe. He had cited Czechoslovakia as the only country lying behind the curtain that was also a democracy. This dual status reflected the reality: there was still a chance. Virtually surrounded by the Soviet bloc, Czechoslovaks could still vote—and vote again; the country's ultimate place remained to be determined.

AMONG THE MOST welcome visitors to the embassy in Belgrade was Jan Masaryk, a man in whom joy seemed in constant competition with sorrow. Around the time of the 1946 election, he came to our living quarters and asked my mother for a sling. "I'll need it," he said. "I don't want to shake hands with Communists." That was a joke he often told among friends. The irony was that he was susceptible to pain and sometimes did have need of a sling. On that particular evening, he accompanied my father to a reception hosted by Tito. A garish display of food and wine was laid out, this in a nation overflowing with hungry children and where little money was set aside for hospitals or schools. Masaryk, who never enjoyed such events, grew agitated. Finally, he asked my father, "Have you a piano at home?"

The two hastily took their leave and returned to our embassy. The foreign minister, having discarded his sling, sat behind the piano and joined my father in singing old Czech songs. In my father's recollection, his companion "soon lost himself in the tunes and in his thoughts." The mood in the room became unusually intimate, given the twenty-three-year difference in ages between the two men and the professional basis of their relationship. To reestablish the natural order, or perhaps to prevent my father from seeing too much, Masaryk turned to him in midsong. "You idiot," he said, "make up your mind whether you are a tenor or a bass. I can do all sorts of things, but I can't change your voice."

My mother had a circle of friends, old and new, with whom she drank thick Turkish coffee and indulged in one of her hobbies—fortune-telling with coffee grounds. The process, called kypellomancy, requires placing the saucer on top of the cup, waiting a few seconds, then turning the cup over to examine the grounds first in it, then in the saucer. The meaning of a particular shape will vary, depending on whether the formation is on the top, right, bottom, or left; drips and clumps have special implications; and to those with the requisite faith, forecasts are guaranteed accurate for forty days.

In addition to predicting pregnancies and the sudden appearance of handsome strangers, my mother played card games with me—usually gin rummy. However, she spent the bulk of her time managing the embassy staff. She had to make sure that we had enough food for ourselves and for entertaining and, to that end, sometimes sent to the country for live lambs; these played noisily around the kitchen until they became dinners that I, for one, refused to eat.

Tito kept a busy schedule of public appearances within his country but attended diplomatic receptions only rarely. Thus when our embassy planned a party on the Czechoslovak national day, my father was not perturbed to learn that the prime minister had declined the invitation.

He *was* surprised when, an hour before the event, Tito's chief waiter appeared with baskets of food, seeking directions to the kitchen.

Tito arrived at five exactly, well ahead of most other guests. It was one of those times when Kathy and I were assigned the job of handing out flowers. We gave the great man a bouquet of white roses (which he later forgot and had to send back for); he thanked us, and we all shook hands. Then, to my mother's immense irritation, the dictator was steered away from all food or drink except that provided by his own cook. My mother steamed about for a while, then gathered her courage, elbowed her way through the crowd, and presented herself to Tito. In her hands, she held a plate of *párky*, the famous (and spicy) Czech sausage that she had prepared herself. To show that it was safe, she cut a sausage in half, popped a forkful into her mouth and offered the other portion to our guest. He smiled, ate, and asked for seconds. The score at the end of the night was Mrs. Korbel, 1; Tito's staff, 0.

On a later occasion, during a diplomatic ceremony, my mother was invited to sit in an anteroom with the wives of two other ambassadors. Suddenly the door opened, and a Yugoslav soldier marched in carrying a silver tray on which there sat a trio of red-velvet boxes; in each there was a ring boasting the appropriate birthstone. The one given to my mother—she was born in May—was an emerald surrounded by fourteen diamonds. We called it "Tito's ring," and when my father first saw it, he growled, "I wonder whose finger they cut off to get this." The ring was eventually given to me, and in 1980, I wore it to Tito's funeral.

IN AUGUST 1946, my father was called away from his duties in Belgrade to join Masaryk and Clementis in representing Czechoslovakia at the postwar peace conference in Paris. He asked if I would like to accompany him; I said no, which I still can't believe except that I was frightened of airplanes and had perhaps been moved around enough at that point.

In Paris my father's chief job was to serve as president of the Economic Commission for the Balkans and Finland. In fulfilling that task, he earned the respect of U.S. diplomats for not behaving like a Stalinist. That may seem like faint praise, but it actually signified a great deal.

The political atmosphere between West and East was deteriorating rapidly, as the Soviets adopted a confrontational approach on nearly every issue. They expected the Slavic-country representatives to fall into line, which they routinely did. This supine behavior was a source of dismay to the United States, which had not yet accepted the division of the globe into two fiercely opposed power blocs. One afternoon in Paris, Secretary of State James Byrnes sat seething while a Soviet speaker denounced American foreign policy in vicious and sarcastic terms; he became furious when he saw two Czech diplomats smiling and applauding the offensive statements. I cannot help thinking how different my life would have been had my father been one of them.

Czechoslovakia's highest priority, sadly, was to amend the draft peace treaty with Budapest in order to authorize the forcible removal of 200,000 ethnic Hungarians from its soil. The country's decision to expel most of its German population could at least be defended on the basis of extreme circumstances. This parallel effort lacked that rationale. After Munich, Hungarian leaders had exploited Prague's weakness in order to reclaim a slice of southern Slovakia; the country had also fought on Germany's side through most of the war. But Slovakia, too, had been a Nazi ally. Czech and Slovak officials, from Beneš on down, often equated Hungarian crimes with those committed by Germany; that was unfair. In truth, the purge had been proposed because it was politically popular among Slovaks and because it would make Czechoslovakia less diverse and hence easier to govern. These reasons were hardly persuasive, and at the Potsdam Conference, neither the U.S. nor the British government had gone along. Instead, the issue had been set aside for consideration in Paris.

Debate began late in the day on August 14, the Hungarian representative speaking first. He painted a depressing picture of the suffering already being visited on his kinsmen in Slovakia, including the loss of property, jobs, schools, pensions, and voting privileges. Although acknowledging that his country had supported Germany during the war, he denied that it had played a meaningful role in the dismemberment of Czechoslovakia or in triggering the conflict. He argued that Hitler's cynical prewar manipulation of minority rights did not justify their elimination, citing, as an example, the need to safeguard Jews. Summing up, the Hungarian urged the conference to avoid hasty action and send, instead, an international team of experts to review the situation. As an added jab, he drew a contrast between the narrow-minded policies of the current Czechoslovak government and the noble ideals of T. G. Masaryk. When, after speaking for three hours, he finished, the conference adjourned for the night.*

The Czechoslovaks had to prepare a response. Clementis was better versed in the issue than others in the delegation but, as a Slovak, might appear biased. My father and his colleagues decided that the answer should be given by our most persuasive orator, Jan Masaryk. The drawback was that Masaryk had not studied the details of the subject and, to the extent he had, voiced private sympathy for the Hungarians. The delegation met with him at 9 p.m. at the Hotel Athenée to underline the points for emphasis the following morning. A committee was constituted to prepare a draft. My father remembered:

At two a.m. Masaryk entered our room. "Well, boys," he said, "let's have a look at what you have produced and what you order me to say." He glanced through our painstakingly prepared text, paused for a second or two, and then said, with a disarming smile, "It's wonderful; you are all political scientists of great caliber; the whole delegation is composed of Talleyrands. But for God's sake don't ask me to use all these highbrow terms. I could not pronounce them. I would blush. Why don't we speak straight from our shoulders?"

He retired to his room and started to write. He finished at five; the text was retyped and mimeographed. At ten Masaryk took the floor. Members of the delegation pulled out their copies in order to follow his words. To their amazement, Masaryk left his text in his pocket and delivered one of his greatest speeches.

---

* While in college in the late 1950s, I went on a date with a boy of Hungarian ancestry. We would have had a good time had he not accused Czechoslovakia of having stolen his country's land after World War I. There was no second date.

This anecdote says more about Masaryk (and my father's admiration for him) than it does about the goal of expelling Hungarians from their homes. The foreign minister did indeed serve up a fine speech, but he neglected to articulate the Czechoslovak position. Instead, he said, "Like my country, I am a very poor hater," and expressed his desire for peace. He asked the delegates to remember that the Hungarians had complained constantly even when, under the republic, they had enjoyed the rights they were now upset about losing. The Czechoslovaks, said Masaryk, had done their best to champion minority protections and been betrayed for their trouble. No one could fairly blame them for being angry.

Surprisingly, the foreign minister stopped there. He made no effort to defend the involuntary expulsion of Hungarians and, as to the statistics cited by the opposing spokesman, said only that "I am not going to deal with them today." When the Czechoslovak amendment came up for decision five weeks later, the Americans asked that it be referred to a subcommittee "for further study," a polite way of killing it. Rather than push the matter to a vote, Masaryk surrendered in what the ordinarily neutral U.S. note taker referred to as "an extremely moving speech." Once again, Masaryk lamented how difficult it was for him to hate.

The peace conference was not a total success for Czechoslovakia, but there were some benefits for the Korbels. My father and Clementis returned from France with a pair of identical black cocker spaniel puppies. We named ours Era. I don't know why—perhaps my parents felt that we were entering a new one.

JAN MASARYK FLEW directly from the peace conference to Long Island to participate in the second session of the United Nations, held at Lake Success, its temporary headquarters. While in New York, he kept company with a lady friend, the American author Marcia Davenport, who appealed to his interest in music (her mother, Alma Gluck, was a famed lyric soprano), his appetite (she was a superb cook), and his intelligence (she had, after all, graduated from Wellesley). The two had been together off and on since their first meeting, at a New York dinner party in 1941, shortly before Pearl Harbor. As a writer, the forty-three-year-old Davenport was best known for her favorably reviewed biography of Mozart and for a novel, *The Valley of Decision*, just made into a movie with Gregory Peck.

She wrote of that time that Masaryk felt torn apart by Cold War politics, the social demands of his position, and the burden of trying to live up to his father's name. She credited her friend with "intuitive diplomatic skill" but acknowledged that he derived no pleasure from the hectic interplay of politics. "Left to himself," she said, "he would have been happy just to play the piano." The couple spent the holidays between 1946 and 1947 at a borrowed farm in Florida, amid a grove of citrus trees and the barnyard chatter of what Masaryk referred to as "dooks." The period gave the foreign minister a rare opportunity to escape from the conflicting advice he had been receiving and from the pressures that were building both around and inside him. One day, in a conversation with Davenport, he vented his contempt for the kind of nationalist cant that had dominated the Paris Peace Conference and that had been so present in nearly every stage of his country's history:

You're no more full-blooded what you think you are than I am. I must be Jewish somewhere, though the presentable story doesn't say so. And you! How the hell do you know who you are?

I don't.

And neither does anyone else who comes as far back as he can tell, from the parts of Europe that were the battlegrounds of the Napoleonic wars. You think you have no Czech ancestry. You're wrong. Some forefather of yours came through there as a conscript in the Russian armies, and if he didn't leave a souvenir on some local *slečna*, then it was the other way round and some Czech in the Austrian army had a bit of fun with some pretty girl in Galicia whom they married off to your great-grandfather. You're like everybody else whose people fled to America in the eighties and nineties—all the villages and synagogues with the family records were burnt up in the pogroms. Nobody knows anything. . . . As for the nobility, with . . . their thousand-year genealogies; there you get into the fun-and-games department. . . . My father was the son of a Slovak coachman and a Moravian housemaid, who were serfs. I can't prove what the blood of their parents was and neither can anybody else.

# 27

# Struggle for a Nation's Soul

Spring 1947. The elections the previous May had given the Communists hope that they could indeed put an end to democracy through democratic means. What better way to answer Western critics than to show that Marxism actually did mirror the popular will? Gottwald insisted that the Russians would abide by their promise not to interfere in the country's internal affairs—but then, so far there had been no need.

To Beneš, the paramount goal was preserving his country. If that meant deferring to the Soviets on foreign policy, it was a burden he would bear. Like T. G. Masaryk before him, he knew that time would be required for political institutions to mature and for party leaders to learn how to submerge their differences for the common good. He expected the months ahead to be a period of testing as candidates prepared for the next round of elections, planned for the spring of 1948. The Communists sought to claim an absolute majority. The democrats were determined to prevent that and to do better themselves.

As so often happens, well-laid political plans had to be adjusted in light of unforeseen economic changes, in this case provoked by the weather. Weeks of hot, rainless days caused panic among farmers, drove food prices up, and created the prospect of a harvest of less than half its normal abundance. Help was needed, so America's announced plan for the reconstruction of all Europe was greeted with enthusiasm. The outlines of the program were described by U.S. secretary of state George Marshall during a commencement speech at Harvard. He put forward not an aid package so much as a generous and coordinated system of loans to help Europe recover its footing. Invitations were sent to capitals in every part of the continent, including the Soviet Union. A preliminary meeting took place at the end of June at which the Russian foreign minister, Molotov, was present along with some one hundred advisers. The French government then asked twenty-two countries to attend a follow-up conference in mid-July. The question of the moment became: would the nations of Central Europe take part in America's grand scheme?

In a cabinet meeting on July 4, Jan Masaryk argued that U.S. loans could help to refloat the economy until farm conditions improved and Czechoslovak industry recovered. He foresaw no diplomatic obstacles; the Poles and Romanians intended to participate, and the Soviets hadn't objected. He suggested that the country send an ambassador to the Paris briefing to find out what the Americans were offering and with what conditions. Even Gottwald agreed with this recommendation; the vote was unanimous.

While the cabinet was deliberating in Prague, my family was vacationing in Slovenia, where Tito and his senior advisers were relaxing as well. In that casual setting, my father compared notes with the Yugoslavs, who said that their country—which was also hurting economically—would send a delegation to the Paris conference. Two days later, my father was informed that the decision had been reversed; Yugoslavia would not participate. Why? According to Tito, Soviet pressure had had nothing to do with the switch; he simply didn't trust the Americans. My father thought the second half of that statement might be plausible, but the first half was plainly false.

By this time Gottwald, Masaryk, and Drtina were in Moscow to consult on a proposed treaty between Czechoslovakia and France. The trio was a fair representation of a divided government: Gottwald, the committed Communist; Drtina, the fervent democrat; and Masaryk, the soulful humanitarian with little taste for confrontation. The meeting began in the middle of the night, as sessions with Russian leaders typically do. Stalin was amiable but unyielding. The Marshall Plan, he declared, was not a program to rebuild Europe but a device to attack him. "If you go to Paris," he warned, "you will prove that you wish to . . . isolate the Soviet Union." Masaryk said that he saw nothing in the plan to harm the USSR and that his own country was highly dependent on imports from the West. "We need financial credits to reinvigorate our industrial base."

The Soviet leader rose and motioned for the others to follow. He pointed to a map of Europe spread out on his desk. "Look at your country and look at Germany," he said. "We are the only ones who can protect you from the resurgence of German power. Why would you want to break your treaty with us, the treaty Beneš made in 1943?" That question, with its thinly veiled threat, made further discussion academic. Postwar Czechoslovakia had nothing to fear from Germany, but the country's people, barely two years after V-E Day, feared little else. No politician could stand against that tide. To salve any hard feelings, Stalin offered to sell the Czechoslovaks a large quantity of desperately needed wheat.

That night in Moscow, Drtina went to the theater while Masaryk retreated glumly to his room. Both men knew that the real issue had little to do with economics and everything to do with power politics. Stalin was determined to keep the United States out of what he considered to be his sphere of control. Neither Czechoslovakia nor any other country in Central or Eastern Europe could participate in Secretary Marshall's plan without defying the Kremlin; that they felt unable to do. Reluctantly, the two men awoke the next morning and called the cabinet in Prague to recommend a reversal of the decision to send a representative to Paris.

Returning home, Masaryk was asked by Marcia Davenport how he had been treated by Stalin. "Oh, he's very gracious," came the reply. "He'd kill me if he could. But very gracious."

AMERICAN DIPLOMATS OFTEN expressed their frustration that Beneš and Masaryk, though genuine democrats, made little visible effort to wriggle free of the Soviet hook. Beneš countered that the United States had made the job more difficult by siding with Hungary at the Paris Peace Conference. Masaryk claimed that the only goal that mattered was preventing the Soviets from interfering in Czechoslovakia's internal affairs. Why should Washington care if his government cast anti-U.S. votes at the United Nations? Such votes rarely affected the outcome, while his country would need time if it were to outlast the Communists and regain its status as an independent democracy. He expressed sorrow that, due to budget cuts, Ambassador Laurence Steinhardt was not making a more vigorous effort through aid, cultural exchanges, and propaganda to compete with the Soviets for popular affection.

The State Department did not agree that Czechoslovakia's hostile voting pattern was meaningless, nor was it impressed by the government's knuckling under with respect to the Marshall Plan. Cables from Steinhardt show an embassy primarily concerned with curbing anti-American press coverage and securing compensation for U.S. investors who had a financial stake in nationalized properties. The ambassador opposed providing economic assistance because it might help the Communists and because he thought that a hard line would cause the "Czechos," as he called them, to recognize their need for the West more fully. Steinhardt acknowledged Czechoslovakia's vulnerability to a long list of Soviet pressure points, including control of strategic ports, media dominance, influence within the trade unions, and the fact that the country was almost surrounded by Communist regimes. But instead of developing a plan to bolster the moderates, the embassy was content to sit on the sidelines and snipe.

This lack of initiative was doubly regrettable because Steinhardt had considerable clout in Washington. Once a successful Wall Street lawyer, his generous financial contributions had paved the way to a second career as a diplomat, where he had acquired a reputation as a hard-driving troubleshooter. His attitude toward the Czechoslovaks, however, was condescending; he described them as "little people, inclined to double-talk [and] more adept in opposition than when . . . in charge." To his credit, he made two useful suggestions: that the United States—like the USSR—establish a consulate in Bratislava; and that it publish the messages between Eisenhower and Soviet military leaders prior to Prague's liberation, thus showing that it was at Russian insistence that U.S. troops had remained in Plzeň. The Truman administration responded to these ideas with inexcusable tardiness. The Bratislava consulate did not begin operating until March 1948, after the Communist coup. The exculpatory military documents were released in May 1949—far too late to make a difference.

MY BROTHER, JOHN (officially Jan), was born in Belgrade on January 15, 1947. He was a handsome child with a round, ruddy face and, when very young, longish hair. In the spirit of true confession, I can now admit that I used his baby picture in my high school yearbook, because whatever pictures there might have been of me had failed to survive all the packing and unpacking of our family luggage.

In the spring, I went with my father to Czechoslovakia, where he would participate in the twentieth anniversary of his high school graduation. We traveled by car, just the two of us. I loved having him all to myself, listening to stories about his school days and how he had courted my mother; I didn't mind that the trip seemed to last forever. It was the first time that I had seen Kyšperk and Kostelec nad Orlicí, the villages where my parents had been born, enjoyed their childhoods,

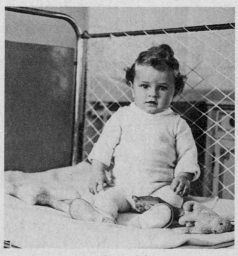

*John Korbel*

and fallen in love. I was surprised to see how small the towns were, even compared to Walton-on-Thames. We visited the house in which my father had grown up, his old high school, and the stationery store where he had purchased notebooks and pencils long before. There was also a candy shop whose sign proclaimed proudly, "Serving Kostelec and the Whole Vicinity," which was not, in truth, much of a boast. We stayed in Kostelec at the house of a family friend on the same street where my mother had lived; our host offered me a glass of goat's milk that I was too polite to refuse—good training for the afternoon in 1998 when, as secretary of state, I would be treated in Mongolia to a bowl of fermented mare's milk.

My time in Belgrade felt like an adventure but was, on occasion, a lonely one. The Jankovics, whom we had known before the war, were the only Yugoslav family with whom we spent time regularly. They had a little boy, Nidza, who was a few months older than my sister, then four. Mr. Jankovic was a journalist who was lively company and who helped my father to stay informed about what was going on in Belgrade. We took weekend hikes with them to Kalemegdan, the massive ancient fort perched on a bluff where the Sava flows into the Danube. The Jankovics came to our house for Saint Nicholas Day and our Christmas, and we went with them to the Greek Orthodox holiday services. Those celebrations took place despite the fact that, in Tito's Yugoslavia, there were no decorations in the streets, no carols on the radio, no days off for workers, or any official acknowledgment of the season. In Czechoslovakia, the Communists did not yet feel strong enough to kill Christmas; in Yugoslavia, they had already tried.

To the extent that I had playmates in Belgrade, they were from within the diplomatic community. I went swimming at the home of the British ambassador and developed my first crush on the son of a French diplomat. He was much taller than I and very good-looking. We didn't see each other again for fifty years, at which point he had shrunk to my size and we both had wrinkles.

During the summers, my family spent July in the Croatian beach town of Opatija, where I must confess that we stayed in the Hotel Moscow, so named to honor Stalin. We didn't go out on boats because

*The author with Nidza Jankovic and Kathy Korbelová, Belgrade*

the Adriatic had been mined during the war and no one was sure whether all the explosives had been removed. In August, we went to Bled, Slovenia, where we lived by a lake. There I made friends with a boy, who also disappeared from my life for fifty years. When I was in government, I received a picture of the two of us in Bled accompanied by a note saying that he had grown up to become the coroner of Jacksonville, Florida.

Back in Belgrade, I was always supervised even though I had reached the age of ten. My behavior was, in Goldilocks fashion, neither perfectly good nor perfectly bad, keeping me out of serious trouble except for once when I was at a party that ran much later than expected. My parents, in the dark concerning my whereabouts, were frantic, and when finally I arrived home, my father was angrier than I had ever seen—sentencing me to my room for three days except for studying and piano. Throughout my unjust ordeal, he kept a stern expression on his face; my mother, meanwhile, smuggled in raspberries.

In June 1947, my father was awarded a medal, presented by Jan

Masaryk, for his contribution to the liberation of the Czechoslovak Republic. By that time he was deeply engaged in the challenge of preserving its freedom. From our perspective in Belgrade, there seemed ample cause for worry. In March, the United States had promulgated the Truman Doctrine, which vowed to aid countries threatened by armed subversion, thus spurring an influx of military assistance to Turkey and Greece. For several years, Stalin had provoked the West without getting much of a response. Now Truman was making some moves of his own, and the Kremlin seemed likely to push back.

Politically, my father considered himself "a man of the Left." He was a democrat to the core but also believed that governments should be active in helping the disadvantaged. This was such a part of his identity that years later, when I was about to wed, he insisted playfully that we begin our walk down the aisle with our left feet. However, he was never tempted by the siren call of communism. His skepticism deepened in Belgrade, where a close-up view convinced him that the Soviet system had serious design problems. First, the economics didn't work because people needed incentives to be productive. That explained why perfectly good Yugoslav grapes and Albanian oranges were allowed to rot on the way to market; there was no reward for efficiency. Second, Communist leaders insisted that class warfare provided the answer to every question, even to the exclusion of such other factors as religion and national feeling. Finally, Communists were overly dogmatic, lacking the kind of intellectual creativity that my father prized. They were trained not to think for themselves but to memorize and, like parrots, repeat only what they were taught. This led directly to the kind of excesses that plagued any one-party system: centralized control of every institution, indoctrination of the young, and the elevation of a single collective goal above every other value.

My parents had been raised in a tradition that emphasized curiosity and humanist thinking. One of their favorite writers was Karel Čapek, who popularized the word "robot" and whose work made fun of precisely the kind of automaton-like behavior that communism encouraged. In Čapek's view:

The strangest and least human element of communism is its weird gloominess . . . there is no middle temperature between the freezing bourgeoisie and the revolutionary fire. . . . [For them] the world contains no lunch or dinner; it is either the moldy bread of the poor or the gorging of the overlords.

My father worried that Stalinists throughout Europe had their eye on Czechoslovakia. A senior Yugoslav army officer told him, "I do not agree with the policy of your government . . . you have too many parties. . . . [In my country, the Communists] lead in parliament, in the army, in public administration, on the collective farms, in industry—everywhere. As they act on behalf of the nation . . . it is a dictatorship of democracy." My father saw how that peculiar system worked when he tried to persuade the government-controlled Yugoslav press to report on events in Czechoslovakia. He thought it part of his job to promote an awareness of what his country was accomplishing and so had his staff transmit a weekly summary of information to the local news agency. When this approach bore no fruit, he complained to the minister of information, who apologized for the neglect and promised to expand coverage. Several weeks later, the minister returned and, with a grin, handed my father a package; inside was a fat file of clippings and quotations—all expressing scorn toward the Czechoslovak government.

DURING THE WAR, Beneš had sought to persuade the West that Stalin could be trusted and that, over time, the Soviet Union would begin to change. In mid-1947, his memoirs were published and became a best seller. Typical of Beneš, he included words of praise for Moscow, which irritated the West, and for the West, which angered Moscow. The president had not lost his sunny outlook nor the hope that his country's mediation might prevent relations between the two sides from deteriorating. He had, however, become less certain of his analysis. Late the previous year, he had undertaken what he confided to the U.S. ambassador, Steinhardt, was a "major fight" to purge Soviet agents and spies from his Defense Ministry. By the end of 1947, he had concluded that Stalin was not mellowing and that the Communists were unlikely to evolve into just another party. This didn't mean that a Marxist takeover was inevitable; it did mean that the democrats would have to engineer a revival at the ballot box. The elections of May 1948 would be crucial.

Beneš himself had only a diminishing level of energy for the struggle. In July, he had suffered a stroke, and he would limp for the rest of his life. That, plus the symptoms of arteriosclerosis he had exhibited for some time, contributed to personality changes that would make him less decisive and more stubborn. In keeping with the practice of the era, information about the president's condition was withheld from the public.

As the two sides continued their maneuvering, the Communists had multiple advantages: superior organization, clear objectives, control of most major ministries, and the unambiguous backing of the Soviet Union. Most important, they had the power of intimidation. Whether one was a cabinet minister or a village clerk, a Communist in good standing had protection; if trouble arose, the alarm would spread and the party activists would mobilize. Democrats pleaded with their countrymen to open their eyes, to see that the Communists who had bragged about opposing fascism were now aping its techniques. Portraits of Stalin had been posted where pictures of Hitler had previously been; the hammer and sickle had replaced the swastika. The Communists, like the Nazis, were manipulating the press, smearing political rivals, demanding total loyalty from their members, and threatening anyone who stood in their way.

Still, that autumn there were positive signs. A Communist-backed proposal to increase taxes was defeated in parliament. In nationwide balloting for student leadership positions, the Communists finished a well-beaten third. Gottwald's own surveys showed his party losing ground, and as for the cultural battle, democrats were winning hands down. Western films, books, magazines, and newspapers were far more popular than their Eastern counterparts. More young people were learning English than Russian. Travelers to Paris and London returned home loaded down with clothes, radios, and household goods that could not be obtained in local stores; 80 percent of the country's trade was with the West. The events that brought the country together were those that celebrated local artists, honored veterans, or showed off the athletic skills of the nation's youth; it did not seem like a society ripe for a workers' revolution. Steinhardt reported to Washington:

> As far as can be judged from constant observations of the people's reactions since May 1945, they have no particular liking for Soviet methods. They regard the Soviet alliance as an unpleasant necessity. They continue to prefer western business methods and . . . standards. They are still rather skeptical about the postwar nationalization of industry. They have no real liking for Marxist doctrines which in any case are not openly advocated by the Czechoslovak communists.

Two dramatic incidents further undermined the Communist position. On September 10, bombs hidden in boxes marked "Perfume" were delivered to the offices of three democratic cabinet members: Drtina, Masaryk, and Petr Zenkl, a deputy prime minister and former mayor of Prague. None of the explosives ignited, but the subsequent investigation turned into a media free-for-all as Communists scrambled—despite damning evidence—to steer the inquiry away from their own bumbling functionaries, among them Gottwald's son-in-law.

The second development was a rebellion among the Social Democrats. For two years, Fierlinger had kept his party subservient to Moscow. This was fine with some, but others desired a more independent voice or, at minimum, a less craven one. Fierlinger was renowned for being obsequious to the powerful and rude to everybody else. In November 1947, party leaders assembled in Brno for their annual meeting. Ignoring the Communist threats, they voted to replace the incumbent with a more conventional career politician. Fierlinger's defeat transformed the electoral arithmetic. If the Communists could not count on the Social Democrats, their ability to win a controlling majority in parliament would be very much in question.

These setbacks added to Gottwald's mounting frustration. When he traveled to neighboring countries, his Communist colleagues reminded him that while they wielded absolute power, he was forced to pander to public opinion and to defer on many occasions to Beneš, who remained a better-liked and more internationally prominent figure. Unlike Tito in Yugoslavia, Gottwald was not a war hero; there were no children singing songs about his bravery. His position was made more difficult by his incumbency; as prime minister, he was poorly positioned to denounce the government or to demand change. Beneš had done little to oppose him on social and economic policy; Masaryk, aside from the occasional sarcastic comment, had done nothing to give offense in foreign affairs. The Communists had few planks with which to build a platform for their campaign. But the darkest cloud on the horizon was Stalin. Not only was the Soviet leader still unhappy about the back-and-forth over the Marshall Plan, but he also disliked Gottwald's references to a special Czechoslovak path to socialism and was in no mood to hear discouraging news about the Communists' electoral prospects. If the balloting went poorly, Gottwald would not only be defeated; he would almost surely be shot.

I WAS TEN years old when my parents concluded that our governess had taught me all she could and that it was time for me to receive a more comprehensive education. I was too young to enroll at the gymnasium (high school) back in Prague, so they proposed to send me to a Swiss boarding school. I reacted as most ten-year-olds would—with anxiety, tears, and feigned illness. Having heard that Zurich was a center for treating polio, I claimed upon arrival that my legs hurt so much that further travel wasn't possible. My mother, who was not easily deceived, found a doctor who pronounced me well. There was nothing left but to go unwillingly to school, in nearby Chexbres.

The Prealpina Institut pour Jeunes Filles was as horrible and unfair as I expected, at least at first. Upon arrival, I was given to believe that one couldn't obtain anything except by asking for it in French, of which I knew little. Not only would I flunk but I was convinced that I would also starve. But within a month, I began to pick up the language, make friends, and do well in my studies. My room had a view of Lake Geneva; we were allowed into the village to buy chocolate on Saturdays; I was still studying piano and learning to skate and ski. I had fought the decision to send me there but now had no brief for complaint. That didn't stop me from yearning to join my family in Belgrade when the school shut down over winter break. Instead I was sent to its sister institution, where everyone spoke that dreaded language, German, and where I was as bewildered and miserable as I was lonely. The only solace came via the Christmas service, with its festive lights, beautiful music, and the text in the neutral tongue of Latin. Not until years later did I grasp the reason behind my desolate holiday: my parents, as always, were trying to protect me from what had become throughout Central Europe an uncertain and increasingly perilous political situation.

# 28

# A Failure to Communicate

Years ending in eight have an outsize role in Czech history. Charles University was founded in 1348; in 1618, Habsburg emissaries were thrown from the castle window, triggering the Thirty Years' War; in 1848, the first pan-Slav congress convened in Prague; the Czechoslovak Republic was founded in 1918; Munich occurred twenty years later. In its first three months, 1948 would earn a place on the less fondly remembered side of that list of milestones. In January, my father went to Prague for a consultation with Beneš. Having witnessed the cutthroat proclivities of Communist leaders in Yugoslavia, he hoped to find the president fully aware of the danger that democratic forces faced and in possession of a clear strategy to fight back. When he entered Beneš's office in Hradčany Castle, he was greeted by an intellectually alert but ill man. Beneš had been a major world figure for three decades and the leader of his strife-torn land for a dozen years. The stroke (or strokes) he had suffered caused him to drag one leg slightly but did not prevent him from striving to work, as he always had, twice as hard as other men.

For four hours on January 12, two in the morning and two in the afternoon, the president and the ambassador reviewed the world situation, with the former exhibiting his characteristic doubts about the West but now reserving his strongest criticism for the aggressive policies of the Soviet Union. My father finally succeeded in shifting the discussion to his own primary concern: the internal situation in Czechoslovakia.

Was Beneš prepared to defend the Constitution against the Communists? Did he have a plan for uniting the democratic forces? Did he realize how extensively Gottwald's men had infiltrated the army, police, trade unions, media, and even the Foreign Ministry?

Few words could have been more alarming to my father's ears than the Panglossian ones offered by Beneš. "As much as I am pessimistic about international developments," he said, "I am calm about the internal situation. The elections will be held in the spring. The communists will lose and rightly so. People understand their policy and will not be duped. I just don't want them to lose too much. That would arouse Moscow's anger."

In Yugoslavia, my father had seen the brutal pressure that Stalin could place on local leaders. He expressed his fear that the Czechoslovak Communists, faced with the specter of electoral defeat, might try to engineer a coup as the sole way to save their necks. Again, Beneš said not to worry:

> They thought of a putsch in September but abandoned the idea and will not try any more. They found out for themselves that I enjoy certain authority in the nation. And not only that. They know that I have numerous supporters among the working class, even among many Communist workers. They have come to realize that they cannot go against me.

Still anxious, my father asked Beneš to comment, individual by individual, on the loyalty of senior defense and military officials. The president vouched for most and was astonished to be told that the commander of the air force was a Communist. When my father raised doubts about General Svoboda, the minister of defense, Beneš replied that he was a reliable man. "Don't be worried, Ambassador," Beneš said as the meeting drew to a close, "return to Belgrade and carry on."

On that same trip, my father joined Gottwald for lunch at his villa. The conversation inevitably turned to a comparison between the situations in Yugoslavia and Czechoslovakia. Perhaps unwisely, my father could not resist taunting his host. "The communists in Belgrade don't think you know what you're doing," he said. "They say that you're nothing but a slow coach and that you're going to lose." Fueled by a combination of anger and brandy, Gottwald shot back, "I'll show them how we will win. And it will not be with the comical ballot as they do in Belgrade."

EVERY CABINET MEETING from mid-January on was marked by bitter arguments. With an election but months away, the rise in partisanship was natural, but Czechoslovak democracy had become like a rundown car with overused shock absorbers; every bump was felt, and the next jolt might be the last. Still the car kept rolling and the potholes kept appearing. The democrats demanded that Communists be prosecuted for trying to blow up three of their ministers; the Communists accused the democrats of plotting to expel them from government. Each side warned that the elections would be unfair because of the underhanded tactics of the other; both declared indignantly how unjust—when directed against them—such allegations were. As January rushed headlong into February, the evenly divided ministers fought over tax and economic policy, the pace of nationalization, and the wages of civil servants. The only respite came courtesy of a commission established to clarify Slovakia's status within the country. According to its report, "Czechoslovakia constitutes a State, one and indivisible, formed of two inseparable nations of equal rights." This was sufficiently confusing that no one knew how to argue about it.

Hubert Ripka was one of the people closest to Beneš. He was experienced and smart, knew everyone in government, and was liked by most. He had served in a variety of positions and at that point was both a democratic party leader and minister of trade. My father considered him among the finest men in Prague. In a conventional political setting, Ripka would have been an effective advocate and leader, but in the Czechoslovakia of 1948 he was a flounder swimming with sharks. On February 9, he met with Gottwald in an effort to ease the anger that threatened to deadlock the government and split the country. Instead, after a few polite exchanges, the session became a shouting match as each repeatedly interrupted the other. Gottwald accused Ripka of opposing socialism, shielding traitors, and conspiring to create, as had the Nazis, an anti-Bolshevik coalition. Ripka claimed that Gottwald's agenda, also in imitation of the Nazis, was a totalitarian state.

The next cabinet meeting, four days later, brought the government to the brink. The democrats introduced evidence in the form of a lengthy, statistics-laden report from Drtina that the Communists were attempting to establish a stronghold on the police, possibly in preparation for a coup. While the meeting was in session, word arrived that the interior minister had ordered the demotion and replacement by Communists of eight divisional police commanders—the officers authorized to distribute arms and ammunition. Drtina moved immediately to reinstate the police and to halt further hiring and firing pending a cabinet inquiry. The Social Democrats, some of whose members were among the police targeted for dismissal, joined the other non-Communist parties in supporting the motion; its approval was an embarrassment to Gottwald and escalated tensions even further. The newspapers were rife with allegations of treason, and on Sunday, February 22, thousands of trade union representatives, 90 percent of whom were Communists, were scheduled to assemble in Prague.

Ripka was convinced that the Communists had a plan and that he knew what it was. His informants had advised him that Gottwald intended to unveil a more radical economic program designed to attract support from the gathering of union delegates and probably also from Social Democrats. Ripka feared that such a maneuver would trigger an explosion of populist fervor and divert attention from the police controversy. On February 16, he proposed to the other democratic ministers that they resign as a group, thereby precipitating a showdown in advance of the Sunday mass meeting:

> It is the only way to counteract the plan of the communists. . . .
> If it is on this question that we bring about a crisis, the Social
> Democrats cannot disassociate themselves from us. Once the
> crisis is upon us we shall without doubt have to hold immediate

elections. If the date of the elections is moved forward, the communists will no longer have the necessary time to gain control of the police and the army.

Two days later, Ripka and Zenkl met at Hradčany to inform Beneš of their strategy and seek his approval. That was crucial because the president had the authority, if the government collapsed, to order new elections. Even in Ripka's account, the discussion was not business-like but rambling and confused. Beneš agreed that the Communists had an obligation to obey the cabinet's instructions with respect to the police. He also stated, when asked, that a cabinet in which the democratic parties had been excluded would be unacceptable. He did not say—because no one deemed it necessary to inquire—exactly what he would do if, as planned, the democratic ministers all resigned. Ripka thought he had a clear signal to proceed as intended, but it is far from certain that Beneš shared that understanding. The president told Ripka to "stand firm" and "avoid blunders," but that does not mean that he had focused on the incendiary consequences of what the democratic leaders were about to do.

Had he done so, he might have pointed out that the plan could not succeed without full cooperation from the Social Democrats, for in their absence the ministers lacked the majority required to bring down the government. The Social Democrats had been supportive on the police controversy but had rebuffed Ripka when he broached the idea of a group resignation. Such a dramatic move would amount to a declaration of political war, something they could not endorse without ripping apart the fabric of their organization.

The democratic ministers went ahead nevertheless, submitting their resignations on Friday, February 20, hoping to catch the Communists off guard. They did not. Gottwald immediately began to mobilize his network of loyalists. The venom of the Communist-controlled media was concentrated on the twelve ministers, who were allegedly taking orders from financial interests abroad and who had resigned in an effort to sabotage democracy and block the popular will.

The next morning, standing in front of the massive statue of Jan

Hus in Old Town Square, Gottwald called upon the president to take the ministers at their word and replace them with a National Unity Front made up of "good" Czechs and Slovaks. Gottwald proceeded to Hradčany to repeat the demand. Beneš refused, instead calling on party leaders from all sides to broker a solution. The democrats were upset that he had not demanded that the remainder of the cabinet resign. Since that alternative had not been raised with him in advance and dismissing the cabinet was, in any event, beyond his constitutional authority, the complaint is hard to understand. In fact, Ripka's failure to gain the endorsement of the Social Democrats meant that a majority of the cabinet remained in place; the government had not fallen. Gottwald was still prime minister; the president had little choice but to deal with him.

Amid all the political wheeling and dealing, the business of Cold War diplomacy ground on. My father returned from Belgrade to Prague to participate in meetings with Yugoslav and Polish representatives that had been convened—as part of Gottwald's grand design—primarily for the purpose of denouncing the West. Receptions were held, and speakers were applauded for their criticisms of British imperialism and U.S. hegemony. To underline those points, a joint statement was proposed. My father detested the whole show but could do little to influence the proceedings. Masaryk, who as foreign minister should have been at center stage, withdrew entirely, either feigning illness or genuinely sick. He asked my father to follow him to his flat, which was on the third floor of the Foreign Ministry. "I met him in his bedroom," said my father. "He was lying in bed, as he often did . . . to escape . . . unwelcome visitors. He told me, 'I can't put my signature to such a document. Half of my life I spent among Americans and British, every bit of my soul is with them, and now I am asked to sign a declaration against them. I just cannot do it. Try to change it somehow.' " My father sighed; before leaving, he peered out the windows into the darkened courtyard below. He would never speak to his boss again.

THAT SUNDAY, BENEŠ was resting at his retreat in Sezimovo Ústí, fifty miles from the capital; Masaryk was still in bed; other democratic leaders were scattered about, accepting an honorary award here, visiting a women's conference there, attending a ski championship in the Tatras, or giving speeches that urged their followers to remain calm. Gottwald, meanwhile, was in Prague addressing thousands of cheering, expectant, "Internationale"-singing representatives of the working class. By that time the full weight of his mobilization order was being felt. Around the country, party leaders were distributing guns to their militias and mobilizing agents in the security forces to confront rivals wherever they could be intimidated, detained, arrested, or beaten up. Radios and newspapers called on the rank and file to rally behind Gottwald and his demand for a new government. Workers were ordered to begin purging non–party members from their factories. Thousands of telegrams were sent to Beneš, insisting that he accept the resignations of the ministers or risk civil war. Throughout the weekend, Gottwald was in contact with his agents at the Interior, Defense, and other ministries; he also had a direct line to the Soviet Embassy, whose deputy foreign minister had suddenly arrived in the country. When Masaryk asked him why he was there, Mr. Zorin replied, "To supervise the delivery of wheat."

On Monday afternoon, Beneš met with Ripka and three of his senior colleagues, Drtina, Stránský, and Zenkl—for the first time since their resignations. The men had known one another for decades; Zenkl had spent the war as a prisoner in Buchenwald, but the others had been with Beneš in London. They were the president's allies and, to the extent he would allow it, his friends. Yet at that crucial moment they did not communicate very well. Beneš did assure the ministers that he would neither accept their resignations nor agree to a list of replacements without their approval. Beyond that, he had nothing to offer. He had pressed Gottwald to seek a negotiated solution, but the Communists would not bargain, insisting instead that the ministers who had resigned were traitors. "And how did you respond to that?" the four democrats wanted to know. Beneš replied, "I did not react; it's up to you to defend yourselves. As far as I am concerned I must remain above the fight, above the parties."

That night, ten thousand students staged a march in support of democracy. Singing and chanting, they moved as one along the winding streets to the castle. The president received them and pledged to remain true to the spirit of T. G. Masaryk. It was an inspiring moment but also the only public demonstration on behalf of liberal government in the entire crisis. Ripka and his aides were convinced that they had the Constitution and the majority of people on their side; they did not, however, have a strategy for proving it.

February 25, 1948, was the day when the rule of law was mugged on the streets of Prague. Democratic leaders on their way to work were prevented from entering their offices; some had their homes searched or were handcuffed and shoved into jail. The last independent newspapers and radio stations were taken over, trashed, or closed down. Fierlinger, accompanied by police and armed thugs, reasserted control over the party that had dumped him. The Communist unions called for a nationwide one-hour strike; workers who failed to participate were tossed from their jobs. Everywhere the demands were the same: out with the old government, in with the new. Gottwald met with Beneš once more and was once again urged by the president to negotiate with the democrats; he refused.

By that time Gottwald had put together a proposed new cabinet that included Communists and some token representatives of the democratic parties. At 11 a.m., he presented the slate to Beneš with a request for its immediate acceptance. According to Eduard Táborský, who was in the room, Gottwald brandished a second list as well—of democratic supporters earmarked for arrest and possible execution if the president refused to sign. Beneš promised an answer later in the day.

My father described what happened next:

At 4 p.m. Gottwald drove to Hradčany Castle for the president's answer. Then, a few minutes later, he drove back to Saint Wenceslas Square. He had a paper in his hands. It was the list of a new government, signed by the president of the Republic. His [Gottwald's] head was covered by a Russian sheepskin cap. Two hundred thousand mobilized workers awaited him. Police and workers' militia mingled with them. He announced the constitution of a new government and read the list. He expressed gratitude to President Beneš for respecting the will and wish of the people.

The mob accompanied Gottwald's every word with frantic applause and thunderous shouts. Somewhere close to the president's castle a few thousand university students were gathering again to march to his residence. The police fired on them. The deposed ministers listened in their homes, surrounded by the police, to Gottwald's address. He was obviously drunk, drunk with alcohol and with success. The day was bitterly cold. Gray skies obscured the sun. In Czechoslovakia, democracy was dead.

# 29

# The Fall

Jan Masaryk lived in a private flat in the northeast corner of Černín Palace. The long, narrow apartment, modest compared with its surroundings, could be reached by private lift. The sitting room had space for a couch and several armchairs, a writing table, and bookshelves. The radio, which is still there, was three and a half feet high and stood across from the brass bed on the same side as the door to the bath. The outside wall, punctuated by four windows, each tall and rectangular, overlooked an inner courtyard thirty feet below. Numerous passageways—some concealed and meant for servants—led into the adjacent hall. An uninvited visitor, once inside the palace, could easily have found his way to the foreign minister's door.

Because he was not a member of a political party, Masaryk had played no direct role in the February crisis. He hadn't been consulted about Ripka's plan, hadn't resigned, and may never have been asked to do so. He was not a political strategist and had been worn down, in any case, by bronchial problems. The morning after the coup, he sent a note to Marcia Davenport telling her that he would remain in the government for the time being and that, despite the shocking events, "This is not the end."

Masaryk didn't say much about the democratic ministers' resignation strategy except to concede privately that it had been a mistake. He was surely right about that. By resigning, the ministers had given Gottwald the chance to seize power through what many would see as constitutional means. He had not had to rely on Soviet troops or public threats, prevailing instead through a combination of police subversion, political gamesmanship, and well-timed mob action. Even he had been surprised by how easy it had been. "I knew I'd get them in the end," he told Masaryk giddily, "but I never thought they'd hand me their asses to kick on a platter."

Quite possibly, the resignation plan simply accelerated a takeover that would have occurred in any case; Gottwald would surely have found some other pretext for causing trouble. But by exposing their adversary's plot to radicalize the police, the democratic ministers had put the Communists into a tight spot. If Gottwald had been forced to move out of desperation, he might have been the one to make mistakes; the Social Democrats might have sided with the other democratic parties; Beneš might have had the confidence to rally the nation; and the army, whose loyalties were split, might have come down on the right side. Like the plethora of what-ifs that arose after Munich, these cannot be answered with certainty.

As it was, Czechoslovakia's place in the Cold War solar system was now fixed. Shortly after the takeover, the minister of education decreed that a portrait of Stalin be displayed in every classroom and that the Soviet national anthem follow each rendition of "Where Is My Home?" The new justice minister was Gottwald's son-in-law, under whose direction the legal system became an arm of the Communist Party, democratic activists were hauled into court on false charges, and a string of labor and reeducation camps was established. To stop prominent democrats from escaping, the country's borders were sealed. Men such as Ripka and Drtina were shadowed by police and had their phone lines cut and mail intercepted; visitors to their homes could count on having their names and addresses recorded in the guides' little black books.

Governments in the West condemned the coup in indignant terms. Gottwald responded that he did not need a lesson in democracy from the perpetrators of Munich.

In England in 1940, my family had lived in the same apartment house as Drtina, the deposed justice minister. Our friend had always been an optimist, even during the Nazi occupation, when he had stayed on in Prague for more than a year to help organize the resistance. Together our families had weathered the Blitz and looked forward to the war's end and with it a second chance to fulfill the dream of Czechoslovak democracy. On the night of February 28, three days after the coup, Drtina tried to commit suicide but did not do a good job of it. Jumping from the window of his third-floor apartment house, he was seriously injured, taken to the hospital, and thrown into jail, where he would remain for a dozen years.*

Inside the government there remained but two top-level personalities: Beneš, a president without power, and Masaryk, a foreign minister without a government he wished to represent. Eleven years earlier, when T. G. Masaryk had been on his deathbed, he had asked Jan to help Beneš: "You know much of the world better than he does. Stand by him always. Promise me that you will never leave him alone." That was one explanation for Masaryk's fidelity, and it struck deep. As he told Marcia Davenport:

> [Beneš] was . . . a martyr for my father's sake. . . . Every time that any political difficulty confronted him, Beneš stood in the breach. For any mistake, he took the blame. He was the whipping boy. He got down in the ditch and did the political spadework that left my father clear of it all, to remain the saint that the people thought him and the saint he really was. For this I will be loyal to Beneš until I die.

The president no longer lived in the castle. In the wake of the February 25 disaster, he moved to his private estate in Sezimovo Ústí, a town where, coincidentally, Jan Hus had taken refuge more than five

---

* From his release in 1960 until his death twenty years later, Drtina did all he could to aid the revival of Czechoslovak freedom, writing a memoir that was published in the West and—in 1977—bravely adding his name to Charter 77, a protest that served as a precursor of the Velvet Revolution.

hundred years before. Why hadn't Beneš resigned? He told Masaryk that Gottwald had threatened to unleash massive violence that would kill thousands and destroy the country. The Communists insisted that the president remain so that they could portray their coup as legal. No name in Europe was more closely associated with democracy than Beneš. The old man saw room for only one gesture, and that was to leave the castle and all it represented.

What about Masaryk; why didn't he resign? Surely his affection toward Beneš was one reason; moreover, his job was a source of protection both for others and for him. He told Steinhardt that he had intervened on behalf of more than two hundred people, either shielding them from arrest or helping them escape. If he had a plan for his own future at that stage, it was to navigate a way out. After all, what kind of foreign minister is not allowed to travel? There is never a shortage of international conferences. Perhaps from exile he could start again. This meant, however, that he must try to convince the Communists that they had nothing to fear, that he would continue playing the good soldier Švejk—the role he had adopted, albeit with a shrug, not a smile, since Beneš had first linked Czechoslovakia's fate to Stalin.

This fragile strategy was undermined when, on March 2, the ambassador to Washington, Juraj Slavík, resigned and in a dramatic statement denounced the Gottwald government. Masaryk and Slavík were friends; the Communists were sure to hold the resignation against the foreign minister, and they did. From that day, in addition to his longtime bodyguard, he was shadowed by two rough-looking figures from the Interior Ministry. Masaryk warned Davenport to leave the country; he feared that she would be arrested, perhaps on charges of being a U.S. spy. Reluctantly she agreed and made plane reservations for the seventh of March. The night before, he visited her apartment. She later wrote:

> He came at half-past eight. He looked absolutely ghastly. All those days he had an exhausted, claylike pallor, but that evening he was even more grey of face. He had come from Sezimovo Ústí where he had lunched with Beneš and spent the afternoon. He

told me nothing of what had happened there. . . . I saw only that
he was distraught. He muttered, "Beneš . . ."

Masaryk was unable to finish the sentence. The couple sat for a time
and did not talk much. He told her that, upon her arrival in London,
she should stay close to her hotel. Soon, within a few days, she would
hear from him. She was to tell their English friend Bruce Lockhart
of his intention to escape. The hour grew late, and Masaryk rose and
struggled into the old cinnamon-colored overcoat that he had been
wearing on those wintry days. They said their farewells; she listened
as he trudged down the stairway, then watched from a window as his
guards fell in beside him on his short journey back up the slope to
Černín Palace.

GOTTWALD'S POLITICAL NETWORK had tentacles long enough to reach
the Czechoslovak legation in Belgrade. The day before the coup, my father's deputy, Arnošt Karpišek, had presented him with a draft telegram
signed by members of the embassy's newly formed action committee,
pledging their support for Stalin and the Communist Party. Karpišek
asked the ambassador to forward the telegram to Prague and to include
his own name at the bottom to make it official. Red-faced and furious,
my father crumpled the paper, threw it away, and reminded Karpišek
that the embassy's sole loyalties were to Beneš and the Constitution.

When word arrived that Gottwald had come out on top in his street
fight with the democratic ministers, my father's first instinct was to find
help. He contacted Charles Peake, the British ambassador in Belgrade,
and asked to meet with him privately. Both men took precautions to
avoid being overseen. My father said that our family might be forced
to seek asylum in Great Britain; he did not think that he could, nor
would he want, to remain with the Communists in power. He didn't
fear Gottwald but worried that more brutal elements might soon take
control. In his visit to Prague earlier that month, he had told the prime
minister that the country might be better served in Belgrade by a Communist. Gottwald replied that a UN commission was being formed to
find a solution to the violent dispute involving Kashmir, a resource-rich

province claimed by both India and Pakistan. Czechoslovakia was one of the commission's members. Gottwald, seconded by Masaryk, suggested that my father might be a good person to represent the country on that panel.

That night the British ambassador sent a "Most Immediate—Top Secret" cable to London:

> [Korbel] and his family are in great distress. He tells me that he is now being closely watched and followed when he leaves the Embassy. . . . Since I took up this post eighteen months ago, I have observed that he has been uncompromisingly pro-British in season and out of season, and he has never failed to provide me with any information . . . he thought would be of use to my government. . . . I have found him in every way decent, honest, respectable, and I have no hesitation in recommending him to you as a particularly deserving case.

Peake's telegram was treated with urgency. "What have we done?" wrote Foreign Secretary Ernest Bevin across the top of the cable. Within days, the British government agreed to issue visas for our immediate family. However, the United Nations was still debating the terms and conditions under which the Kashmir commission would operate, and my father worried that Gottwald's offer would be withdrawn. He took solace in the fact that Jan Masaryk remained as foreign minister.

ON MARCH 7, Czechoslovakia's new government sponsored a celebration of the ninety-eighth anniversary of T. G. Masaryk's birth. The speakers, most of whom were Communists, told the familiar stories of his exploits and made the absurd claim that, had the elder Masaryk still been alive, he would have cheered the recent developments. Following the speeches, a group of cabinet ministers proceeded to Lány, where they gratified official photographers by posing at the founder's grave site.

One minister did not arrive until after the others had left.

There is no doubt that Jan Masaryk visited his father's grave that afternoon. Less certain is whether he did so in the company of a private secretary, as the secretary later asserted; of his niece, as the niece later claimed; or only of his bodyguard, as the bodyguard said was the case. The foreign minister may have remained for only five minutes, or he may have stayed for an hour. He may have engaged in soulful reflections resulting in a momentous decision—or he might simply have performed a filial duty.

Masaryk knew by then that Davenport was safely en route to London; he had sent a friend to the airport to make sure she departed without incident. The night before, he had been unable to put into words how much his meeting earlier that day had depressed him. He had gone to Sezimovo Ústí to ask what Beneš intended and what he should do. Did the president have a plan? Was there anything further the foreign minister could do to honor the pledge he had made to his father? The old president had not liked the inquiries; he had become excited and angry, telling Masaryk that he did not care what he did, that

*The Masaryks, son and father*

he should solve his own problems. The situation was impossible, came the reply. Jan said he could not continue; he intended to leave.

At least that is one account of the meeting. A second highlights instead a discussion of Drtina's suicide attempt, which Masaryk is said to have dismissed as the kind of thing "a servant girl would do." "Suicide," he reportedly declared, "doesn't absolve anybody of his responsibilities. It's a very poor escape." In that version, there is no confrontation with Beneš but instead a suggestion by Masaryk to a third party, Dr. Oskar Klinger, that he and Klinger leave the country together. The doctor who treated both Masaryk and Beneš was the fourth person present at the March 6 meeting, along with the two principals and Mrs. Benešová. This second description of what transpired was given by Klinger to the Czech-born English journalist Henry Brandon. The first, also originating with him, was given to Davenport. Curiously, the two accounts neither directly contradict each other nor overlap.

On Tuesday, Masaryk had cause to see Beneš once more. The newly designated Polish ambassador had arrived in Prague and wished to present his credentials. Masaryk and Clementis accompanied him to Sezimovo Ústí for a brief meeting with the president. Jan stayed behind to talk with Beneš in his study; he left about fifteen minutes later in relatively high spirits, telling jokes to the secretaries, exhibiting his charm.

Back in his flat, Masaryk, who was still nagged by bronchitis and high blood pressure, napped for two hours. Upon awakening, he conducted routine business and reviewed his schedule for the coming day. He was slotted to attend the reopening of parliament and a meeting of the Polish-Czechoslovak Friendship Society, for which he would have to draft a short speech. According to what his secretary would later tell police, Masaryk also planned to leave that night for a two-week stay in Gräfenberg, a Moravian spa famed for its natural healing techniques. There was no mention of how he was to get there.

After the secretary left, the butler brought in a dinner of roast chicken, potatoes, and salad. When the dishes were cleared, Masaryk asked that the bedroom window be opened and that two bottles of mineral water and one of beer be placed on his table. "Don't forget," he said, "wake me at 8:30!"

For the next hour or two, the foreign minister continued with his paperwork, filled an ashtray with cigarette stubs, and wrote his 126-word speech including the line, "We . . . look with open eyes towards the future." He may also have read, for there were familiar books close to hand: *The Good Soldier Švejk* and his father's Bible. Then he took two tablets of Seconal and retired for the night.

AROUND SUNRISE, JAN MASARYK's body was found in the ministry courtyard, several yards out from the wall. He was partly clothed, his unmarked face a mask of fright. Far above, the bathroom window gaped open; inside, the night table—which contained a loaded revolver—was overturned. The contents of the medicine chest were scattered and smashed; there were shards of broken glass in the bath, one soiled pillow in the tub, a second beneath the sink. Closet doors were open, as were dresser drawers. A search of the rooms revealed Masaryk's freshly drafted speech, written in pencil, but no farewell note. Some witnesses claimed that the Bible was closed, others that it was open to a suggestive and freshly underlined passage from Saint Paul: "Those that are

*The bath in Masaryk's apartment*

Christ's have crucified their own flesh." A forensic examination uncovered traces of paint beneath his fingernails, a long scratch on his abdomen, and two sleeping pills half dissolved in his stomach. His heels were shattered and the bones splintered; someone had placed the fragments in a small pile. Officials rushed to secure the scene. Within hours, any mystery about what had happened had apparently been resolved. The Communist authorities declared that Jan Masaryk had been driven to take his own life because of Western criticisms. The time of death had been in the early hours of March 10.

# 30

# Sands Through the Hourglass

March 12, 1948: The queue of mourners stretched for two miles or more up the steep hill to Černín Palace. They filed by four abreast, students and factory workers, teachers and farmers, to pay final respects to their Honza, the irrepressible Jan. Candle flames kept watch at each corner of the open casket; flowers were everywhere; so, too, were the secret police.

March 13: The nation's largest funeral since that of the founder, T. G. Masaryk. Mammoth crowds gathered along sidewalks and staircases to look on as the procession made its way down the castle heights, across the bridge, past storefronts draped with black bunting, to Wenceslas Square, and up the steps to the cavernous National Museum. The public's grief could not have been more heartfelt; still, official hypocrisy would rule the day. The ten and a half years between the funerals of Masaryk father and son had been marked by war, occupation, renewal, and disintegration; a series of trials and transitions leading—where? In 1937, when the elder man had died, the dream of a democratic and humane Czechoslovakia still lived; now that vision had been distorted into something dark and cold.

For Klement Gottwald the ceremony was less a solemn rite than a coronation. No one inside the country dared point to the irony that he, of all people, would preside on this occasion. Perhaps only Jan could have found apt words, for the Communist had come both to praise Masaryk and to bury democracy. Yet Gottwald might have

*Funeral procession, March 13, 1948*

felt a tinge of unease. The murder, if murder it had been, was not (or probably not) his doing. A month earlier, he had been on the edge of bungling away his party's position, wasting years of preparation. Due to his adversaries' mistakes and Masaryk's death, the path to power was now clear—but Gottwald was both boss and underling. He had often expressed his belief that communism was compatible with Czech nationalism and that his country could have a workers' revolution distinct from any other. Stalin had grown impatient with such talk, and if Soviet agents could murder Masaryk and call it suicide (attributed officially to "insomnia and nervous disorder"), they could do the same to Gottwald. But they would not have to. When Moscow played the tune, he would dance.

Among the officials present was Vlado Clementis, the successor to Jan as foreign minister. It would be his duty to speak at the interment ceremony in Lány. Clementis had been one of two cabinet ministers summoned to Masaryk's living quarters when the body was discovered. What had he thought when he entered the apartment and found it in disarray? He had searched for a note and seen nothing except the speech Jan had prepared and some personal keepsakes that he would later send to Marcia Davenport. What did he think when he looked at the narrow bathroom opening through which the fleshy sixty-year-old had supposedly squeezed, ignoring the more accessible window in the bedroom? Had the man really chosen to end his life in the middle of the night while barefoot and in a mismatched set of pajamas? Had he decided to throw himself out of a window barely ten days after Drtina had, with embarrassing results, done the same? Why, if death was his decision, had he not used his revolver or quadrupled his dose of sleeping pills? Was Clementis troubled by the speed and sloppiness of the so-called police investigation? Did he have doubts about the official version of events? There can be no question that he mourned Masaryk's death. His idea of communism was different from that being dictated by Moscow. This was a tribute to Clementis's character and (as will shortly be described) a reason for his own demise.

The first reaction of many of Jan's friends and acquaintances had been to accept that he had committed suicide. They didn't know—for they were not told—of the countervailing evidence but were well aware of Masaryk's tendency toward gloom. My father learned of the death while on an excursion in Yugoslavia with a group of Czechoslovak sightseers. His inclination, based on his most recent encounters with Jan, was to think that a man so careworn might well have been driven to take his own life. My mother felt otherwise; she knew of their friend's aversion to pain and thought that even if he had decided to kill himself, he would never have done so in the manner described.

Reporting to Washington on the afternoon of the tragedy,

Ambassador Steinhardt speculated that Jan had reached a breaking point and could not go on allowing the Communists to exploit his family name. He suggested that the foreign minister had been weighed down by depression and that the visit to his father's grave had been to explain what he had decided to do—perhaps even to ask permission. He might have hoped that his death would be seen as the kind of eloquent protest he had failed to voice publicly. "In his desperation," mused Steinhardt, "he appears to have turned to suicide as the only means of expressing his disapproval." Bruce Lockhart, the English diplomat whom Davenport had been asked to contact in London, held a similar view:

> What he thought or felt [at the grave site] no one will know, but of one thing I am sure. The knowledge that his father's birthday was being celebrated hypocritically and for purely opportunistic reasons by the men who were undoing his work must have been agony to Jan, and I think it probable that during that lonely vigil he made his decision. I do not doubt that he had made his plans to escape. I also do not doubt that when the time for action came he preferred the simpler way.

By this point in their long history, Czechs had grown accustomed to official explanations that they did not for a minute take seriously. Many citizens, perhaps most, suspected that Masaryk had been murdered. Less than a month after the alleged suicide, Steinhardt too was having second thoughts. "I cannot escape the feeling that the repeated rumors . . . might have some basis," he wrote. He was puzzled, in particular, by the absence of a farewell message. "Masaryk was a showman and knew the value of such a statement. Nor do I believe that there was . . . [one] which has been suppressed or destroyed, for Masaryk was too shrewd and knew too well what was going on not to have made certain that at least a copy would be in the possession of Marcia Davenport or myself."

From almost the moment Jan began falling, the battle of perceptions was launched. The government produced a fifty-page illustrated tribute to the deceased hero, recounting his career and repeating the theory that he had jumped because of harsh criticism from his supposed friends in the West. In England, Dr. Klinger, Masaryk's physician, told the *New York Times* that Jan had arranged for a plane to whisk them away on the morning of his death, so that together they might begin a new campaign against communism. No evidence for that notion has been found, nor has a second part of Klinger's story been corroborated—that Masaryk became embroiled in a shootout, killing four of his would-be assassins before losing his own life. Klinger supposedly had an informant, one of his patients, who claimed to have seen coffins being ferried from the ministry that night.

I believe that Masaryk was murdered, probably by Stalin's agents. I can't prove this and would hardly be shocked should conclusive evidence one day surface to the contrary. But the Soviets had a motive, especially if they thought Masaryk was on the verge of flight. He may have been overheard discussing plans to leave, whether on wiretaps in his apartment or when meeting with Beneš in the president's study. The Communists could not realistically put Jan into prison. They could hardly fire him and still claim, for public purposes, that they had his support. Suicide, blamed on the West, was the ideal solution. Also pointing to murder was the foreign minister's businesslike behavior on his last night, the signs of fear and struggle in his bedroom and bath, the lack of a professional investigation, the government's rush to judgment, the absence of any last words, the half-digested sleeping pills, and the fact that he had taken time to draft a speech for the following day.

There is another reason, little discussed, why I find it hard to believe that T. G. Masaryk's son would kill himself. The senior Masaryk's earliest book was a study of suicide, in which that phenomenon is depicted as a symptom of social and spiritual loss, a judgment that one's life has no meaning, a negative verdict on the world. Such a judgment could not have been further from Jan's inheritance. Would a child so conscious of his father's opinion willfully disregard it at a defining moment in his life?

Officially, the Masaryk case was reopened three times: in 1968, when

Communist control relaxed for the period known as Prague Spring; in
1993, after the return of democracy (via Václav Havel and the Velvet
Revolution); and in 2003. The first two inquiries were inconclusive; in
the third, the state prosecutor ruled that Masaryk had been murdered,
based primarily on expert testimony about the position in which the
body had been discovered. Investigators argued that the foreign minister must have been pushed; they came to no conclusion as to who might
have done the pushing.

AT T. G. MASARYK's funeral, Beneš had delivered the main oration. At
Jan's, he refused to speak. He would not share the dais with Gottwald
and had only at the last moment consented to attend. Wrapped in a
heavy coat, he sat slumped in a seat next to Hana.

Following the ceremony, Beneš would leave his estate in Sezimovo
Ústí but one more time, in April, to mark the six hundredth anniversary of the founding of Charles University. There he would summon
a final measure of eloquence in support of freedom "of belief, science,
thought, and vocation . . . founded on man's respect for man." Not until
June did he formally resign, departing the office four years younger than
T. G. Masaryk had entered it. His successor, no surprise, was Gottwald,
to whom—out of an excess of courtesy (or perhaps fear)—he addressed
a congratulatory note.

In his final months, the circle of aides and friends that had long assisted Beneš melted away. Aside from his doctors, there was the always
steady Hana, a personal secretary, and such visitors as were willing to
run the gauntlet of gun-toting Communist guards. Weary and lacking
hobbies, the former president wandered about his gardens or sat in his
chair without papers, lost in thought. He sometimes listened to broadcasts from the Voice of America, but Hana insisted that he do so only
on the second floor, beyond the hearing of his security detail. Politics
had filled his life, but the hourglass was running out.

But Beneš still cared about his reputation. From exile, Ripka and some
of the other democratic ministers had been quick to tell their side of the
February story, blaming the president for accepting their resignations,

failing to arrest Gottwald, and losing control of the military and police. Above all, they criticized him for placing the nation's fate in Soviet hands.

On August 19, Beneš called on his last remaining reserves of energy to fight back, telling an interviewer:

> They are accusing me of disappointing them. But I am accusing them of disappointing me. . . . When Gottwald filled Old Town Square with armed, bloodthirsty militia, I expected a counter-rally on Saint Wenceslas Square. . . . I believed that the demonstration of unarmed students would be a signal for a general uprising. When however, nobody made a move, I would not allow Gottwald's hordes, who were spoiling for a fight, to perpetrate a wholesale massacre on the defenseless Prague population.

At the time of this conversation, Beneš nurtured a hope of escaping the isolation of Sezimovo Ústí; he had spoken to Hana about moving to an apartment in the capital, near the Foreign Ministry, where he had presided for so many years. But the day after the interview, his health took a sharp downward turn. For a few days he lost his voice, then rallied for a time, then began again to weaken. He lapsed into a coma and, on September 3, 1948, breathed his last.

Beneš had neither T. G. Masaryk's intellectual range nor Jan's easy way with people. He was not a war hero or a gifted politician. My father, especially in his last book (on the meaning of Czechoslovak history), was among those who faulted him for failing to fight after Munich and for his lack of effective leadership against the Communists. But in my view, much depends on the standard against which Beneš is judged. Those anticipating a second T. G. Masaryk were right to be disappointed; on that scale, he did not measure up. He was too much the lawyer and analyst, striving to gauge the course of events while lacking the boldness and charisma to shape them. He worked consistently within the confines of the democratic and humane values that Masaryk championed but rarely pushed back against the flow of public opinion. If the majority wanted to kick out the Germans and

Hungarians, he would lead the effort. If the people were drawn toward socialism, he would help to nationalize the economy. If the consensus held that Stalin was their liberator, so be it. Popular opinion was a fact, one of many that Beneš took into account when calculating his next move. When he did challenge the public mood, it was to cool the boiling pot, as after Munich, when this Czech Sancho Panza tried to save his country from what he thought a quixotic response.

Tomáš Masaryk, by contrast, was the rare leader who taught as he led. Even as a relatively young man, he exposed patriotic—yet fraudulent—documents, fought anti-Semitism, espoused women's rights, promoted public health, and emphasized the responsibilities of democratic citizens. He spoke instinctively to the decency of his listeners, seeking not to prove how astute he was but to bring out the best in those who cared to listen. Decades later, Václav Havel would do the same when striving, as president, to heal the deep wounds that remained between the Czech and German populations and to elevate public debate into a discussion of ethics and mutual responsibility. In sum, Beneš was a less commanding figure than T. G. Masaryk and a less compelling moral arbiter than Havel. As criticisms go, these damn but faintly.

Measured against the other European leaders of his day, and especially considering his health problems toward the end, Edvard Beneš was a man of lasting stature. Early in his career, his diplomatic genius helped to create the Czechoslovak Republic and contributed much to the senior Masaryk's success and reputation. As president, he performed miracles in holding the government in exile together and realizing its goals. After the war, he gave his country a better chance than others in the region to preserve its freedom. In later years, Jan Masaryk was the only one to rival his popularity yet the (slightly) younger man would not have been a successful president. He was a moody, volatile, compassionate jester who never took anything quite as seriously as Beneš took everything. Jan Masaryk complemented his boss; he could not have replaced him.

Between 1937 and 1948, the team of Beneš and Masaryk was matched first against Hitler and Ribbentrop, then against Stalin and Molotov.

History's narrative tells us that, in both cases, the more powerful duo prevailed—at least for a time; but history's judgment suggests that Beneš and Masaryk were the kind of leaders we might wish to see again.

WHEN MY PARENTS returned to Prague for Masaryk's funeral, Dáša was there to greet them. At my parents' hotel, she learned that our family would soon be leaving Yugoslavia as my father began a new assignment. Perhaps we would be returning to London; would she like to come, too? My cousin was torn; the great-aunt with whom she had stayed after we went to Belgrade had herself left—to join family members in England. Dáša had then been "planted" with her aunt Krista, of whom she was not overly fond. As with the Nazis years earlier, no one knew in advance exactly what life in a Communist-controlled Czechoslovakia would be like. The Cold War had been given a name (by George Orwell), and Walter Lippmann had already written a book about it, but life behind the Iron Curtain was still in the process of being defined. Dáša had a boyfriend, Vladimir Šima, and wished to complete her studies at Charles University. Now twenty years old, she decided to stay in Prague.

Alas, her life, like those of so many others, would be knocked askew by politics. In January 1949, she was summoned by security officials for questioning about the activities of my father and about her own views on the people's revolution. Her affirmations of political indifference were not enough to save her from being expelled and virtually disowned by her aunt, described later by my cousin as "a Communist beast." Dáša was so upset that she went to her fiancé's apartment and threw her student book with all her identification papers into the fire—from which her future mother-in-law was just able to retrieve them.

Dáša and Vladimir married; she became an accountant, translator, and journalist and he a military construction engineer. Theirs would not be an easy life, but they built their own family and persevered. My mother did her best to help by sending them the deed to her parents' Czechoslovak property, which was later sold to raise money. I find it touching that, out of respect for their heritage, Dáša's grandchildren developed a special passion for helping young refugees from Bosnia, the Caucasus, and Asia.

BEFORE DEPARTING PRAGUE, my father met with Clementis to confirm that the government's offer of the UN position still stood. My parents then returned to Belgrade, where the political waters remained turbulent. Even as the Communists were celebrating victory in Czechoslovakia, signs of a falling-out were rising to the surface in Yugoslavia. Tito had an oversize ego and did not like being told what to do—even by Stalin. Officials in Washington were unaware of the extent of his wrath until my mother paid a farewell visit to the family of Andrija Hebrang, a prominent local politician with close ties to Moscow. She discovered the house empty except for a frightened maid, who told her that the entire family had been arrested. Apparently, Tito had become convinced that the Soviets were grooming Hebrang as a replacement for him—a possibility he refused to accept. The U.S. Embassy included a report on my mother's house call in a top secret cable to Washington. A month later, Yugoslavia was expelled from the Soviet bloc and the historic rivalry between Stalin and Tito burst into the open.

I did not see my family until May 1948, when they came to Switzerland so that my father could confer in Geneva with UN officials. The rest of us went with him, and although the trip itself was uneventful, we did love watching the peacocks parade in front of the organization's European headquarters. For Kathy, the highlight was an introduction—courtesy of yours truly—to the wonders of bubble gum.

After Geneva and our rendezvous with the peacocks, my father returned to Belgrade, while I stayed on to finish school in Chexbres and my mother, Kathy, and John went to London. When I joined them, we moved into a dark basement apartment, memorable only because the bathtub was in the kitchen. Before that, they were hosted by Eduard Goldstücker, the scholar with whom we had lived previously in Walton-on-Thames and who had since been promoted to second in command at our country's embassy. Naturally, as a Communist, his reaction to the February events was far sunnier than that of my parents. To him the change in leadership meant a chance to prove

*The author, age ten, Switzerland*

that the ideology in which he believed could deliver on its promise of social justice. He was soon rewarded for his loyalty with a post he coveted: ambassador to the newly created state of Israel. For a time, the two countries enjoyed warm relations; Israel needed weapons and training, especially for its fledgling air force; the Czechoslovaks provided them willingly and at a fair price. The friendship soured, however, when Israel's leaders refused to align their nation's strategic interests with those of Moscow.

Sadly, the fates of Clementis and Goldstücker were not as either envisioned. Both would be caught in the web of the Stalinist show trials that terrorized Czechoslovakia in the early 1950s. Moscow was determined to prevent its Central European satellites from imitating Tito's independent line. Its chosen method was to push the governments of Hungary, Romania, and Czechoslovakia to make an example of selected government officials, whether or not they were actually guilty of revisionist thinking. Gottwald, who feared being purged himself, cooperated by offering up more than a dozen colleagues, including Clementis, who was arrested and later hanged, and Goldstücker, who was sentenced to life in prison.* The specific charges varied, but the general allegation was that the suspects had been plotting to betray communism to the West.

The period also marked a frenzy of activity on the part of the secret police, whose young investigators were urged by the Soviets to find conspiracies wherever they could—especially among the men and women who had been active in the London-based exile community during the war. By the new logic, anyone who had considered the West a true ally in the fight against Hitler should be judged a traitor to the proletariat. One was either a Communist or a spy; there were no neutrals. Among the supporters of Beneš who were investigated in absentia and at length was Josef Korbel—a badge of honor.

GOTTWALD'S TIME ATOP the power pyramid in Prague was brief, for the role of Soviet puppet proved an exhausting one. In March 1953, a few days after attending Stalin's funeral in Moscow, he died of a burst artery brought on by heart disease and several decades of alcohol abuse. The story of his burial site is worth recounting for what it says about the tangled web of Czech history.

In the nineteenth century, Czech Protestants sought to place a statue of the one-eyed Hussite general Jan Žižka alongside the other major monuments of the city. Conceived in 1882, the project was delayed by Catholics, who had no wish to honor a Protestant hero, and by cubists, who favored a design more abstract than the conventional man on a horse. When, in 1913, a competition was held, three artists were declared to have won second place but none first. Eventually, a brilliant young sculptor of the Rodin school was selected and began work

---

* When Goldstücker was released in 1955, he was, in the words of a friend, "so small and skinny that he looked like a small boy." Undaunted, he resumed his academic career. As chair of the Writers' Union he successfully championed the idea that Franz Kafka should be honored instead of dismissed as a "decadent bourgeois" writer. He also played a leading role in Prague Spring. I renewed my acquaintance with the old ambassador in the early 1970s, when I interviewed him in England for my dissertation. Until he passed away in 2000 at the age of eighty-seven, Goldstücker continued to defend Communist beliefs, arguing that the principles were right even if the implementation was not.

on what would become the world's largest bronze equestrian statue: seventy-three feet high. Construction was then suspended because Czech legislators felt that Žižka's horse looked suspiciously Austrian and because church leaders thought the general should be shown holding a Bible instead of a sword. By the time the sculptor had completed the mold, the Nazis had invaded and begun searching for the work, which was hastily cut into pieces and concealed in various locations around Prague.

Finally assembled after the war, the statue was dedicated on July 14, 1950. Because the government by then was Communist, the Hussite rebellion now had to be reinvented as an early demonstration of secular class warfare, with Žižka glorified as the first Marxist. Suitable bas-reliefs celebrating the proletariat were added. Thus it seemed only right to his admirers that when Gottwald died, his remains should be committed to the mausoleum behind Žižka's statue—where the two could commune, one great people's warrior to another. The task of embalming, however, was entrusted to amateurs. Within a decade, the decomposing corpse had to be removed and cremated.

MY FATHER'S UN job was not a permanent position, even though the quarrel over Kashmir—still unresolved six decades later—might make it seem so. The commission had been established in January, and its first three members were selected the following month. Two more were added in April; in July, the full panel, with my father as chairman, departed for Pakistan. Its members spent most of the summer shuttling between there and India.

Through the last half of 1948, my father wrote us cheery letters about the exotic landscape and wildlife of the subcontinent, including the monkeys that came into his hotel room. He was aware, however, that he was working on borrowed time. The Czechoslovak Foreign Ministry suspected strongly—this according to recently released documents from the secret police—that he had no intention of returning to Prague. Yet he also had three young children and little in the way of accumulated assets. It was natural that his thoughts for the future would turn to America, where the United Nations was based and where opportunities exceeded those of anywhere else. At year's end, the commission was due in New York to submit its report. He would use the time there to inquire whether a suitable position might be available in the UN Secretariat. He would also arrange for his family to meet him.

On the evening of November 5, my mother, Kathy, John, and I made our way to Southampton, where we boarded the SS *America* and crossed the Channel to France, where we spent the night. The next morning, after breakfast, we resumed our journey westward, chasing the sun. Greeted by the Statue of Liberty, we arrived in New York Harbor shortly after 10 a.m.; it was, fittingly enough, Armistice Day. Shortly before Christmas, my father joined us, crossing the Atlantic on the *Queen Mary*. Although his hope for a job at the United Nations was not realized, his request that our family be granted political asylum was supported strongly by the U.S. and British diplomats who knew him and by the Czechoslovak democratic exiles with whom he had served in times of war and peace. For several months my parents anxiously waited; the application was approved on the first day of June 1949. So began the further adventures of my family—in what Antonín Dvořák had referred to in his famous symphony as the New World.

> **Czech Diplomat Asks For Sanctuary in U. S.**
>
> Special to THE NEW YORK TIMES.
>
> LAKE SUCCESS, Feb. 14—Dr. Josef Korbel, the diplomat who was dismissed by the Czechoslovak Communist regime from his job on the United Nations Kashmir Commission, asked the United States today to give political asylum to himself and his family.

# The Next Chapter

Few sentiments are expressed more often than gratitude for the sacrifices made by earlier generations; so be it, originality isn't everything. I truly am indebted to my parents for the love and protection they provided and for the inheritance I received—including a commitment to freedom and an understanding that its survival cannot be presumed. I am grateful, too, for my father's example; without it, I would have had neither the passion for public affairs that has driven me throughout my life nor the confidence to insist that my voice be heard. As should be evident by now, he was not someone who was content sitting in an armchair reading about world events. He had a desire to know every detail, to delve into the motivations of leaders and countries, to learn their history, to sample the opinions of all he met, and to search for solutions that accorded with his high standards. Many children rebel against their parents; I wanted to make my father proud and to act as I thought he might in comparable circumstances—whether as a diplomat, teacher, or citizen.

After our arrival in the United States, my father began a second career as a professor at the University of Denver, where the school of international affairs today bears his name. Upon his death in 1977, the school published a memorial book of essays on Czechoslovak history and heritage. The volume included tributes from former students lauding Professor Korbel's "passion for learning [and] devotion to truth." In 2011, the Czech Foreign Ministry also honored him, in this case with a film, *A Man with a Pipe: A Documentary on the Life of Josef Korbel.*

The film had its premiere in the fall of 2011, while I was in Prague for the dedication of a statue of Woodrow Wilson. An earlier version of the statue, built in the 1920s, had stood in front of the train station until it was ordered destroyed by Heydrich. The new one is paired in Washington, D.C., with a comparable memorial to Tomáš Masaryk, highlighting the deep historic ties that bind Czechs and Slovaks to the United States, a bond that is also part of my inheritance. For years, on the Fourth of July, my mother called to ask if we were watching the parades and fireworks and singing patriotic songs. She was proof—as are her children and millions of other immigrants—that allegiance to country can spread from one land to another. In *A Man with a Pipe*, my brother observed that although my father had been seen as intellectual and my mother more a creature of temperament, she had often been the more levelheaded of the two. In sum, we miss them as we loved them, equally and always.

I FEEL AN obligation I can never repay to those who helped me learn more about my family and what they experienced. The Holocaust has yielded many moving accounts from people who survived—whether in concentration camps or in hiding—and also from those whose diaries lived on even if they did not. The stories are important in their own right but even more so because they give us a better idea of the histories we will never hear from the millions who lacked the means, the strength, or the opportunity to commit their thoughts to paper. The members of my family who were murdered by gun, gas, or disease left behind but a limited quantity of letters. Part of my goal in writing this book has been to learn more. For that, I thank the remarkable people who lived alongside my relatives in Terezín and the many since who have dedicated themselves to honoring the dead. Remembrance is the least we owe.

Looking back at the stories that fill these pages, I am struck, too, by the magnitude of our debt to the men and women who fought and won World War II and who created the institutions that would contain and ultimately defeat communism. Foremost among these institutions was the North Atlantic Treaty Organization (NATO), whose origins can be traced directly to the shock of Jan Masaryk's fall from the Foreign Ministry window on the night of March 9, 1948. His death erased any lingering hope that collaboration between the Soviet Union and the West—so essential during the war—could survive even in diluted form. The European winter that began with Hitler's occupation of Prague resumed a decade later with the loss of democracy's favorite son. We know today what was then "behind the mountains." NATO would prove equal to its responsibilities; the West would stand firm; and the Iron Curtain would one day be torn down from both sides by a revolution that liberated Poland over the course of ten years, Hungary in ten months, East Germany in ten weeks, Czechoslovakia in ten days, and Romania in ten hours. Twice in my lifetime, Central Europe lost and then regained its freedom; that is cause for celebration—and also for vigilance. NATO's job is far from over.

STILL, COMPARED WITH our parents and grandparents, we live in a transformed world. Due in large measure to technology, the means of diplomacy have been revolutionized, the geopolitical center of gravity has shifted from the West toward the East and South, and new threats to international security have arisen. Fortunately, the role of our principal World War II adversary has turned right side up. In the decades after Hitler, the German people lifted their country to greatness in the best way, as a bulwark of democracy, a good neighbor, and a model in protecting human rights. It is an irony of our age that the United States now asks its ally in Berlin to be more, not less, assertive on the international stage. Ironic too that in 2011, on the seventy-second anniversary of the German invasion of Prague, that country's ambassador called to ask me whether I would accept an award (the Federal Cross of Merit) for service on behalf of U.S.-German relations. I said, "Yes, of course, I would be honored," thinking that by now even my mother would approve.

The marriage between Czechs and Slovaks survived wars both hot and cold, but, on December 31, 1992, the two split peacefully through

what was called the Velvet Divorce. Like my parents, I had always embraced the idea of a united Czechoslovakia, but perhaps it was never
truly meant to be. The majority of Slovaks earnestly desired their own
state—a sentiment that Czech nationalists might regret but could
hardly fail to understand.

All this is not to say that the new era is free from echoes of the old. The
lessons of the Second World War and its aftermath have been learned
at best imperfectly. Minor irritations are frequently enough to rekindle
medieval resentments involving the Slavic peoples and their neighbors,
or between East and West. In Moscow, authorities have pushed to replace monuments of Stalin that had been toppled and to teach students
what is referred to as "positive history," that is, a thoroughly Russocen-
tric version of landmark events. This doctrine promotes the idea that
Stalin almost single-handedly won World War II while British and
U.S. leaders sought cravenly to engage Hitler in a separate peace. Few
choices have proved more damaging to the future than teaching children to resent the past. In Europe, politics remain tarnished by extrem-
ist parties, some transparently anti-Semitic or anti-Muslim, that place
national identity above a commitment to democratic values. Left-wing
organizations also survive; in the Czech Republic, the Communist
Party is the third largest, in Russia, the second.

People everywhere, including the United States, are still prone to
accept stereotypes, eager to believe what we want to believe (for example, on global warming) and anxious to wait while others take the
lead—seeking in vain to avoid both responsibility and risk. When trou-
bles arise among faraway people, we remain tempted to hide behind the
principle of national sovereignty, to "mind our own business" when it
is convenient, and to think of democracy as a suit to be worn in fine
weather but left in the closet when clouds threaten.

Just as extreme nationalism, bigotry, and racism remain very much
a part of contemporary life, so too do torture, ethnic cleansing, and
genocide. In one of the many boxes in my garage, I came across a quo-
tation attributed to Otokar Březina, a nineteenth-century Moravian
poet: "It is no longer possible," he asserted, "to strangle one's brethren
unheard. Somebody will always hear the cry of agony and let it fly from

mouth to mouth throughout the land like a hurricane that blows the holy fires into flame." Unfortunately, news of genocide can outpace the wind and still not prompt timely action to save lives. Well-intentioned people have been striving for generations to find an effective guarantee against atrocities, but we are not there yet.

MY FATHER'S UNPUBLISHED novel concludes with a reminder, shared between the protagonist, Peter, and a friend: "The main thing is to remain oneself, under any circumstances; that was and is our common purpose." Peter, when alone, repeats the mantra, as if seeking a source of certainty in a world where supposed absolutes have lost their meaning: "The main thing is to remain oneself."

Upon first reading, I had to wonder what my father had meant by that sentence. Was he referring in some oblique way to our family's Jewish heritage? I am sure now that he was not. Writing about the period after the war, a time in which he saw his countrymen divided against themselves, religious and even racial identity were probably not the first subjects on his mind. To him, "being ourselves" meant living up to the humanitarian values that had been championed in the first Czechoslovak Republic. The ethos of T. G. Masaryk is what burned most brightly in my father's intellect and soul. In that sense, the idea of "being ourselves" is not confining, as categories of nation and creed inherently are. In fact, the very belief that "being ourselves" should be an aspiration reflects profound optimism—especially after the events that shook Europe and the world between 1937 and 1948.

It would be good news indeed if people were behaving unnaturally when, under the stress of wartime conditions, they exhibited more cruelty than compassion and more cowardice than courage; or if those who rushed to salute Hitler and Stalin had first been twisted into something other than "themselves." In saying this, I do not intend to leap into a philosophical or still less a theological discussion of human character. There is no need to go beyond what we know and have seen.

Given the events described in this book, we cannot help but acknowledge the capacity within us for unspeakable cruelty or—to give the virtuous their due—at least some degree of moral cowardice. There is a piece of the traitor within most of us, a slice of collaborator, an aptitude for appeasement, a touch of the unfeeling prison guard. Who among us has not dehumanized others, if not by word or action, then at least in thought? From the maternity ward to the deathbed, all that goes on within our breasts is hardly sweetness and light. Some have concluded from this that what is needed from our leaders is an iron hand, an ideology that explains everything, or a historical grievance that can serve as a center to our lives. Still others study the past and despair that we will ever learn anything, comparing us instead to a laboratory animal on an exercise wheel, always running, never advancing.

If I agreed with this dismal prognosis, I would never have arisen from bed this morning, much less written this book. I prefer the diagnosis of Václav Havel, whose conclusions about human behavior were forged in the smithy of the Cold War. Amid the repression of those years, he discerned two varieties of hope. The first he compared to the longing for "some kind of salvation from the outside." This caused people to wait and do nothing because they had "lost the feeling that there was anything they could do. . . . So they waited [in essence] for Godot. . . . But Godot is an illusion. He is the product of our own helplessness, a patch over a hole in the spirit . . . the hope of people without hope."

"On the other end of the spectrum," said Havel, there are those who insist on "speaking the truth simply because it [is] the right thing to do, without speculating whether it [will] lead somewhere tomorrow, or the day after, or ever." This urge, too, is fully human, every bit as much as the temptation to despair. Such daring, he argued, grows out of the faith that repeating truth makes sense in itself, regardless of whether it is "appreciated, or victorious, or repressed for the hundredth time. At the very least, it [means] that someone [is] not supporting the government of lies."

There are many examples of cruelty and betrayal in this book, but they are not what I will take with me as I move to life's next chapter. In the world where I choose to live, even the coldest winter must yield to agents of spring and the darkest view of human nature must eventually find room for shafts of light.

Let us focus, then, not on the frozen ground but on the green blades rising, on the men and women who met adversity in the right way, with courage and faith. Let us remember those who were drawn closer together by Hitler's bombs, who stood virtually alone in the battle for a continent, finding in the moment of crisis the bravery and strength they had almost forgotten they had. Let us honor the fighters who jumped into hell on Omaha Beach and who battled through the snows of the Ardennes to secure victory over a tyrant. Let us recall the aviators and soldiers who, from exile, fought to restore the honor of their country—and the shopkeepers who hurled pieces of cobbled stone at tanks in a mad effort to reclaim their homeland. Let us salute the quiet English stockbroker who, while others did nothing, single-handedly devised the means for saving the lives of my cousin and hundreds of other innocent children. Let us reflect on the courage of a middle-aged woman navigating the streets of occupied Prague with contraband in her purse, the fate of brave men on her mind, and cyanide in her pocket. Let us recall the boys and girls who had the nerve to write poetry and create works of art, and the adults who cared enough about life to debate philosophy, devote themselves to healing, and share their meager belongings—all in a prison expressly designed to crush their spirit. Let us refresh our minds with the image of Jan Masaryk breaking free from the company of appeasers, fascists, and Communists to tell a joke, pound the piano, and belt out songs at the top of his lungs about wood nymphs and water sprites. Let us imagine the soft voice of a Jewish prisoner singing a requiem beneath the stars while shoveling earth beside a fallen church in Lidice.

"The soul is purified by misfortune and sorrow, as gold by fire." So says the grandmother in Božena Němcová's novel. "Without sorrow there can be no joy."

I have spent a lifetime looking for remedies to all manner of life's problems—personal, social, political, global. I am deeply suspicious of those who offer simple solutions and statements of absolute certainty or who claim full possession of the truth. Yet I have grown equally skeptical of those who suggest that all is too nuanced and complex for us to learn any lessons, that there are so many sides to everything that we can pursue knowledge every day of our lives and still know nothing for sure. I believe we can recognize truth when we see it, just not at first and not without ever relenting in our efforts to learn more. This is because the goal we seek, and the good we hope for, comes not as some final reward but as the hidden companion to our quest. It is not what we find, but the reason we cannot stop looking and striving, that tells us why we are here.

# GUIDE TO PERSONALITIES

## *Bohemia*

KING VÁCLAV (Wenceslas) (d. 935)

CHARLES IV (1316–1378): emperor and builder

JOHN OF NEPOMUK (1345–1393): Catholic martyr

JAN HUS (c. 1371–1415): religious reformer and martyr

JAN ŽIŽKA (1360–1424): Hussite warrior

JAN KOMENSKÝ (Comenius) (1592–1670): educator

JOSEF II (1741–1790): Austrian emperor and reformer

BOŽENA NĚMCOVÁ (1820–1862): novelist and poet

KAREL HAVLÍČEK (1821–1856): journalist

JAN NERUDA (1834–1891): novelist and poet

## *Czechoslovak Republic, Protectorate, and Government in Exile*

EDVARD BENEŠ (1884–1948): foreign minister and president

HANA BENEŠOVÁ (1885–1974): first lady

KAREL ČAPEK (1890–1938): writer

VLADO CLEMENTIS (1902–1952): deputy foreign minister

PROKOP DRTINA (1900–1980): minister of justice

ALOIS ELIÁŠ (1890–1942): prime minister, executed by Nazis

ZDENĚK FIERLINGER (1891–1976): minister to Moscow, prime minister

KARL HERMANN "K. H." FRANK (1898–1946): Sudeten German leader during Nazi occupation of Czechoslovakia

EDUARD GOLDSTÜCKER (1913–2000): scholar, diplomat

KLEMENT GOTTWALD (1896–1953): Communist Party head, prime minister

EMIL HÁCHA (1872–1945): president during Nazi occupation of Czechoslovakia

VÁCLAV HAVEL (1936–2011): writer, revolutionary, president

KONRAD HENLEIN (1898–1945): prewar leader of the Sudeten Germans

JOSEF KORBEL (1909–1977): ambassador to Yugoslavia and Albania

JAN MASARYK (1886–1948): ambassador to Great Britain; foreign minister

TOMÁŠ G. MASARYK (1850–1937): founder and first president

GENERAL FRANTIŠEK MORAVEC (1895–1966): head of intelligence

MARIE MORAVCOVÁ (?–1942): volunteer with the antifascist resistance

GONDA REDLICH (1916–1944): youth leader at Terezín

HUBERT RIPKA (1895–1958): state secretary, minister of trade

EDUARD TÁBORSKÝ (1910–1996): personal secretary to Beneš

JOZEF TISO (1887–1947): president of Slovakia during World War II

## Czechoslovak Parachutists

KAREL ČURDA (1911–1947) (Vrbas)

JOZEF GABČÍK (1912–1942) (Little Ota)

JAN KUBIŠ (1913–1942) (Big Ota)

ADOLF OPÁLKA (1915–1942)

JOSEF VALČÍK (1914–1942) (Zdenda)

## United Kingdom

CLEMENT ATTLEE (1883–1967): postwar prime minister

ALEXANDER CADOGAN (1884–1968): undersecretary in the Foreign Office

NEVILLE CHAMBERLAIN (1869–1940): prewar prime minister

WINSTON CHURCHILL (1874–1965): wartime prime minister

SHIELA GRANT DUFF (1913–2004): journalist

ANTHONY EDEN (1897–1977): wartime foreign secretary

LORD HALIFAX (1881–1959): prewar foreign secretary

BRUCE LOCKHART (1887–1970): liaison to Czechoslovak government in exile

## France

ÉDOUARD DALADIER (1884–1970): president

## Germany

HERMANN GÖRING (1893–1946): commander, air force
REINHARD HEYDRICH (1904–1942): acting reichsprotektor
HEINRICH HIMMLER (1900–1945): chief of security forces
ADOLF HITLER (1889–1945): chancellor
KONSTANTIN VON NEURATH (1873–1956): reichsprotektor

## Soviet Union

VYACHESLAV MOLOTOV (1890–1986): foreign minister
JOSEF STALIN (1878–1953): premier

# TIME LINES

## Czech History

NOVEMBER 8, 1620: Battle of White Mountain

1836–1867: František Palacký's *History of Bohemia* is published

MARCH 7, 1850: Birth of Tomáš Masaryk

MAY 28, 1884: Birth of Edvard Beneš

OCTOBER 28, 1918: Czechoslovakia declares independence

NOVEMBER 11, 1918: Armistice Day, end of World War I

NOVEMBER 14, 1918: Tomáš Masaryk becomes president

JUNE 1, 1925: Jan Masaryk becomes ambassador to Great Britain

## Prelude to War

JANUARY 30, 1933: Hitler comes to power

MAY 16, 1935: Czechoslovak-USSR Treaty signed

MAY 19, 1935: Czechoslovak elections; big gains for German nationalists

DECEMBER 18, 1935: Beneš becomes president

SEPTEMBER 14, 1937: Death of Tomáš Masaryk

SEPTEMBER 15, 1938: First Hitler-Chamberlain meeting, Berchtesgaden

SEPTEMBER 22, 1938: Second Hitler-Chamberlain meeting, Godesberg

SEPTEMBER 30, 1938: Third Hitler-Chamberlain meeting, joined by Daladier and Mussolini, Munich agreement

OCTOBER 1, 1938: German troops enter the Sudetenland

OCTOBER 5, 1938: Beneš resigns

OCTOBER 22, 1938: Beneš enters exile

MARCH 14, 1939: Slovakia declares independence

MARCH 15, 1939: Germany invades what is left of Czecho-Slovakia; declares protectorate of Bohemia and Moravia

AUGUST 23, 1939: Hitler-Stalin pact

## *World War II*

### 1939

SEPTEMBER 1: Germany invades Poland

SEPTEMBER 3: Great Britain and France declare war on Germany; Czechoslovak National Committee created

NOVEMBER 30: The Soviet Union invades Finland

### 1940

APRIL 9: Germany invades Norway and Denmark

MAY 10: Winston Churchill becomes prime minister

MAY 10: Germany invades the Low Countries, then France

JUNE 22: Capitulation of France

JULY 21: The British recognize the provisional Czechoslovak government in exile

AUGUST: The Battle of Britain commences

SEPTEMBER 7: Bombing shifts from coastal areas to London; the Blitz begins

NOVEMBER 13: Beneš moves to Aston Abbotts

### 1941

JUNE 22: Germany invades the USSR

JULY 18: Great Britain and Soviet Union recognize Czechoslovak government in exile

SEPTEMBER 27: Reinhard Heydrich is appointed acting reichsprotektor

DECEMBER 7: Japan attacks Pearl Harbor; the United States enters the war the next day

### 1942

MAY 27: Attack by Czechoslovak assassins on Heydrich

JUNE 4: Heydrich dies

JUNE 10: Destruction of Lidice

JUNE 18: Assassins trapped in church basement; shot or commit suicide

AUGUST 5: Great Britain officially revokes the Munich agreement

## 1943

JANUARY: Churchill and FDR meet in Casablanca

MAY 12: Beneš begins visit to Washington

JULY 10: Allied invasion of Sicily commences

SEPTEMBER 3: Italy surrenders

NOVEMBER 28–December 1: Meeting of Big Three in Tehran

DECEMBER 12: In Moscow, Beneš signs treaty with USSR

## 1944

JUNE 6: Normandy Invasion, D-Day

JUNE 13: Germans begin V-1 ("doodlebug") attacks

AUGUST 1: Warsaw uprising

AUGUST 25: Liberation of Paris

AUGUST 29: Slovak uprising begins

SEPTEMBER 3: Allies take Brussels

SEPTEMBER 12: First V-2 bombs ("gooney birds") launched against Great Britain

DECEMBER 16: Battle of the Ardennes begins

## 1945

FEBRUARY 4–11: The Big Three meet in Yalta

MARCH 11: Beneš flies to Moscow

APRIL 4: Beneš, in Kosice, announces the program of the postwar Czechoslovak government

APRIL 12: Franklin Roosevelt dies

APRIL 25: UN conference begins in San Francisco

APRIL 28: Mussolini is killed

APRIL 30: Hitler commits suicide

MAY 5: Prague uprising begins

MAY 8: V-E Day

## Postwar

1945

MAY 9: Red Army enters Prague

MAY 16: Government in exile returns to Prague

JULY 17–AUGUST 2: Potsdam meeting of the heads of the Allied powers

DECEMBER: U.S. and Soviet troops withdraw from Czechoslovakia

1946

MARCH 5: Churchill's "Iron Curtain" speech

MAY 26: Czechoslovak Communists triumph in parliamentary elections

JULY 29–OCTOBER 15: Paris Peace Conference

1947

JUNE 5: Marshall Plan is announced

JULY 9: Stalin forbids Czechoslovak participation in Marshall Plan

1948:

FEBRUARY 25: Communist coup

MARCH 10: Jan Masaryk found dead

MARCH 13: Jan Masaryk's funeral

JUNE 7: Beneš resigns

SEPTEMBER 3: Beneš dies

1952

DECEMBER 3: Vlado Clementis and other Czechoslovak officials condemned and executed.

## Körbel (Korbel)–Spiegel Chronology

JUNE 7, 1878: Arnošt Körbel is born

SEPTEMBER 20, 1909: Josef Körbel is born

MAY 11, 1910: Anna Spiegelová is born

1933: Josef Körbel completes his doctorate

NOVEMBER 22, 1934: Josef Körbel joins the Czechoslovak Ministry of Foreign Affairs

APRIL 20, 1935: Wedding of Josef Körbel and Anna Spiegelová

JANUARY 1937: Josef Körbel assigned to embassy in Belgrade

MAY 15, 1937: Maria Jana "Madlenka" Körbelová is born

NOVEMBER 1938: Josef Körbel is withdrawn from Belgrade

MARCH 25, 1939: Körbel family escapes Prague

MAY 1939: Korbel family arrives in England

JULY 1, 1939: Dáša Deimlová boards "Winton" train in Prague

SEPTEMBER 1939: First BBC broadcasts by Czechoslovak government in exile

SUMMER 1940: Korbel family moves to Princes House, 52 Kensington Park Road, Notting Hill Gate

MAY 1941: Family moves in briefly with Jan "Honza" Körbel's family

JUNE 11, 1942: Růžena Spiegelová arrives in Terezín; three days later, is sent on transport to the east, probably to Trawniki

JULY 30, 1942: Arnošt and Olga Körbel arrive at Terezín

SEPTEMBER 18, 1942: Arnošt Körbel dies at Terezín

OCTOBER 7, 1942: Kathy Korbelová is born; Madeleine starts kindergarten (Kensington High School for Girls)

NOVEMBER 26, 1942: Rudolf Deiml and Greta and Milena Deimlová arrive at Terezín

FEBRUARY 15, 1943: Greta Deimlová dies of typhoid at Terezín

MAY 1943: Korbel family moves to Walton-on-Thames (shares house with Goldstücker family); Madeleine enrolls in Ingomar school

SEPTEMBER 28, 1944: Rudolf Deiml is transported to Auschwitz

OCTOBER 23, 1944: Olga Körbelová and Milena Deimlová are transported to Auschwitz

MAY 1945: Josef Korbel returns to Prague

JULY 1945: Mandula, Madeleine, and Kathy Korbelová and Dáša Deimlová return to Prague

SEPTEMBER 28, 1945: Korbels arrive in Belgrade

JUNE–AUGUST 1946: Josef Korbel attends Paris Peace Conference

JANUARY 15, 1947: Jan "John" Korbel is born, Belgrade

FEBRUARY 5, 1948: Josef Korbel is asked by Czechoslovak government
to serve as its representative on the UN Commission on Kashmir

MAY 13, 1948: Josef Korbel is officially named to UN Commission on
Kashmir

NOVEMBER 11, 1948: Korbel family (except for Josef) arrives in United
States

DECEMBER 1948: Josef Korbel joins family in United States

JUNE 7, 1949: Korbel family is granted political asylum in the United
States

# NOTES

### Setting Out

4   **"so she is not"**: Božena Němcová, *The Grandmother: A Story of Country Life in Bohemia [1852]* (Chicago: A. C. McClurg, 1892), 231.

6   **"On a high mountain"**: Mandula Korbel, unpublished essay, 1977.

### PART I: BEFORE MARCH 15, 1939
#### *1. An Unwelcome Guest*

15   **"The Czechs may squeal"**: Adolf Hitler, quoted in Callum MacDonald and Jan Kaplan, *Prague: In the Shadow of the Swastika* (London: Quartet Books, 1995), 19.

#### *2. Tales of Bohemia*

17   **"A scholar,"** wrote my father: Josef Korbel, *Twentieth Century Czechoslovakia: The Meaning of Its History* (New York: Columbia University Press, 1977), 5.

18   **"Promised Land"**: Quotations in this section are adapted from the Bohemian chronicle of Cosmas (1045–1125). Secondary sources include Kathy and Joe T. Vosoba, *Tales of the Czechs* (Wilber, Neb.: Nebraska Czechs of Wilber, 1983), and J. M. Lutzow, *The Story of Prague* (London: Dent, 1902).

23   **"This wicked people"**: Pope Pius II, quoted in J. V. Polišenský, *History of Czechoslovakia in Outline* (Prague: Bohemia International, 1991), 48.

26   **"My whole life"**: Jan Ámos Komenský, quoted in Vosoba, *Tales of the Czechs*, 53.

#### *3. The Competition*

31   **"A Czech does not rely"**: Karel Havlíček, "Pan-Slavism Declined," in *From Absolutism to Revolution (1648–1848)*, ed. Herbert H. Rowen (New York: Macmillan, 1963), 289.

31   **"When a Czech owns"**: Ladislav Holy, The *Little Czech and the Great Czech Nation* (Cambridge, England: Cambridge University Press, 1996), 75.

31   **"The Czech skull is impervious"**: Theodor Mommsen, quoted in Tomáš G. Masaryk, *Problem of a Small Nation* (Prague: Trigon Press, 2010), 66.

33  **"If every Czech person"**: Tara Zahra, *Kidnapped Souls: National Indifference and the Battle for the Children in the Bohemian Lands, 1900–1948* (Ithaca, N.Y.: Cornell University Press, 2008), 30.

35  **"Poor Jews"**: Herzl, quoted in Lisa Rothkirchen, *Jews of Bohemia and Moravia* (Lincoln: University of Nebraska Press; Jerusalem: Yad Vashem, 2005), 21.

35  **"all kinds of ways"**: T. G. Masaryk, quoted in Karel Čapek, *Talks with T. G. Masaryk* (North Haven, Conn.: Catbird Press, 1995), 42–43.

37  **"He didn't know"**: Ibid., 143.

38  **"love of nation"**: T. G. Masaryk, *Problem of a Small Nation*, 22.

38  **"Every Sunday," he said**: T. G. Masaryk, quoted in Čapek, *Talks with T. G. Masaryk*, 77.

## 4. The Linden Tree

40  **"artificial state"**: T. G. Masaryk, memorandum for British friends, April 1915, included in R. W. Seton-Watson, *Masaryk in England* (Cambridge, England: Cambridge University Press; New York: Macmillan, 1943), 122–123.

41  **"Death to the enemy!"**: Jaroslav Hašek, *The Good Soldier Švejk* (New York: Penguin, 1973), 213.

42  **"It is up to us"**: Jan Janák, quoted in Victor S. Mamatey, "The Birth of Czechoslovakia: Union of Two Peoples," in *Czechoslovakia: The Heritage of Ages Past: Essays in Memory of Josef Korbel*, ed. Hans Brisch and Ivan Volgyes (New York: East European Quarterly, Columbia University Press, 1979), 81.

42  **"somewhat touchy"**: T. G. Masaryk, *Světová Revoluce: Za Valky a ve Valce* (Prague: Čin-Praha, 1938), 365.

43  **"all branches of the Slav race"**: Secretary of State Robert Lansing, June 24, 1918, quoted in *Foreign Relations of the United States, 1918*, vol. 10, suppl. 1 ( Washington, D.C.: United States Government Printing Office, 1933), 816.

43  **"I was just nine years old"**: J. Korbel, essay written for fiftieth anniversary of Czechoslovak Independence Day, unpublished.

44  **"Beneš and Masaryk were"**: Margaret MacMillan, *Paris 1919: Six Months That Changed the World* (New York: Random House, 2001), 229–230.

44  **"at eleven o'clock"**: Isaiah Bowman, U.S. delegate, quoted in Mary Heimann, *Czechoslovakia: The State That Failed* (New Haven, Conn.: Yale University Press, 2010), 58.

49  **"trousers hung like an accordion"**: J. Korbel, unpublished manuscript.

50  **"I had a very nice childhood"**: M. Korbel, interview by Katie Albright, unpublished.

50  **"Joe was certainly a man"**: M. Korbel, essay, 1977.

52  **"As other European countries went"**: J. Korbel, quoted in Madeleine Albright, *Madam Secretary* (New York: Miramax Books, 2003), 6.

53  **"There he was"**: J. Korbel, speech draft in honor of T. G. Masaryk's one hundredth birthday, February 27, 1950, unpublished.

54  **"captivated by radicalism and socialism"**: Compton Mackenzie, *Dr Beneš* (London: George G. Harrap, 1946), 36–37.

55  **"rip the enemy apart"**: Kliment Voroshilov, quoted in Igor Lukes, *Czechoslovakia Between Stalin and Hitler: The Diplomacy of Edvard Beneš in the 1930s* (New York: Oxford University Press, 1996), 53–55.

56 **"That man for chancellor?"**: President Hindenburg, quoted in Winston Churchill, *The Second World War,* vol. 1: *The Gathering Storm* (London: Houghton-Mifflin, 1948), 62.

### 5. A Favorable Impression

58 **"You heard only"**: Peter Demetz, *Prague in Black and Gold: Scenes from the Life of a European City* (New York: Hill and Wang, 1997), 363.

58 **"the awakener"**: Edvard Beneš, funeral oration for T. G. Masaryk, September 21, 1937, quoted in Compton Mackenzie, *Dr Beneš* (London: George G. Harrap, 1946), 138.

59 **"My satisfaction," he explained**: Masaryk, quoted in Karel Čapek, *Talks with T. G. Masaryk* (North Haven, Conn.: Catbird Press, 1995), 248.

60 **"We love Beneš"**: Josef Korbel, *Tito's Communism* (Denver: University of Denver Press, 1951), 4.

61 **"We shall never be able"**: Adolf Hitler, quoted in J. W. Bruegel, *Czechoslovakia Before Munich* (London: Cambridge University Press, 1973), 160.

62 **"We have lived with the Czech[s]"**: Franz Spina, December 26, 1926, quoted in ibid., 79.

63 **"To judge by his personality"**: R. H. Hadow, British legation in Prague, cable to London, December 27, 1935, quoted in ibid., 137.

63 **"One wonders why Dr. Beneš"**: Hadow, cable to London, January 31, 1936, quoted in ibid., 137.

63 **"He makes a most favorable"**: Vansittart, quoted in ibid., 138–139.

64 **"He goes wherever I do"**: Masaryk, quoted in Sir Robert Bruce Lockhart, *Comes the Reckoning* (London: Putnam, 1947), 61.

65 **"an inexorable first draft"**: Thomas Mann, foreword to Erika Mann, *School for Barbarians* (New York: Modern Age Books, 1938), 6–7.

66 **"the glory of the German nation"**: Ibid.

66 **"education must have the sole object"**: Hitler, quoted in ibid., 20.

67 **"History is replete"**: Churchill, *Great Contemporaries* (New York: W. W. Norton, 1990), 165.

68 **"Those who have met Hitler"**: Ibid., 170.

68 **"I only wish"**: Lloyd George, quoted in Lynne Olson, *Troublesome Young Men: The Churchill Conspiracy of 1940* (New York: Farrar, Straus and Giroux, 2007), 68–69.

69 **"On all these matters"**: Halifax, quoted in Igor Lukes, *Czechoslovakia Between Stalin and Hitler: The Diplomacy of Edvard Beneš in the 1930s* (New York: Oxford University Press, 1996), 82–83.

70 **"It is well"**: Baldwin, address to the House of Commons, November 1933, quoted in Telford Taylor, *Munich: The Price of Peace* (Garden City, N.Y.: Doubleday, 1979), 211.

71 **"It takes time"**: John F. Kennedy, *Why England Slept* (New York: Wilfred Funk, 1940), 5.

### 6. Out from Behind the Mountains

72 **"Before civilization dawned"**: Voskovec and Warich, quoted in Hana Stránská, unpublished manuscript, 1994.

73 **"I'd dedicate the combined power"**: T. G. Masaryk, quoted in Karel Čapek, *Talks with T. G. Masaryk* (North Haven, Conn.: Catbird Press, 1995), 247.

75 **"respect of the human person"**: Beneš, speaking in Liberec, August 19, 1936, quoted in Radomír Luža, *The Transfer of the Sudeten Germans: A Study of Czech-German Relations, 1933–1962* (New York: New York University Press, 1964), 90.

75 **"really wishes to improve"**: Ernst Eisenlohr, German minister in Prague, November 11, 1937, quoted in J. W. Bruegel, *Czechoslovakia Before Munich* (London: Cambridge University Press, 1973), 161.

75 **"has made the internal appeasement"**: Eisenlohr, December 21, 1937, quoted in ibid., 167.

75 **If Hitler, he said**: František Moravec, *Master of Spies: The Memoirs of General František Moravec* (London: Bodley Head, 1975), 117.

75 **"I once saw the president's"**: Compton Mackenzie, *Dr Beneš* (London: George G. Harrap, 1946), 263–264.

76 **"mathematician of politics"**: Josef Korbel, *Twentieth Century Czechoslovakia: The Meaning of Its History* (New York: Columbia University Press, 1977), 129.

76 **"not in a warlike"**: Telford Taylor, *Munich: The Price of Peace* (Garden City, N.Y.: Doubleday, 1979), 368.

77 **The land where Schubert had been born**: Marcia Davenport, "Elegy for Vienna," in Marcia Davenport, *Too Strong for Fantasy* (Pittsburgh: University of Pittsburgh Press, 1967), 245.

77 **"Sagen Sie in Prag"**: Igor Lukes, *Czechoslovakia Between Stalin and Hitler: The Diplomacy of Edvard Beneš in the 1930s* (New York: Oxford University Press, 1996), 124.

77 **"even a boa constrictor"**: Exchange between Halifax and Jan Masaryk, London, March 13, 1938, quoted in ibid., 129.

77 **"We must always demand"**: Henlein, quoted in ibid., 142.

78 **"solve the German problem"**: Directive by Supreme Commander, Wehrmacht, December 21, 1937, quoted in Breugel, *Czechoslovakia Before Munich*, 185.

78 **"It was of utmost importance"**: Harwood L. Childs and John B. Whitton, *Propaganda by Short Wave* (Princeton, N.J.: Princeton University Press, 1942), 37.

79 **To lift his spirits**: Madeleine Jana Korbel, "Zdeněk Fierlinger's Role in the Communization of Czechoslovakia: The Profile of a Fellow Traveler," Wellesley College, May 1959, 24.

## 7. *"We Must Go On Being Cowards"*

81 **"You have only to look"**: Alexander Cadogan, *The Diaries of Sir Alexander Cadogan (1938–1945)*, ed. David Dilks (New York: G. P. Putnam's Sons, 1972), 65.

82 **"His Majesty's Government could not"**: Ibid., 78.

82 **"the German Government"**: Halifax, quoted in J. W. Bruegel, *Czechoslovakia Before Munich* (London: Cambridge University Press, 1973), 199.

83 **"Is it not positively horrible"**: Chamberlain, quoted in Cadogan, *Diaries*, 92.

84 **"What I wonder is"**: Ibid., 70.

84 **"We cannot but feel"**: S. Grant Duff, *Europe and the Czechs* (Harmondsworth, England: Penguin, 1938), 200.

86 **"This misery of the Sudeten Germans"**: Hitler, addressing Nazi Party Congress, Nuremberg, September 12, 1938, quoted in Compton Mackenzie, *Dr Beneš* (London: George G. Harrap, 1946), 12.

87  **"I am in no way willing"**: Hitler, addressing Nazi Party Congress, Nuremberg, September 12, 1938, quoted in Cadogan, *Diaries*, 97.

87  **"not in any way ill-natured"**: Major Reginald Sutton-Pratt, quoted in Igor Lukes, *Czechoslovakia Between Stalin and Hitler: The Diplomacy of Edvard Beneš in the 1930s* (New York: Oxford University Press, 1996), 212.

87  **"The morale of the German soldier"**: Memo of Czechoslovak armed forces chief of staff, General Ludvík Krejčí, to the Supreme State Defense Council, September 9, 1938, quoted in Jiří Doležal and Jan Křen, eds., *Czechoslovakia's Fight* (Prague: Publishing House of the Czechoslovak Academy of Sciences, 1964), 15–17.

88  **"We must warn Hitler"**: Nicolson, quoted in Lynne Olson, *Troublesome Young Men: The Churchill Conspiracy of 1940* (New York: Farrar, Straus and Giroux, 2007), 129.

88  **"None of us can even think"**: Henderson, quoted in Bruegel, *Czechoslovakia Before Munich*, 255.

89  **"cruel, overbearing [and] . . . ruthless"**: Ambassador Joseph Kennedy, cable to Washington, D.C., from U.S. Embassy in London, September 17, 1938, in *Foreign Relations of the United States, 1938*, vol. 1, (Washington, D.C., United States Government Printing Office, 1955), 610.

91  **"It depends solely on you"**: Letter from Czechoslovak patriots, quoted in Josef Korbel, *Twentieth Century Czechoslovakia: The Meaning of Its History* (New York: Columbia University Press, 1977), 131–132.

91  **"It is hence understandable"**: Beneš, quoted in Mackenzie, *Dr Beneš*, 207.

92  **That was "inevitable"**: Alexander Cadogan, *The Diaries of Sir Alexander Cadogan (1938–1945)*, ed. David Dilks (New York: G. P. Putnam's Sons, 1972), 102.

93  **"A week ago"**: Ibid., 103.

93  **"had a narrow mind"**: Chamberlain, in notes from meeting of British cabinet, September 24, 1938, quoted in Bruegel, *Czechoslovakia Before Munich*, 284.

93  **"extremely anxious to secure"**: Cadogan, *Diaries*, 104.

94  **"It was obvious"**: Prokop Drtina, quoted in Korbel, *Twentieth Century Czechoslovakia*, 135.

94  **"The national will manifested"**: Ibid., 126–127.

95  **"Two men stand arrayed"**: Hitler, September 26, 1938, quoted in Mackenzie, *Dr Beneš*, 13.

96  **"I'm wobbling all over"**: Chamberlain, quoted in Telford Taylor, *Munich: The Price of Peace* (Garden City, N.Y.: Doubleday, 1979), 884.

96  **"How horrible, fantastic, incredible"**: Chamberlain, September 27, 1937, quoted in Cadogan, *Diaries*, 108.

96  **"However much we may sympathize"**: Ibid., 108.

99  **"tired, but pleasantly tired"**: Chamberlain, quoted in Lukes, *Czechoslovakia Between Stalin and Hitler*, 254.

99  **"The representatives of the Czechoslovak army"**: Beneš, quoted in Korbel, *Twentieth Century Czechoslovakia*, 139.

100 **"A war—a big European war"**: Beneš, quoted in ibid., 139.

## 8. A Hopeless Task

101 **"Our personal security"**: M. Korbel, essay.

101 **"in her hour of crisis"**: Josef Korbel, *Twentieth Century Czechoslovakia* (New York: Columbia University Press, 1977), 147.

102 **"the valiant ethos"**: Ibid., 148.

102 **"Beneš was wrong to yield"**: Winston Churchill, *The Second World War*, vol. 1: *The Gathering Storm* (London: Houghton-Mifflin, 1948), 272.

104 **"preserved for the exacting tasks"**: George Kennan, *From Prague After Munich, 1938–1940* (Princeton, N.J.: Princeton University Press, 1968), 5.

105 **"Certainly not"**: Keitel, testimony at the Nuremberg trials, quoted in Churchill, *The Gathering Storm*, 286.

105 **"Some day the Czechs will see"**: Chamberlain, letter to archbishop of Canterbury, quoted in Telford Taylor, *Munich: The Price of Peace* (Garden City, N.Y.: Doubleday, 1979), 66.

106 **"Oh Mr. Masaryk"**: Lady Chamberlain, quoted in Harold Nicolson, *The War Years: Diaries and Letters 1939–1945*, ed. Nigel Nicolson (New York: Atheneum, 1967), 354–355.

106 **"We have suffered a total"**: Churchill, address to the House of Commons, October 13, 1938, quoted in Stanislav Fejfar, *A Fighter's Call to Arms: Defending Britain and France Against the Luftwaffe 1940–42*, ed. Norman Franks with Simon Muggleton (London: Grub Street, 2010), 13.

107 **"the first prime minister"**: Dorothy Parker, quoted in Sir Robert Bruce Lockhart, *Comes the Reckoning* (London: Putnam, 1947), 23.

110 **"The question is"**: Hitler, quoted in Victor S. Mamatey and Radomir Luža, eds., *A History of the Czechoslovak Republic 1918–1948* (Princeton, N.J.: Princeton University Press, 1973), 268.

111 **"What would they do"**: Veronika Herman Bromberg, "Tell Me Again So I Won't Forget: My Father's Stories of Survival and Courage During World War II," unpublished manuscript, 29.

113 **"There are several thousand"**: U.S. Embassy, Prague, cable to Department of State, March 19, 1939. *Foreign Relations of the United States*, 1939, vol 5, *Diplomatic Matters* (Washington, D.C., Government Printing Office, 1955), 310.

113 **"To leave Czechoslovakia immediately"**: M. Korbel, unpublished essay.

## PART II: APRIL 1939–APRIL 1942
### 9. Starting Over

118 **"common butchery"**: New York mayor Fiorello La Guardia, quoted in Edvard Beneš, *Memoirs of Dr Edvard Beneš* (Boston: Houghton Mifflin, 1954), 61.

118 **"I'll be back"**: Franklin Roosevelt, quoted in "Roosevelt War Talk Begins and Roosevelt Peace Call Ends a Fateful Week of Power Politics," *Life*, April 24, 1939, 20.

118 **"the future of American youth"**: Walter Winchell, Walter Lippmann, and David Lawrence, quoted in "The Nation's Columnists Divide in Great Debate on American War & Peace," ibid., 24–25.

118 **"prevent the hideous consequences"**: Ibid.

118 **"a second peace conference"**: Ibid.

119 **"political commitments"**: 1936 Democratic Party platform, Democratic Party Book, 1936, 24.

119 **"I wonder"**: Eleanor Roosevelt, quoted in "The Nation's Columnists Divide in Great Debate on American War & Peace," *Life*, April 24, 1939, 24–25.

120 **"I don't know how things"**: Churchill, quoted in Harold Nicolson, *The War Years: Diaries and Letters 1939–1945*, ed. Nigel Nicolson (New York: Atheneum, 1967), 166.

120 **"Here is a key"**: Korbel, "Portrait of J. Masaryk," unpublished manuscript.

120 **"clever as a bag of monkeys"**: Shiela Grant Duff, *The Parting of Ways: A Personal Account of the Thirties* (London: Peter Owen, 1982), 129.

121 **"You cannot hope"**: Quoted in ibid., 72.

122 **"a man called Zid"**: J. Korbel, letter to Hubert Ripka, June 21, 1939.

122 **"I wasn't allowed"**: Nicholas Winton, quoted in Mark Jonathan Harris and Deborah Oppenheimer, *Into the Arms of Strangers: Stories of the Kindertransport* (New York: Bloomsbury, 2000), 151.

124 **"We took her over"**: J. Korbel, postscript on a letter from Dáša Deimlová to her parents, July 2, 1939.

125 **"If the worst came to the worst"**: Jan Struther, *Mrs. Miniver* (New York: Harcourt, Brace, 1940), 148.

127 **"Of course it's all a game"**: Stalin, quoted in Ian Kershaw, *Fateful Choices: Ten Decisions That Changed the World, 1940–1941* (New York: Penguin, 2007), 258.

## 10. Occupation and Resistance

129 **"Prague children beg for food"**: Anecdote adapted from George Kennan, *From Prague After Munich, 1938–1940* (Princeton, N.J.: Princeton University Press, 1968), 117.

130 **"I am looking forward"**: Emil Hácha, quoted in Jan Drabek, "V Krajina: Hero of European Resistance and Canadian Wilderness," unpublished manuscript, 31.

131 **"one of humanity's oldest"**: Kennan, *From Prague After Munich*, x.

133 **"If German authority"**: Ibid., 235.

138 **"With great fanfare"**: Tahra Zahra, *Kidnapped Souls: National Indifference and the Battle for the Children in the Bohemian Lands 1900–1948* (Ithaca, N.Y.: Cornell University Press, 2008), 203.

139 **"friends went shopping"**: Miroslav Karny, "The Genocide of the Czech Jews," ed. David P. Stern, 2009, www.phy6.org/outreach/Jewish/TerezinBook.htm.

140 **"Milena cried a lot"**: Greta Deimlová, letter to Dáša Deimlová, July 6, 1939.

## 11. The Lamps Go Out

144 **"The hour of retribution"**: J. Masaryk, Czech-language BBC broadcast, September 8, 1939.

145 **"We huddled around"**: Radomír Luža and Christina Vella, *The Hitler Kiss: A Memoir of the Czech Resistance* (Baton Rouge: Louisiana State University Press, 2002), 29.

147 **"sat down on air"**: BBC Written Archives Centre, Caversham Park.

148 **"We shall fight"**: Alexander Cadogan, *The Diaries of Sir Alexander Cadogan (1938–1945)*, ed. David Dilks (New York: G. P. Putnam's Sons, 1972), 214.

148 **"People call me defeatist"**: Lloyd George, quoted in Lynne Olson, *Troublesome Young Men: The Churchill Conspiracy of 1940* (New York: Farrar, Straus and Giroux, 2007), 258.

148 **One refugee compared**: Sonia Tomara, "Nazi-Red Animosity Described Along Tense Frontier Border in Poland," *New York Herald Tribune*, November 20, 1939, in *Reporting World War II*, part 1: *American Journalism 1938–1944* (New York: Library of America, 1995), 30.

149 **"ten times as confident"**: Chamberlain, address to the House of Commons, London, April 4, 1940, quoted in Olson, *Troublesome Young Men*, 276.

150 **"When Winston was born"**: Baldwin, quoted in Ian Kershaw, *Fateful Choices: Ten Decisions That Changed the World, 1940–1941* (New York: Penguin, 2007), 21.

151 **"The flat countryside"**: Rommel, quoted in John Carey, ed., *Eyewitness to History* (New York: Avon Books, 1987), 529.

151 **"an awful day"**: Cadogan, *Diaries*, 283.

152 **"I trust you realize"**: Churchill, quoted in Kershaw, *Fateful Choices*, 209.

152 **"holding the bag"**: Kennedy, quoted in ibid., 210.

152 **"drunken French soldiers"**: Jan Stránský, *East Wind over Prague* (New York: Random House, 1950), 10.

153 **"on those terrible encumbered"**: Ibid.

153 **"position of B.E.F. quite awful"**: Cadogan, *Diaries*, 290–291.

153 **"From the margin of the sea"**: John Charles Austin, quoted in Carey, ed., *Eyewitness to History*, 532.

153 **"blood, toil, tears, and sweat"**: Churchill, address to the House of Commons, May 13, 1940, http://churchill-society-london.org.uk/SpchIndx.html.

153 **"fight on the seas"**: Churchill, address to the House of Commons, June 4, 1940, ibid.

154 **"The Battle of Britain"**: Churchill, address to the House of Commons, June 18, 1940, ibid.

### 12. The Irresistible Force

157 **"charming . . . a pleasure"**: Prokop Drtina, *Československo Můj Osud* (Prague: Melantrich, 1991), 564.

157 **"Madlenka is very cute"**: Dáša Deimlová, letter to her parents, January 1940.

157 **"When it rains"**: Dáša Deimlová, letter to her parents, July 9, 1939.

158 **"We were living"**: M. Korbel, unpublished essay.

160 **"pretty awful"**: Cadogan, *Diaries*, 273.

163 **"That is where the officers"**: Renata A. Kauders, "From Prague to Denver: Sketches from My Life," unpublished manuscript, 38.

164 **"the unbounded French drive"**: Hitler, quoted in Hans Kohn, ed., *The Modern World (1848–Present)* (New York: Macmillan, 1968), 223.

164 **"the war is over, Hermann"**: Hitler, quoted in Marcel Jullian, *The Battle of Britain, July–September 1940* (New York: Orion, 1967), 6.

165 **"Since, despite its desperate"**: Hitler, July 16, 1940, quoted in ibid., 24.

165 **"I feel it is my duty"**: Hitler, July 19, 1940, quoted in ibid., 35–36.

165 **"It is a pity!"**: Joyce, quoted in J. A. Cole, *Lord Haw-Haw: The Full Story* (New York: Farrar, Straus and Giroux, 1964), 164.

166 **"German parachutists"**: quoted in Jullian, *The Battle of Britain*, 78.

167 **"Madam, there is no honey"**: Philip Ziegler, *London at War, 1939–1945* (New York: Alfred A. Knopf, 1995), 159.

### 13. Fire in the Sky

169 **"From Reichsmarshal Göring"**: Göring, August 13, 1940, quoted in Richard Hough and Denis Richards, *The Battle of Britain* (New York: W. W. Norton, 1989), 154.

172   **A Luftwaffe pilot**: William Shirer, "Berlin After a Year of War: September 1940," in *Reporting World War II*, part 1: *American Journalism 1938–1944* (New York: Library of America, 1995), 121–123.

173   **"Since they attack our cities"**: Hitler, September 4, 1940, quoted in Hough and Richards, *The Battle of Britain*, 244.

173   **"The British have been asking"**: Hitler, September 4, 1940, quoted in J. A. Cole, *Lord Haw-Haw: The Full Story* (New York: Farrar, Straus and Giroux, 1964), 171.

174   **"Suddenly we were gaping"**: Desmond Flower, quoted in John Carey, ed., *Eyewitness to History* (New York: Avon Books, 1987), 537–538.

177   **"I'm so sorry"**: Roosevelt, quoted in Hough and Richards, *The Battle of Britain*, 294.

178   **"generally a very pleasant"**: Mrs. Orlow Tollett, interview by Libby Cook and Sonia Knight, London, early 2011.

178   **"The pub had a direct hit"**: Tollett, interview for HistoryTalk, Reminiscence at Home, London Care Connections, David Welsh, coordinator, May 20, 2004. Interview manuscript forwarded courtesy of Isobel Czarska.

178   **"Don't be afraid"**: Anecdote based on the recollections of the author's cousin Dáša (Deimlová) Simová.

179   **"The whizz of a flying bomb"**: Prokop Drtina, *Československo Můj Osud* (Prague: Melantrich, 1991), 573.

179   **"I have become"**: Hess, quoted in Stanislav Fejfar, *A Fighter's Call to Arms: Defending Britain and France Against the Luftwaffe 1940–42*, ed. Norman Franks with Simon Muggleton (London: Grub Street, 2010), 91–92.

179   **"We were flying"**: Fejfar, quoted in ibid., 94.

180   **"Stanislav did not take"**: Mrs. Fejfar, quoted in ibid., 14.

## 14. The Alliance Comes Together

182   **"a man would not say"**: Roosevelt, press conference, Washington, D.C., December 17, 1940, quoted in Ian Kershaw, *Fateful Choices: Ten Decisions That Changed the World, 1940–1941* (New York: Penguin, 2007), 227.

183   **"They're sustained in part"**: Edward R. Murrow, "The London Blitz, September 1940," in *Reporting World War II*, part 1: *American Journalism 1938–1944* (New York: Library of America, 1995), 87, 93.

183   **"There is a tremendous vitality"**: James R. Reston, "Nazi Fliers Foiled by London's Smoke," *New York Times*, October 23, 1940.

183   **"Please pass the marmalade"**: Brian Meredith, radio broadcast, July 4, 1940, quoted in Harwood L. Childs and John B. Whitton, *Propaganda by Short Wave* (Princeton, N.J.: Princeton University Press, 1942), 134.

184   **"It is industrial England"**: J. B. Priestley, radio broadcast, quoted in ibid., 117.

185   **"I suppose you wish to know"**: Hopkins, January 1941, quoted in Alexander Cadogan, *The Diaries of Alexander Cadogan (1938–1948)*, ed. David Dilks (New York: G. P. Putnam's Sons, 1972), 348.

187   **"at last an Englishman"**: Beneš, quoted in Vít Smetana, *In the Shadow of Munich: British Policy Towards Czechoslovakia from the Endorsement to the Renunciation of the Munich Agreement (1938–1942)* (Prague: Charles University, Karolinum Press, 2008), 217.

188   **"I see no reason"**: Churchill, memo to Eden, April 20, 1941, quoted in ibid., 217.

194 **"Any man or state"**: Churchill, radio broadcast, June 22, 1941, quoted in Cadogan, *Diaries*, 389.

196 **"heroic vanguard of mankind"**: Josef Korbel, *The Communist Subversion of Czechoslovakia (1938–1948): The Failure of Coexistence* (Princeton, N.J.: Princeton University Press, 1959), 56.

196 **"You who labor in factories"**: J. Masaryk, radio broadcast, September 1941, quoted in Jan Masaryk, *Jan Masaryk: Speaking to My Country* (London: Lincolns-Prager, 1944), 122.

## 15. The Crown of Wenceslas

201 **"Not a single German"**: Heydrich, October 2, 1941, quoted in Jiří Doležal and Jan Křen, eds., *Czechoslovakia's Fight* (Prague: Publishing House of the Czechoslovak Academy of Sciences, 1964), 60–66.

202 **"If we give these gourmands"**: Hitler, quoted in Vojtech Mastny, *The Czechs Under Nazi Rule: The Failure of National Resistance, 1939–1942* (New York: Columbia University Press, 1971), 206.

202 **"always remained a happy"**: Bormann, June 7, 1942, quoted in *Reinhard Heydrich: The Ideal National Socialist* (Lincoln, Nebr.: Preuss Press, 2004), 48.

203 **"The Jew," wrote Heydrich**: Heydrich, quoted in ibid., 40.

203 **"I too am a Zionist"**: Eichmann, quoted in Cara de Silva, ed., *In Memory's Kitchen: A Legacy from the Women of Terezin* (Northvale, N.J.: Jason Aronson, 1996), xxxviii n.

207 **"Theresienstadt will allow us"**: Eichmann, quoted in George E. Berkley, *Hitler's Gift: The Story of Theresienstadt* (Boston: Branden Books, 1995), 58.

207 **"a spectacular action"**: Beneš, quoted in František Moravec, *Master of Spies: The Memoirs of General František Moravec* (London: Bodley Head, 1975), 210.

209 **"One of them seemed"**: Táborský, quoted in Callum MacDonald, *The Killing of Reinhard Heydrich, the SS "Butcher of Prague"* (New York: Da Capo, 1989), 125.

209 **"they were both ordinary chaps"**: Vladimir Skacha, quoted in Miraslav Ivanov, *Target: Heydrich* (New York: Macmillan, 1972), 51.

## PART III: MAY 1942–APRIL 1945
### 16. Day of the Assassins

213 **"If you need anything"**: Marie Moravcová, quoted by Marie Soukupová, quoted in Miraslav Ivanov, *Target: Heydrich* (New York: Macmillan, 1972), 68.

215 **"You see that wooden crate, Ata?"**: František Spinka, quoted in ibid., 117.

215 **"In this situation"**: Beneš, radio broadcast of May 15, 1942, quoted in ibid., 198.

216 **"BBC broadcasts call too much"**: Memo dated May 21, 1942, unsigned and on blank paper, with the name "Korbel" scribbled in the upper-right corner, obtained by the author from the Czech Foreign Ministry.

218 **"in a bad way"**: František Sitta, quoted in Ivanov, *Target: Heydrich*, 174.

218 **"The shots which sounded"**: J. Korbel, Czechoslovak-language broadcast from London, May 30, 1942.

218 **"From certain indications"**: Masaryk, interview with NBC radio, New York City, June 15, 1942, Czechoslovak Sources and Documents, *Speeches of Jan Masaryk in America* (New York: Czechoslovak Information Service, 1942), 71–72.

219  **"traitor or a quisling"**: J. Korbel, broadcast from London, May 30, 1942.
219  **"The leaders of today's Germany"**: J. Korbel, broadcast from London, May 27, 1942.
219  **"Can Korbel be told"**: June 1942, BBC Written Archives Centre, Caversham Park.
221  **"We didn't find any traitors"**: Gestapo agent, quoted by Josef Chalupsky, quoted in Ivanov, *Target: Heydrich*, 260.
222  **"right through the whole city"**: Tereza Kasperová, quoted in ibid., 215.
228  **"if future generations"**: Knox, quoted in Radomír Luža, *The Transfer of the Sudeten Germans: A Study of Czech-German Relations, 1933–1962* (New York: New York University Press, 1964), 236 n.

## 17. Auguries of Genocide

229  **"This is the most crucial year"**: J. Masaryk, addressing the women's division of the American Jewish Congress, New York, April 28, 1942, quoted in Czechoslovak Sources and Documents, *Speeches of Jan Masaryk in America* (New York: Czechoslovak Information Service, 1942), 56.
229  **"periodic ethical and moral blackouts"**: J. Masaryk, radio broadcast, Columbia Broadcasting Company, New York, November 5, 1941, quoted in ibid., 18.
229  **"Until the war is over"**: Masaryk, address at Rollins College, Winter Park, Fla., February 4, 1942, quoted in ibid., 22.
230  **"Every Sudeten German"**: Beneš, quoted in Compton Mackenzie, *Dr Beneš* (London: George G. Harrap, 1946), 293.
230  **"President Beneš has found"**: Lockhart, memo to Halifax, October 7, 1940, quoted in Vít Smetana, *In the Shadow of Munich: British Policy Towards Czechoslovakia from the Endorsement to the Renunciation of the Munich Agreement (1938–1942)* (Prague: Charles University, Karolinum Press, 2008), 276.
231  **"puffed-up gangster"**: J. Masaryk, quotations from Jan Masaryk, *Jan Masaryk: Speaking to My Country* (London: Lincolns-Prager, 1944), 19–21.
231  **"Masaryk entered the hall"**: J. Korbel, "Portrait of J. Masaryk," unpublished manuscript.
232  **"because my charwoman"**: Sir Robert Bruce Lockhart, *Jan Masaryk: A Personal Memoir* (Norwich, England: Putnam, 1956), vii–viii.
232  **"My dear children"**: J. Korbel, "Portrait of J. Masaryk," unpublished manuscript.
234  **"a stamp album"**: Dáša Deimlová, letter to her parents, January 1940.
234  **"Daddy is always at home"**: Greta Deimlová, letter to Dáša Deimlová, August 8, 1940.
235  **"To our own Jews"**: Quoted in Lisa Rothkirchen, *Jews of Bohemia and Moravia* (Lincoln: University of Nebraska Press; Jerusalem: Yad Vashem, 2005), 184.
235  **"the reasons of higher interests"**: Beneš, quoted in Jan Láníček, "The Czechoslovak Service of the BBC and the Jews During World War II," paper distributed at "Ties That Bind," London conference commemorating the seventieth anniversary of the Czechoslovak government in exile, September 2010, 8.
236  **"the principal Nazi slaughterhouse"**: Eden, debate in the House of Commons, London, vol. 385, December 17, 1942, 2082, www.ww2talk.com/forum/holocaust/41529-mr-eden-commons-dec-1942-a.html.
236  **"eyewitness stuff"**: Murrow, "A Horror Beyond What Imagination Can Grasp,"

in *Reporting World War II*, part 1: *American Journalism 1938–1944* (New York: Library of America, 1995), 453.

236 **"All the help and relief"**: Strańský, quoted in Láníček, "The Czechoslovak Service of the BBC and the Jews During World War II," 18.

236 **"every crime"**: Beneš, quoted in Benjamin Frommer, *National Cleansing: Retribution Against Nazi Collaborators in Postwar Czechoslovakia* (Cambridge, England: Cambridge University Press, 2005), 40.

## 18. Terezín

241 **"I have to get used to"**: Olga Körbelová, letter to Greta Deimlová, July 22, 1942.

241 **"Father will have a heavy heart"**: Ibid.

241 **"We have had visitors"**: Olga Körbelová, letter to Greta Deimlová, July 29, 1942.

244 **"After they looted"**: Gerty Spies, *My Years in Theresienstadt: How One Woman Survived the Holocaust* (Amherst, N.Y.: Prometheus Books, 1997), 59–61.

249 **"most people couldn't carry"**: Letter from Hana Malka to Michael Dobbs, January 14, 1998.

250 **"Lilka's sister has died"**: Helga Weissová, quoted in Hannelore Brenner, *The Girls of Room 28: Friendship, Hope and Survival in Theresienstadt* (New York: Schocken Books, 2009), 46.

251 **"Wash your hands"**: Ibid., 48.

251 **"You didn't have to"**: Weissová, quoted in ibid., 154.

253 **"I hereby declare"**: Petr Ginz, diary, February 16, 1944, quoted in Alexandra Zapruder, ed., *Salvaged Pages: Young Writers' Diaries of the Holocaust* (New Haven, Conn.: Yale University Press, 2002), 169.

253 **"Manchuria is not"**: Ginz, quoted in Marie Rút Křížková, Kurt Jiří Kotouč, and Zdeněk Ornest, eds., *We Are Children Just the Same: Vedem, the Secret Magazine by the Boys of Terezín* (Philadelphia: Jewish Publication Society, 1995), 135.

254 **"Is a man who is given"**: Egon Redlich, *The Terezín Diary of Gonda Redlich*, ed. Saul S. Friedman (Lexington: University Press of Kentucky, 1992), 134.

255 **"Although I was not bound"**: Vera Schiff, *Theresienstadt: The Town the Nazis Gave to the Jews* (Toronto: Lugus, 1996), 74.

## 19. The Bridge Too Far

256 **"Hitler and his lot"**: Clara Emily Milburn, *Mrs. Milburn's Diaries: An Englishwoman's Day-to-Day Reflections 1939–45*, ed. Peter Donnelly (New York: Schocken Books, 1980), 167.

256 **"Our cause is internationally assured"**: František Moravec, *Master of Spies: The Memoirs of General František Moravec* (London: Bodley Head, 1975), 229.

259 **"right interesting"**: U.S. Department of State, IDS Special Report no. 574, July 6, 1943.

260 **"Roosevelt esteems your sound advice"**: Hopkins, May 18, 1943, quoted in Eduard Táborský, *President Edvard Beneš: Between East and West 1938–1948* (Stanford, Calif.: Hoover Institution Press, Stanford University, 1981), 129.

261 **"I passed . . . demolished hamlets"**: Edvard Beneš, *Memoirs of Dr Edvard Beneš* (Boston: Houghton Mifflin, 1954), 279.

262 **"Masaryk refused to accept"**: Beneš, quoted in Táborský, *President Edvard Beneš*, 135.

262 **"better than yesterday"**: Stalin, quoted in ibid., 167.

262 **"I consider all our negotiations"**: Beneš, quoted in Josef Korbel, *The Communist Subversion of Czechoslovakia (1938–1948): The Failure of Coexistence* (Princeton, N.J.: Princeton University Press, 1959), 85.

262 **"a new Soviet Union"**: Beneš, *Memoirs of Dr Edvard Beneš*, 262.

264 **"survive the war better"**: Ibid., 274.

## 20. Cried-out Eyes

267 **"The Cleaning Service"**: George E. Berkley, *Hitler's Gift: The Story of Theresienstadt* (Boston: Branden Books, 1995), 132.

267 **"Performances multiplied"**: Gerty Spies, *My Years in Theresienstadt: How One Woman Survived the Holocaust* (Amherst, N.Y.: Prometheus Books, 1997), 79–81.

268 **"Never become a mere number"**: Rabbi Baeck, quoted in Berkley, *Hitler's Gift*, 156.

274 **"Since yesterday"**: Berkley, *Hitler's Gift*, 176.

274 **"This Jewish city"**: ICRC report, quoted in ibid., 177–8.

277 **"'Now then, gentlemen'"**: Marie Rút Křížková, Kurt Jiří Kotouč, and Zdeněk Ornest, eds., *We Are Children Just the Same: Vedem, the Secret Magazine by the Boys of Terezín* (Philadelphia: Jewish Publication Society, 1995), 128.

277 **"Even the kings of Egypt"**: Egon Redlich, *The Terezín Diary of Gonda Redlich*, ed. Saul S. Friedman (Lexington: University Press of Kentucky, 1992), 160.

278 **"As far as Dresden"**: Jiří Barbier, letter to Dáša Deimlová, November 11, 1946.

278 **"We had to get out"**: Ibid.

279 **"With that we parted"**: Ibid.

280 **"October 23, 1944"**: Alice Ehrmann, in Alexandra Zapruder, ed., *Salvaged Pages: Young Writers' Diaries of the Holocaust* (New Haven, Conn.: Yale University Press, 2002), 404.

281 **"I've had enough"**: Eichmann, quoted in Berkley, *Hitler's Gift*, 225.

283 **"Twelve meters long"**: František Kraus, "But Lidice Is in Europe!" in *Art from the Ashes*, ed. Lawrence L. Langer (New York: Oxford University Press, 1995), 66.

284 **"I sink back"**: Ibid., 69.

## 21. Doodlebugs and Gooney Birds

290 **"the light-hearted bulldog view"**: Alexander Cadogan, *The Diaries of Sir Alexander Cadogan (1938–1945)*, ed. David Dilks (New York: G. P. Putnam's Sons, 1972), 647.

290 **"One can see"**: Harold Nicolson, *The War Years: Diaries and Letters 1939–1945*, ed. Nigel Nicolson (New York: Atheneum, 1967), 394.

293 **"Poland in this war"**: Beneš, quoted in Compton Mackenzie, *Dr Beneš* (London: George G. Harrap, 1946), 295.

295 **"People say to me"**: Nicolson, *The War Years*, 394, 464.

296 **"Lenin said that people"**: Masaryk, quoted in Sir Robert Bruce Lockhart, *Jan Masaryk: A Personal Memoir* (Norwich, England: Putnam, 1956), viii–ix.

296 **"What I want"**: Hitler, quoted in Wayne Biddle, *Dark Side of the Moon: Werner von Braun, the Third Reich, and the Space Race* (New York: W. W. Norton, 2009), 120.

297 **"Every time one goes off"**: George Orwell, quoted in Philip Ziegler, *London at War, 1939–1945* (New York: Alfred A. Knopf, 1995), 298.

### 22. Hitler's End

300 **"By rushing out"**: Dwight D. Eisenhower, message to the Allied Expeditionary Force, December 21, 1944, quoted in Forrest C. Pogue, *The Supreme Command: United States Army in World War II, The European Theater of Operations, Office of the Chief of Military History* (Washington, D.C.: U.S. Department of the Army, 1954), 380.

301 **"You, the veterans"**: Remarks by the author at fiftieth-anniversary commemoration of the Battle of the Bulge, Bastogne, Belgium, December 16, 1994.

302 **"a silly old man"**: Alexander Cadogan, *The Diaries of Sir Alexander Cadogan (1938–1945)*, ed. David Dilks (New York: G. P. Putnam's Sons, 1972), 706.

302 **"rather disturbing"**: Ibid., 716.

303 **"free and sovereign"**: Churchill, quoted in ibid., 716.

303 **"poor Chamberlain believed"**: Ibid.

304 **"the offending passage"**: J. Korbel, letter to the BBC, BBC Written Archives Centre, Caversham Park.

304 **"We want a strong"**: J. Masaryk, New Year's address, December 31, 1944.

304 **"What can one do"**: J. Masaryk, quoted in Stephen C. Schlesinger, *Act of Creation: The Founding of the United Nations* (Boulder, Colo.: Westview Press, 2003), 133.

305 **a man who, while sitting**: This image was inspired by a line in Mario Cuomo's address to the 1984 Democratic National Convention, "ever since Franklin Roosevelt lifted himself from his wheelchair to lift this nation from its knees," San Francisco, California, July 16, 1984.

306 **"Apparently, Hitler has croaked"**: Eva Ginzová, in Alexandra Zapruder, ed., *Salvaged Pages: Young Writers' Diaries of the Holocaust* (New Haven, Conn.: Yale University Press, 2002), 188.

### PART IV: MAY 1945–NOVEMBER 1948
### 23. No Angels

311 **"The source of the tragedy"**: quote from *Katolické Noviny*, April 26, 1942, cited in an article by Y. Jelinek, "The Vatican, the Catholic Church, the Catholics, and the Persecution of the Jews during World War II: The Case of Slovakia," included in B. Vago and G. L. Moss, eds., *Jews and Non-Jews in Eastern Europe, 1918–1945* (New York: John Wiley & Sons; Jerusalem: Israel Universities Press, 1974), 226.

314 **"Our soldiers will be going"**: Stalin, quoted in Jan Stránský, *East Wind over Prague* (New York: Random House, 1950), 30.

315 **"They have liberated us"**: J. Korbel, unpublished manuscript.

315 **"In our view"**: Eden, quoted in Alexander Cadogan, *The Diaries of Sir Alexander Cadogan (1938–1945)*, ed. David Dilks (New York: G. P. Putnam's Sons, 1972), 735.

316 **"Come help us everyone!"**: Heda Margolius Kovály, *Under a Cruel Star: A Life in Prague 1941–1968* (New York: Holmes & Meier, 1997), 40.

317 **"are fighting unexpectedly well"**: Teleprinter message to commander of Waffen SS from SS General Pückler, May 5, 1945, Prague, quoted in Jiří Doležal and Jan

Křen, eds., *Czechoslovakia's Fight* (Prague: Publishing House of the Czechoslovak Academy of Sciences, 1964), 109.

317    **"the Soviet forces"**: Exchange between Eisenhower and Soviet High Command, quoted in "Anniversary of Liberation of Czechoslovakia," State Department Bulletin, May 22, 1949, 666.

318    **"stand firm and strike"**: Proclamation of the Czech National Council, May 7, 1945, quoted in Doležal and Křen, eds., *Czechoslovakia's Fight*, 111–112.

319    **"People streamed into the streets"**: Kovály, *Under a Cruel Star*, 44.


## 24. Unpatched

325    **"were deep, always emotional"**: Josef Korbel, *The Communist Subversion of Czechoslovakia (1938–1948): The Failure of Coexistence* (Princeton, N.J.: Princeton University Press, 1959), 123.

326    **"frail stature [with] red cheeks"**: Hana Stránská, unpublished manuscript, 1994, chap. 5, p. 1.

328    **"The plane was about to land"**: J. Korbel, unpublished manuscript.


## 25. A World Big Enough to Keep Us Apart

331    **"Our people," he declared**: Beneš, May 9, 1945, quoted in Kálman Janics, "1945: The Year of Peace," in *Czechoslovak Policy and the Hungarian Minority*, www.hungarianhistory.com/lib/jani/jani111.htm.

332    **"is the historic task"**: Prokop Drtina, *Československo Můj Osud* (Prague: Melantrich, 1991), 62–63.

332    **"Now we will definitely"**: Gottwald, June 23, 1945, quoted in Janics, *Czechoslovak Policy*, www.hungarianhistory.com/lib/jani/jani111.htm.

332    **"In Nový Bydžov"**: Benjamin Frommer, *National Cleansing: Retribution Against Nazi Collaborators in Postwar Czechoslovakia* (Cambridge, England: Cambridge University Press, 2005), 43.

333    **"arm-in-arm with dirndled"**: Hana Stránská, unpublished manuscript, 1994, chap. 15.

334    **"The war is over"**: Ibid.

336    **"in order to remain hidden"**: Tahra Zahra, *Kidnapped Souls: National Indifference and the Battle for the Children in the Bohemian Lands 1900–1948* (Ithaca, N.Y.: Cornell University Press, 2008), 257.

339    **According to the Czechoslovak government's**: Memo from Jan Masaryk to the U.S. ambassador in Prague, October 24, 1945, quoted in *Foreign Relations of the United States, 1945*, vol. 2: *General Economic and Political Matters* (Washington, D.C.: United States Government Printing Office, 1976), 1298.

339    **"sometimes accompanied by excesses"**: Josef Korbel, *The Communist Subversion of Czechoslovakia (1938–1948): The Failure of Coexistence* (Princeton, N.J.: Princeton University Press, 1959), 138.

339    **"The disease of violence and evil"**: Havel, remarks at dinner in honor of German Chancellor Helmut Kohl, Prague, February 27, 1992.

343    **"I think you would do"**: Čurda, quoted in František Moravec, *Master of Spies: The Memoirs of General František Moravec* (London: Bodley Head, 1975), 221.

## 26. A Precarious Balance

344 **"The Yalta line was meant"**: Author's interview with Havel, October 27, 2010.

345 **"victory over one's smallness"**: Heda Margolius Kovály, *Under a Cruel Star: A Life in Prague 1941–1968* (New York: Holmes & Meier, 1997), 62.

345 **"to rebuild the very foundations"**: Gottwald, July 9, 1945, quoted in Josef Korbel, *The Communist Subversion of Czechoslovakia (1938–1948): The Failure of Coexistence* (Princeton, N.J.: Princeton University Press, 1959), 135.

346 **"That window used to be"**: J. Korbel, unpublished manuscript.

347 **"Don't write down anything"**: Beneš, quoted in Josef Korbel, *Tito's Communism* (Denver: University of Denver Press, 1951), 18.

348 **"Just look at it"**: Tito, quoted in ibid., 72.

351 **"not a Communist"**: Card entry, November 25, 1947, secret police document file no. 019952, Security Services Archive, Prague.

356 **"You idiot," he said**: J. Masaryk, cited in J. Korbel, unpublished speech manuscript.

359 **"At two a.m. Masaryk entered"**: Ibid.

360 **"Like my country"**: J. Masaryk, address at Paris Peace Conference, August 15, 1946, quoted in *Foreign Relations of the United States, 1946*, vol. 3 (Washington, D.C.: United States Government Printing Office, 1976), 225.

360 **"an extremely moving speech"**: Description of statement by J. Masaryk at Paris Peace Conference, September 23, 1946, in ibid., 527.

361 **"Left to himself"**: Marcia Davenport, *Too Strong for Fantasy* (Pittsburgh: University of Pittsburgh Press, 1967), 122.

361 **"You're no more full-blooded"**: Masaryk, quoted in ibid., 325.

## 27. Struggle for a Nation's Soul

365 **"If you go to Paris"**: Stalin, quoted in Hubert Ripka, *Czechoslovakia Enslaved* (London: Victor Gollancz, 1950), 67.

365 **"Oh, he's very gracious"**: J. Masaryk, quoted in Marcia Davenport, *Too Strong for Fantasy* (Pittsburgh: University of Pittsburgh Press, 1967), 405.

366 **"little people, inclined"**: Steinhardt, cable to State Department, April 30, 1948.

371 **"The strangest and least human"**: Čapek, "Why I Am Not a Communist," *Přítomnost*, December 2, 1924, translated by Martin Pokorny, capek.misto.cz/english/communist.html.

371 **"I do not agree"**: Josef Korbel, *Tito's Communism* (Denver: University of Denver Press, 1951), 124–125.

373 **"As far as can be judged"**: Steinhardt, cable to State Department, November 3, 1947.

## 28. A Failure to Communicate

377 **"As much as I am pessimistic"**: Beneš, quoted in Josef Korbel, *The Communist Subversion of Czechoslovakia (1938–1948): The Failure of Coexistence* (Princeton, N.J.: Princeton University Press, 1959), 198.

377 **"They thought of a putsch"**: Beneš, quoted in ibid., 199.

378  **"The communists in Belgrade"**: Josef Korbel, *Tito's Communism* (Denver: University of Denver Press, 1951), 306.

379  **"It is the only way"**: Hubert Ripka, *Czechoslovakia Enslaved* (London: Victor Gollancz, 1950), 203.

381  **"I met him"**: J. Korbel, unpublished speech manuscript.

382  **"And how did you respond"**: Korbel, *The Communist Subversion of Czechoslovakia*, 229.

383  **"At 4 p.m. Gottwald drove"**: Ibid., 235.

### 29. The Fall

385  **"This is not the end"**: J. Masaryk, quoted in Marcia Davenport, *Too Strong for Fantasy* (Pittsburgh: University of Pittsburgh Press, 1967), 419.

386  **"I knew I'd get them"**: Gottwald, quoted in Claire Sterling, *The Masaryk Case* (New York: Harper & Row, 1968), 88.

387  **"You know much of the world"**: J. Masaryk, quoted in Davenport, *Too Strong for Fantasy*, 365–366.

387  **"[Beneš] was . . . a martyr"**: J. Masaryk, quoted in ibid., 366.

388  **"He came at half-past eight"**: Ibid., 426.

390  **"[Korbel] and his family"**: Cable from Ambassador Charles Peake, Belgrade, to British Foreign Office, February 25, 1948.

392  **"a servant girl would do"**: J. Masaryk, quoted in O. Henry Brandon, "Was Masaryk Murdered?" *Saturday Evening Post*, August 12, 1948, 38.

### 30. Sands Through the Hourglass

398  **"In his desperation"**: Cable from Steinhardt to Washington, March 10, 1948, included in *Foreign Relations of the United States, 1948*, vol. 4, *East Europe and the Soviet Union* (Washington, D.C.: United States Government Printing Office, 1976), 743.

398  **"What he thought or felt"**: Sir Robert Bruce Lockhart, *Jan Masaryk: A Personal Memoir* (Norwich, England: Putnam, 1956), 78–79.

398  **"I cannot escape"**: Steinhardt, letter to Harold C. Vedeler, April 7, 1948, in *Foreign Relations of the United States, 1948*, vol. 4, *East Europe and the Soviet Union*, 743.

400  **"of belief, science"**: Beneš, address at Charles University, Prague, April 7, 1948, quoted in *Chronology of International Events and Documents* 4, no. 7 (March 18–April 8, 1948), 236.

401  **"They are accusing me"**: Eduard Táborský, *President Edvard Beneš: Between East and West, 1938–1948* (Stanford, Calif.: Hoover Institution Press, Stanford University, 1981), 228–229.

### The Next Chapter

409  **"passion for learning"**: James B. Bruce, "In Memoriam: Josef Korbel," in *Czechoslovakia: The Heritage of Ages Past, Essay in Memory of Josef Korbel*, ed. Hans Brich and Ivan Volgyes (Boulder, Colo.: East European Quarterly, distributed by Columbia University Press, New York, 1979), 7.

411   **a revolution that liberated Poland**: Credit for this formulation of the ten years
       through ten days belongs to the peerless European historian Timothy Garton Ash.
413   **"The main thing"**: J. Korbel, unpublished manuscript.
414   **"some kind of salvation"**: Havel, address to the Academy of Humanities and Po-
       litical Sciences, Paris, October 27, 1992, in Havel, *The Art of the Impossible: Politics
       and Morality in Practice* (New York: Alfred A. Knopf, 1997), 103–104.

# ACKNOWLEDGMENTS

Except for my memoir, completed in 2003, this has been my most personal book, one that I both needed to write and almost feared to attempt because of the feelings it would set loose. From the outset, the challenge has been to blend an account of my family's history with that of the encompassing era. Because of my age, I was a less than ideal witness, especially before and in the early years of the war. I was handicapped further by the fact that, for the first six decades of my life, I had been ignorant of my family's Jewish heritage and of the tragedy that had befallen so many of my relatives. I had a lot of catching up to do. Piecing together the facts, then placing them within the context of the times, has required the help of many hands.

The process began with my family; I am deeply grateful to my sister, Kathy, brother, John, and sister-in-law, Pamela, for the efforts they have made to trace the pattern of past events. Our experience in learning has been a shared one, as are many of our memories. I thank them also for being the first to review and suggest improvements to initial drafts. I appreciate, as well, the warm support of my daughters, Anne, Alice, and Katie, and of their families. Of all my roles in life, mother and grandmother are my favorites.

Working on this book gave me an opportunity to draw closer to my cousin Alena in London and, until her death, cousin Dáša in Prague. In October 2011, I met for the first time yet another cousin, Pedro Mahler, who was born and raised in Brazil. Pedro's grandmother was an older sister of my paternal grandfather, Arnošt Körbel; his father was among the Czech exiles who served with the British Royal Air Force during the war.

In studying the history of Czechs and Slovaks, including the story of my father's career, I have benefited from the assistance of the Czech Foreign Ministry, especially Jiří Schneider, Jiří Kuděla, Martina Tauberová, Ivan Dubovický, Tomáš Pernicky, and Robert Janas—and also Jan Havranek of the Ministry of Defense. Daniel Herman and Pavel Zacek of the Institute for the Study of Totalitarian Regimes were extraordinarily helpful in providing information and access to documents. I thank Dr. Oldřich Tůma of the Institute for Contemporary History, for taking time to meet with me and to share his thoughts about the men with whom my father served; and also the young scholar Tomáš Bouška, for his insights.

The experience of the Jewish community in the period before and during the war is a searing and scarring one. Tomáš Kraus, my dear friend, played host once again to Kathy and me during our researches and shared with us the unforgettable story of his father at Terezín. I am grateful, as well, to his assistant, Alena Ortenová, for her help and to the staff at the Federation of Jewish Communities for providing us, at short notice, with a meal when we most needed it. Dr. Vojtěch Blodig and Jan Munk were our expert guides at the Terezín Memorial and Museum. Thanks are also due to the many writers, artists, and archivists who have worked to preserve the memory of those who passed through the gates of that prison camp during the war; it is because of their efforts that I was able to understand more fully the story of my family.

Words are insufficient to capture all I owe to Václav Havel, who felt like family to me, and whose encouragement, thoughtful observations—and childhood drawings—lent depth and color to this volume. If this complicated book has a single message, it is to heed the wisdom of this matchless man.

In England, special thanks are due to Isobel Alicia Czarska, who welcomed me into her apartment at Princes House and who helped me to learn more about the building where some of my earliest memories were born. Thanks also to Libby Cook and Sonia Knight, friends and neighbors of Isobel, for agreeing to interview Mrs. Orlow Tollett at her nursing home; and a word of appreciation and respect to Mrs. Tollett herself for sharing her experiences. I was saddened to learn of her death in November 2011 at the age of 103 years and nine months.

Research, of course, is how a book like this begins, but transforming that knowledge into a presentable text requires a heavy dose of creativity, countless hours of work, and a strong team.

Bill Woodward played a major role in research and, as he has done on my previous books, served as a partner in the writing. Repeatedly, he urged me to rethink basic assumptions—just as I challenged his inability to pronounce Czech words. Elaine Shocas, another longtime partner, provided essential help in reviewing drafts and offering strategic advice. As always, she supplied a calm voice and a steady hand.

When not writing brilliant books (on swords, the sun, and, coming soon, historians) Richard Cohen has found time to edit each of my volumes. I'm not sure that it's quite normal to love one's editor, but Richard's incredible storehouse of talent and wit render him both

*The author in front of Princes House, 2010*

invaluable and delightful. When he stops editing, I will stop writing, but not before.

Lauren Griffith devoted enormous energy to research, fact-checking, and finding just the right photographs to accompany the text. Her excellent judgment, diligence, organizational skill, and humor made an indispensable contribution.

This may not be the best of times for companies that publish books with actual bindings and paper and such, but Tim Duggan of Harper-Collins is a superb captain, even in this era of rapid technological change. Special thanks, also, to Emily Cunningham, who has devoted many hours to this project, and to the entire HC team, including Brian Murray, Michael Morrison, Jonathan Burnham, Kathy Schneider, Tina Andreadis, Beth Harper, and Fritz Metsch; I am grateful for their ongoing faith and guidance.

Bob Barnett and Deneen Howell, my counselors, are the best in the world at what they do. Bob has been a booster of this book from the beginning; I doubt very much that I would have climbed this mountain without his push.

Like most books, this went through numerous drafts. I am indebted to those who took time to review parts of one or more and for their suggestions, including Ambassadors Wendy Sherman, Jiří Kuděla, Michael Žantovský, and Martin Palous; also Daniel Herman, Evelyn Lieberman, and Alan Fleischmann. I wish I could claim otherwise, but any errors are mine.

Assembling the photos to accompany the text was a labor of love but still a labor. In addition to Lauren Griffith and Elaine Shocas, I am indebted to my family—including my cousins Alena, Dáša, and Pedro; also the journalist Michael Dobbs, who generously shared pictures and other materials gathered in the course of his own research; Jakub Hauser and Michaela Sidenberg of the Jewish Museum in Prague; Martina Šiknerová of the Terezín Memorial; Daniel Palmieri and Fania Khan Mohammad of the International Committee of the Red Cross; Andrej Sumbera, for the image of the Wenceslas crown; Marcela Spacková of the National Gallery in Prague; Robin Blackwood, for running around with me in London, camera in hand; and the effervescent Jan Kaplan, who possesses a stunning archive of photos related to Czech and Slovak history. Early on, we went looking for someone to draw just the right maps; our prayers were answered by the young and talented Laura Lee.

Working on a book demands huge chunks of time, something that can detract from other obligations. I am privileged to work every day among an understanding assemblage at the AlbrightStonebridge Group, including my colleagues Sandy Berger, Tony Harrington, Jim O'Brien, Anne Fauvre, Jen Friedman, Wyatt King, Sarah Lincoln, Matt McGrath, and Fariba Yassaee. Particular thanks go to Suzy George for her keen insights and help in managing the many elements of this enterprise; to Mica Carmio, who had a special interest in the subject matter; to Erin Cochran, who worked closely with Harper-Collins and provided thoughtful comments throughout the process; and to Juliana Gendelman and Robyn Lee, who devoted many hours of their time to helping me make more efficient use of mine.

Finally, I want to express my appreciation to others who assisted during the course of my research or who were kind enough to share their own family remembrances, including Jan Drabek and Veronika Herman Bromberg. The unpublished memoirs of Renata Kauders and Hana Stránská also find a place in this category. I am grateful to Helen Epstein for sending me the text of Jan Masaryk's wartime radio addresses; she was also the English-language translator of Heda Margolius Kovály's superb history of midcentury Czechoslovakia, *Under a Cruel Star*. Rachelle Horowitz offered the kind of candid advice that only she can give. Leslie Thompson supplied much help in the early stages of gathering material. Anne Furlong of Sacred Heart Church in Berkhamsted confirmed information concerning my baptism. Helen Fedor of the Library of Congress was prompt and comprehensive in responding to my inquiries. Jeff Walden of the BBC Written Archives Centre, Caversham Park, provided access to documents related to my father's broadcasting career. Stephen Plotkin of the Kennedy Library kindly responded to questions regarding Ambassador Joseph Kennedy. Finally, I am pleased to thank Petr Vitek, the owner of the Hotel Sax, not only for furnishing lovely accommodations in Malá Strana but for personally driving me to every appointment.

# CREDITS

Except as noted below, all images are courtesy of the author. Grateful acknowledgment is made to the following institutions and individuals for permission to reproduce images in their possession:

Jan Kaplan Archive (pages 14, 133, 145, 318, 319, 334); CTK PHOTO (pages 16 [Martin Štěrba, René Fluger], 24, 38, 162, 167, 199, 206, 208, 209, 217, 221, 260, 263, 284, 327, 348, 393 [Michael Kamaryt], 396); Time & Life Pictures/Getty Images (pages 19, 22); Edu-art Prague/Andrej Šumbera (page 21); Alena Korbel (pages 51, 240); Bundesarchiv, Bild (pages 98 [183-R69173], 100 [183-H13116]); Associated Press (pages 103, 176, 188, 302); Dáša Šimová (page 123, 273); Václav Havel (page 136); Pedro Mahler (pages 157, 163, 234, 238, 282); National Archives (page 175 [306-NT-901F-2743V]); Jewish Museum in Prague Photo Archive collections (pages 243, 252); Terezín Memorial (pages 244 [Franktišek Mořic Nágl, prison sleeping quarters, PT 6728, © Alexandra Strnadová], 245 [the interior of the crematory, FAPT 6283], 267 [a poster for Hans Kráza's children's opera, *Brundibár*, PT4010, Hermann's Collection, © Zuzana Dvořaková]); Yad Vashem Photo Archive (page 270); ICRC (page 273); and Robin Blackwood (page 447).

All maps by Laura Lee.

Cover photograph (regiment of Nazi soldiers marches down Charles Bridge): CTK PHOTO.

Endpaper photograph (Charles Bridge): Frank Chmura/Getty Images.

# INDEX

and Masaryk's death, 395–97
meetings to organize the government, 310, 345
in Moscow, 263, 295, 310, 364
and Munich agreement, 264, 386
and Soviet Union, 263, 353, 354, 363, 364, 397, 405
and Stalinist purges, 405
Gottwald, Marta, 351
*Grandmother, The* (Němcová), 4, 137, 415
Grant Duff, Shiela, *Europe and the Czechs*, 121–22
Great Britain, *see* England
Great Depression, 56, 62
Great Moravian Empire, 26–27
Great War, *see* World War I
Greece, 189, 370
Grenfell-Baines, Milena, 123, 124
Gunther, John, 183

Haakon VII, king of Norway, 195
Habsburg empire, 23–24, 40, 376
Hácha, Emil:
and crown of Wenceslas, 198, *199*
and German invasion, 110–11
and German occupation, 15, 112, 129–30, 195, 202
and Heydrich assassination, 219
as president, 108, 129, 130, 160, 189, 227
Hachenburg, Hanuš, "Terezín," 211
Haile Selassie, emperor of Ethiopia, 195
Halifax, Edward Wood, Lord, 68–69, 77, 82, 84, 89, 90, 96, 127, 150, 165
*Hangmen Also Die!* (movie), 228
Hašek, Jaroslav, 41
Havel, Miloš, 137
Havel, Václav, 9–10, 344, 414
childhood drawing by, 136, *136*
on collective guilt, 339–40, 402
death of, 9
on liberation of Prague, 320–21
and Velvet Revolution, 2, 400
Havlíček, Karel, 31, 35
Haw-Haw, William Joyce, Lord, 165–66
Hebrang, Andrija, 404
Heine, Heinrich, 248

Henderson, Sir Nevile, 88
Henlein, Konrad, *103*
flight of, 87
and German invasion of Sudetenland, 108
and Masaryk, 64
pro-Nazi stance of, 62–63, 64, 74–75, 77, 83, 84–85, 88, 109, 343
propaganda campaign instigated by, 75, 83, 86
and Slovak separatists, 109
and Sudeten German Heimat Front, 62–63, 74–75
suicide of, 343
war aims of, 75, 77–78, 83, 86
Henry V, king of England, 69
Herman, Daniel, 8
Herodotus, 165n
Herzl, Theodor, 35
Hess, Alexander, 179
Heydrich, Reinhard, 199–203, 410
as acting protektor, 199, 200–203
assassination of, 207–10, 216–19, 226–28
brutality of, 200–201, 230
and crown of Wenceslas, 198, *199*, 202
and Czech underground, 200–201, 202, 213, 214, 217, 226
and final solution to Jewish question, 203, 204–5
reprisals for death of, 220–23, 226–27, 228, 239
and Terezín, 205
Hilsner, Leopold, 37
Himmler, Heinrich, 199, 200, 203, 218, 269, 270, 271, 275, 276
Hindenburg, Paul von, 56
Hiroshima, Japan, 353
Hitler, Adolf:
and Beneš, 75, 231, 402
and Chamberlain, 88–91, 92–93, 125, 194, 264, 303
and Czechoslovak Republic, 2, 3, 15, 82–89, 92–95, 97, 101, *103*, 104, 109, 202, 358, 406, 411
death of, 306, 316
on education, 66

# BOOKS BY MADELEINE ALBRIGHT

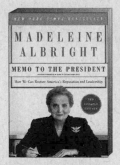